Environmental Justice in North America

Emphasizing the voices of activists, this book's diverse contributors examine communities' common experiences with environmental injustice, how they organize to address it, and the ways in which their campaigns intersect with related movements such as Black Lives Matter and Indigenous sovereignty.

The global COVID-19 pandemic exposed the ways in which BIPOC (Black, Indigenous, People of Color) communities and white working-class communities have suffered disproportionately from the crisis due to sustained exposure to toxic land, air, and water, creating a new urgency for addressing underlying conditions of systemic racism and poverty in North America. In addition to exploring the historical roots of the Environmental Justice movement in the 1980s and 1990s, the volume offers coverage of recent events such as the DAPL pipeline controversy, the Flint water crisis, and the rise of climate justice. The collection incorporates the experiences of rural and urban communities, Alaska Natives, Native Hawaiians, Puerto Ricans, and Indigenous peoples in the U.S., Canada, and Mexico.

The chapters offer instructors, undergraduate and graduate students, and general readers a range of accessible case studies that create opportunities for comparative and intersectional analysis across geographical and ethnic boundaries.

Paul C. Rosier is Professor of History and Director of the Albert Lepage Center for History in the Public Interest at Villanova University. He is the author of multiple books and articles on Native American History and Environmental History and co-editor of two essay collections.

Themes in Environmental History

Themes in Environmental History is a series of books aimed at 2nd and 3rd year undergraduate students and postgraduate students in the fields of history and environmental studies. The collection covers key areas of environmental history from across the globe, running from 500 CE to the present day. These books bring together chapters on the historiography of the field and the new research that is being done to move the field forward, making engaging reading for students. Topics covered are varied and expansive and emphasize the importance of looking back at environmental history to date to understand where we are today.

Water in North American Environmental History
Martin V. Melosi

Disease and the Environment in the Medieval and Early Modern Worlds
Edited by Lori Jones

Energy in the Early Modern Home
Material Cultures of Domestic Energy Consumption in Europe, 1450–1850
Edited by Wout Saelens, Bruno Blondé & Wouter Ryckbosch

Environmental Justice in North America
Edited by Paul C. Rosier

For more information about this series, please visit: https://www.routledge.com/Themes-in-Environmental-History/book-series/TIEH

Environmental Justice in North America

Paul C. Rosier

Routledge
Taylor & Francis Group

NEW YORK AND LONDON

Designed cover image: Protesters in Warren County, North
Carolina try to block trucks from dumping PCB-contaminated
soil in their community in September 1982, helping to launch the
environmental justice movement. © Jenny Labalme.

First published 2024
by Routledge
605 Third Avenue, New York, NY 10158

and by Routledge
4 Park Square, Milton Park, Abingdon, Oxon, OX14 4RN

*Routledge is an imprint of the Taylor & Francis Group, an informa
business*

© 2024 selection and editorial matter, Paul C. Rosier; individual
chapters, the contributors

Library of Congress Cataloging-in-Publication Data
Names: Rosier, Paul C., editor.
Title: Environmental justice in North America / edited by Paul C. Rosier.
Description: New York, NY : Routledge Taylor & Francis, 2024. | Series:
Themes in environmental history | Includes bibliographical references and
index.
Identifiers: LCCN 2023023488 (print) | LCCN 2023023489 (ebook) |
ISBN 9781032102474 (hardback) | ISBN 9781032080376 (paperback) |
ISBN 9781003214380 (ebook) | ISBN 9781000986396 (adobe pdf) |
ISBN 9781000986426 (epub)
Subjects: LCSH: Environmental justice--North America. | Environmental
sociology--North America. | Indians of North America--Social conditions. |
Indians of North America--Government relations. | Environmental
policy--Social aspects--North America.
Classification: LCC GE240.N7 E68 2024 (print) | LCC GE240.N7 (ebook) |
DDC 363.70089/97--dc23/eng/20230608
LC record available at https://lccn.loc.gov/2023023488
LC ebook record available at https://lccn.loc.gov/2023023489

ISBN: 978-1-032-10247-4 (hbk)
ISBN: 978-1-032-08037-6 (pbk)
ISBN: 978-1-003-21438-0 (ebk)

DOI: 10.4324/9781003214380

Contents

Introduction

Paul C. Rosier

America is a thousand Flints.[1]

(Andrew R. Highsmith)

In 1976, when I was 17, I embarked on a canoe trip from Lake Temagami in central Ontario to the James Bay, the southern part of the Hudson Bay in Quebec, Canada. It was a 700-mile, 40-day trip that took me through beautiful country, much of it Cree country. I lugged fishing gear across 3-mile portages in anticipation of catching and eating fish from the 300-mile long Harricana River, considered one of the five wildest rivers in the world by *National Geographic*. But when I arrived at the headwaters of the Harricana I was told by local Cree that no one should eat the fish because of mercury contamination leaching from paper mills. I thought about this experience years later when I came across a poem by a Cree writer, Margaret Sam-Cromarty, which I used in an essay for a 2006 volume I co-edited, *Echoes from the Poisoned Well: Global Memories of Environmental Injustice*. Sam-Cromarty wrote: "The river cod; so excellent in taste; This rich fish is my fondest memory of my tradition.... It's full of mercury they said; I try not to cry." The Cree would also suffer a different form of environmental injustice when part of their ancestral territory was flooded by the massive James Bay Project, a series of hydroelectric dams that would produce energy principally for New York State.[2]

Allow me one other personal story that has shaped my understanding of and commitment to environmental justice (EJ). After getting a teaching job at Villanova University in 1999, my family moved to a small town located a few miles from Chester, Pennsylvania, a predominately Black city targeted for large waste storage and incineration projects by white corporate officials backed by white state politicians. Chester became the dumping ground for not only Pennsylvania's trash but also that of surrounding states.[3] I participated in various events in support of the EJ group Chester Residents for Quality Living (or CIRQL), including a Zoom session with

DOI: 10.4324/9781003214380-1

county officials of the solid waste authority responsible for renewing the permits of companies burning trash and medical waste in Chester. Near the end of the session, a Swarthmore College student put the issue into proper perspective. He asked the officials, "are you willing to accept the fact that these industries are killing people, that if you renew these permits, people will die?" It was a moment of clarity amid a lot of regulatory detail, distilling the matter of EJ down to its core: some citizens die as a result of political and economic decisions, and those citizens are typically people of color who have fewer financial and political resources with which to fight for the health of their communities, where they live, work, play, and pray.

In a documentary about the Chester story called *Laid to Waste*, CIRQL members traveled to a local corporate office to confront those responsible for bringing polluting industries to Chester. When told that the headquarters was in Pittsburgh, CIRQL President Zulene Mayfield said, "Oh, we'll go to Pittsburgh, even if we have to hold a bake sale to get there." And when told that it was a legal issue, she replied, "it's not a legal issue, it's a moral issue."[4]

EJ history is a form of moral accounting, an economic accounting of the medical and financial costs of industrial pollution across the axes of race, class, and gender, and a social accounting of the extraordinary effort people with limited means have made and continue to make to identify the sources of that pollution and travel to distant corporate and government offices in order to protect their families and their communities. What expenses and measures would the Cree have to take to confront the source of mercury poisoning? Or CIRQL members to confront the source of their multiple pollution vectors?

The Cree's story is just one of thousands of environmental crises Indigenous peoples in North America have experienced since the advent of settler colonialism; and Chester is one of thousands of communities in North America which has borne the brunt of what environmental historian Carolyn Merchant called the "malignant side-effects" of industrial capitalism. The Cree and Chester residents experienced daily what Rachel Carson argued in her 1962 book, *Silent Spring*, that when humans poison nature, they poison themselves. But who suffers from those poisons is at the heart of EJ history. Dr. Robert Bullard, a leading scholar of that history, has argued that "America is segregated and so is pollution. Race and class still matter and map closely with pollution, unequal protection, and vulnerability. Today, zip code is still the most potent predictor of an individual's health and well-being. Individuals who physically live on the 'wrong side of the tracks' are subjected to elevated environmental health threats and more than their fair share of preventable diseases."[5] The global pandemic of 2020–2022 further exposed the ways in which Black, Indigenous, people of color (BIPOC), poor and working-class communities in

Figures

Editor's Note

All royalties received by the editor from the sale of this book will be shared with the Environmental Justice and Climate Justice organizations featured in this volume. This book is dedicated to their efforts, past, present, and future.

–Paul C. Rosier

North America suffered disproportionately from COVID-19 due to sustained exposure to toxic land, air and water, creating a new urgency for addressing underlying conditions of systemic environmental racism and poverty.[6]

Emphasizing the voices of activists, young and old, this volume's diverse contributors document these communities' common experiences with environmental injustice, the ways in which they organized to confront it on a local, national, and international level, and how their campaigns intersected with other movements seeking racial and economic justice, such as Indigenous sovereignty and Black Lives Matter. For example, when New York City police officers were choking Eric Garner to death in 2014, he had said repeatedly "I can't breathe." "I can't breathe" became a powerful phrase that speaks beyond police brutality to structural environmental racism in BIPOC, poor and working-class communities suffering from extraordinary rates of asthma caused by airborne particulate matter. Indeed, Eric Garner's daughter, Erica, died from an asthma-induced heart attack at the age of 27.[7]

The volume's collection of essays explores the historical roots of the environmental justice movement of the 1980s and 1990s and offers coverage of 21st-century events such as the Dakota Access Pipeline (DAPL) controversy, the Flint water crisis, and the rise of climate justice (CJ) in accessible case studies that speak to the enduring problem of environmental racism and classism. Written by a set of authors that reflects the diversity of communities examined in the book, the chapters provide in-depth discussions of each topic while creating opportunities for comparative and intersectional analysis across chronological, geographical, and ethnic boundaries.

The individual chapters are unified by a coherent theme of EJ movement actors' economic, political, and social struggles to mediate the toxic impacts of industrialization that disproportionately affect their communities. Existing texts addressing EJ have focused on one ethnic group (Black, Native American, or Latinx[8]), one geographical area (United States or global), or one case study (Love Canal or Flint). And with few exceptions, these texts have ignored the histories of EJ or CJ activism in Alaska, Canada, Hawai'i, and Mexico. This volume's analysis of the class, race, and gender dynamics of EJ movements will cross international and ethnic boundaries and incorporate the neglected experiences of Alaska Natives, Native Hawaiians, Mexicans, Puerto Ricans, and Indigenous peoples in Canada.

This introductory chapter is designed to provide the reader with a historical framework in which to situate the 11 chapters, introducing key events, people, themes, and theories of 20th-century and early 21st-century EJ and CJ campaigns. It focuses on the evolution of the EJ movement in the United States and Canada following the crisis of polychlorinated biphenyl (PCB) contamination in the predominately Black communities of Warren County, North Carolina, which generated new phrases such as "environmental racism" and "environmental justice" for understanding the intersection of environmental and civil rights activism; and how that movement coalesced at the 1991 First People of Color Environmental Leadership Summit, which brought together BIPOC communities from the United States and beyond to create 17 Principles of EJ that animated the multi-racial EJ movement into the 21st century. Finally, it closes with a short summary of each chapter to provide an overview of their central themes.

Some EJ scholars use the term "long Environmental Justice movement" to describe the range of environmental injustices emerging from settler colonialism and global capitalism, tracing those injustices to Europeans' conquest of Indigenous people and to their use of enslaved African people in agricultural and extractive operations. Indeed, Indigenous peoples throughout North America contended with a variety of environmental crises, from the spread of smallpox and other diseases to the expropriation of land and other resources, including water, gold, coal, and uranium. After slavery ended, Black Americans contended with racism that pushed them into the dirtiest parts of factories and other industrial enterprises. To cite one example among many, during the 1930s, several thousand Black workers migrated North to West Virginia to work on a tunnel construction project. Roughly 500 of them died from the lung disease silicosis as a result; the company that hired them had negotiated a discount rate with the local undertaker, knowing that many of the workers would die.[9]

The environmental injustices experienced by BIPOC and working-class communities would increase in scope and scale after World War II during a rapid expansion of the economy in the United States and in Canada, triggering campaigns by activists shaped and strengthened by the Indigenous treaty rights, Black civil rights, environmental, and feminist movements of the post-war era.

During the 1960s and 1970s, Native groups in Washington, Wisconsin, and other states protested pollution that endangered their access to treaty-protected fish stocks. Native people in Arizona, Montana, and elsewhere confronted the degradation uranium and coal mines caused to their water and air. In the Southwest, Navajo families were devastated by radiation-related cancers; one family was left with just 3 members of an original 27.[10] Of 150 Navajo who worked in the Shiprock, New Mexico, uranium mine "133 had either died of radiation-induced lung cancer or had contracted cancer and severe respiratory ailments such as fibrosis."[11] Native communities in New Mexico, Washington State, and South Dakota also faced the dangers of nuclear waste during the 1960s and 1970s.[12]

Black activists and writers confronted the problems of race and pollution in their communities, both rural and urban. In his 1970 essay entitled

"Black Survival in our Polluted Cities," Wilbur L. Thomas, Jr. asked several questions that lie at the heart of the EJ movement: did the new discourse of "saving the environment" ignore Black communities? Are Black residents "disproportionally exposed to greater environmental hazards than non-Blacks?" And if so, how will Black citizens protect their communities from growing environmental hazards? Thomas concluded that Black Americans "must become aware of the types of problems existing in our community such as lead poisoning, air pollution, the lingering effect of DDT, and other similar pesticides."[13] In a seminal legal case addressing environmental injustices in Black communities, residents of the middle-class Houston neighborhood filed the first environmental racism lawsuit in the 1979 case *Bean v. Southwestern Waste* to protest a planned municipal landfill.[14]

In California, Latinx communities warned of the dangers of pesticides that permeated the fields in which they worked, their homes, and their playgrounds. Cesar Chavez, representing predominately Latinx farmworkers in the United Farm Workers of America, sought to expand the boycott of grapes in light of increasing evidence that pesticides sprayed on the grapes constituted a grave health crisis for workers and their families. Chavez argued that "we will not tolerate the systematic poisoning of our people.... [W]e will be damned ... if we will permit human beings to sustain permanent damage to their health from economic poisons."[15]

Leaders of working-class communities raised similar questions as the evidence of toxic waste's impact on their health became increasingly clear. A crisis in Love Canal, New York, a predominately white working-class community in upstate New York, brought the issue of toxic chemicals to the nation's attention in the late 1970s, revealing the legacy of chemical dumping in America's residential neighborhoods and leading to the passage of the 1980 Superfund Act, which forced the federal government to begin cleaning up the thousands of toxic waste sites littering the country. The Love Canal case also provided evidence that race played an important part in environmental injustice, as Black renters were denied compensation offered to white homeowners of Love Canal.[16]

Across the border from New York in Ontario, Canada, mercury discharges from paper companies began contaminating the land and bodies of the Asubpeeschoseewagong (Grassy Narrows) First Nation and the Wabaseemoong (Whitedog) First Nation in the 1960s, leading to generational health crises. Indigenous people in Canada, which comprise Metis, First Nations, and Inuit, represent roughly 5 percent of Canada's population. But, like BIPOC communities in the United States, they fared worse in categories such as income, education, and health and suffered disproportionately from pollution, resource extraction, and degradation of water supplies, among other environmental problems. First Nations also have suffered tremendous health problems from uranium mining in the North-west Territories, as have Indigenous people in central and northern Alberta from oil and gas resource extraction projects.[17] The Lubicon Lake people of northern Alberta, a Cree First Nations community about 300 miles north of Edmonton, faced economic and ecological devastation from resource extraction projects; between 1979 and 1983, energy companies drilled over 400 oil and gas wells without consulting Lubicon leaders, leading to contaminated landscapes and water supplies.[18]

The Warren County Protests of 1982

As the examples cited above indicate, EJ campaigns have been waged for decades. Historian Josiah Rector situates the origins of the broader EJ movement in the 1976 "Working for Environmental and Economic Justice and Jobs" conference held at the Black Lake Walter and May Reuther Education Center in Michigan. The conference was attended by over 350 people from roughly 140 labor, civil rights, and environmental organizations and by representatives of Black, Native, and Latinx communities. Emphasizing the intersectionality of civil rights and economic and environmental policies, the conference also offered workshops on alternative energy, pollution control, and other environmental topics. Three years later, the Urban Environment Conference's (UEC) City Care conference held in Detroit galvanized civil rights, environmental, and labor activists to confront environmental problems in low-income communities of color, especially in southern states such as Alabama and Tennessee.[19]

Most EJ scholars contend that citizen protests in Warren County, North Carolina, in 1982 launched the national environmental movement in the United States. Warren County saw the first series of protests that resulted in mass arrests and generated extensive press coverage; national and international media outlets captured images of Black children, men, and women linking arms with white ministers and other supporters before hundreds of them were taken to jail in buses (see Figure I.1). The protests began after the State of North Carolina decided to create a landfill for soil contaminated with PCB in Shocco, a small town in Warren County, the population of which was 75 percent Black and which ranked 97th in GDP among the state's 100 counties. Alarmed by this toxic threat, on September 15, 1982, 130 protestors marched for 2 miles to the landfill, where 55 of them tried to block the first of thousands of dump trucks from dropping their load of PCB-laced soil. Protestors sang traditional civil rights songs and new ones such as "A Warren County PCB Protest Song": "Our good old earth we've got to guard and share; We've got to keep her safe and free from care; And that means standing up for what is right. We'll fight the poison with all our might."[20]

Figure I.1 One of several marches protesting the dumping of PCBs in Warren County, North Carolina, in September 1982. © Jenny Labalme.

The Warren County protests, then, married two strains of activism, civil rights, and environmentalism. But it was a different kind of environmentalism, one which defined environment as where people live, work, play, and pray, not just where people vacation, hike, or view wildlife. The environment, activists argued, started with the human body, increasingly exposed to a barrage of chemicals that flooded urban and rural ecosystems, contaminated the food chain, and undermined community health and sustainability.[21] The Warren County protests also generated a new vocabulary to describe this crisis: "environmental racism" and "environmental justice."

Several reports which emerged in the mid-1980s furthered the use of these phrases and provided concrete evidence that the problem of toxic and nuclear wastes was borne disproportionately by BIPOC communities. The first, the so-called Cerrell Report, was compiled by the consulting firm Cerrell Associates for the California Waste Management Board, which was struggling to find suitable locations for the state's growing quantity of hazardous wastes. Although the report did not mention race as a criterion, it concluded that potential "sites can be suggested partly on the basis of neighborhoods least likely to express opposition—older, conservative, and lower socioeconomic neighborhoods."[22] The Waste Management Board used the Cerrell Report to justify siting a Chemical Waste Management toxic waste incinerator in Kettleman City where 95 percent of the farmworker community was Latinx, most of them Spanish-speaking. Alarmed by this potential exposure, residents formed El Pueblo para el Aire y Agua Limpio (People for Clean Air and Water) to oppose the project. "The Cerrell report fit us to a T," said one of the El Pueblo leaders. Another stated, "We felt we were being targeted, that Chem Waste as a corporation was targeting these communities on purpose because their ethnic make-up would make people least likely to protest."[23] Chem Waste was wrong in this case, as Kettleman City residents banded together to attend siting meetings, overcome language barriers thrown up by government and corporate officials, and succeed in blunting Chem Waste's proposal, which it withdrew in 1993 after a five-year struggle.[24]

Charles Lee, a Chinese American researcher, followed the Kettleman City case. And after he read a 1983 U.S. General Accounting Office (GAO) report which showed that three out of four major landfills in the South were surrounded by Black communities,[25] Lee suspected that environmental racism was widespread and sought to replicate the GAO study on a national scale. With the support of the United Church of Christ's Commission for Racial Justice, Lee and Rev. Benjamin Chavis helped to author the comprehensive report "Toxic Wastes and Race in the United States: A National Report on the Racial and Socio-Economic Characteristics of Communities with Hazardous Waste Sites" in 1987. The report examined commercial hazardous waste sites and roughly 20,000 abandoned "uncontrolled toxic waste sites," making clear the role of race in siting hazardous waste sites in BIPOC and poor communities by quantifying the impact on Black, Latinx, Native Americans, Asian Americans, Pacific Islanders, and poor white communities. It concluded that "three out of every five Black and Hispanic Americans lived in communities with uncontrolled toxic waste sites" and "approximately half of all Asian/Pacific Islanders and American Indians lived in communities with uncontrolled toxic waste sites."[26] BIPOC communities faced the burden of living near existing sites and were targeted for new commercial hazardous waste sites as well as incinerators of medical and toxic waste. This data and the new discourse of environmental racism and EJ galvanized the communities facing these toxic threats, each waging their own EJ campaigns in isolation from groups fighting similar battles.

Richard Moore, an EJ activist in Albuquerque, New Mexico, had been fighting toxic waste dumping in his Latinx community for decades. He attended a meeting held by Rev. Chavis and Charles Lee where they discussed the findings of their landmark 1987 report *Toxic Wastes*. Moore commented after the meeting that he and his fellow activists "were seeing basically our own community in front of our eyes, but the communities would be in North Carolina or Mississippi or South Carolina or Alabama. And they were using environmental justice language."[27] Similarly, the Native activist Tom Goldtooth (Diné and Dakota) found the language of environmental racism and EJ coming from the report and the national cases it documented reinforcing his work on environmental problems in Native communities: "This is what I'm experiencing here with the work I'm doing," he said. "Racism rears its ugly head when it comes to protecting Native people."[28]

The First People of Color Environmental Leadership Summit, 1991

Moore and Goldtooth would join hundreds of other representatives of EJ groups who traveled to DC for the First People of Color Environmental Leadership Summit, from every state in the United States and from Canada, Central America, Puerto Rico, and the Marshall Islands.[29] The Summit co-chairs included Rev. Benjamin Chavis, Executive Director of United Church of Christ Commission for Racial Justice; Gail Small, Executive Director of the Northern Cheyenne Tribe's Native Action organization, and Rev. Syngman Rhee, president of the National Council of Churches. The National Planning Committee included Damu Smith from Greenpeace; Donna Chavis, a Lumbee Nation elder from North Carolina; and Charles Lee, who helped to write the 1987 *Toxic Wastes and Race* report.

Summit organizers invited representatives of so-called mainstream environmental groups such as Sierra Club and Greenpeace, but EJ activists resented these groups' failure to support their campaigns as environmental issues rather than public health issues. A year before the Summit, members of the Southwest Organizing Project based in Arizona sent a letter to the "Big Ten" American environmental groups criticizing them for not including people of color in their organizations, focusing on endangered species and forests rather than people, promoting environmental policies that neglected the interests of people of color communities, and accepting funding from the companies responsible for much of the nation's environmental degradation. The letter concluded that "people of color in the United States and throughout the world are clearly endangered species. Issues of environmental destruction are issues of our immediate and long-term survival."[30]

These themes of endangerment and survival percolated through some of the early statements made at the Summit. The opening declaration set the urgent tone, describing the Summit as a "multiracial, multicultural convergence of existing local and regional grassroots movements and struggles which are already underway by people of color which are actively resisting various forms of environmental genocide."[31] During the opening Plenary Session on October 24, co-chair Benjamin Chavis defined the delegates' "common struggle to prevent the destruction of our peoples and our communities, and to rescue the environment from the clutches of persons and institutions gone mad with racism and greed." Chavis focused on the environmental violence suffered by children: "In Chicago, our children are dying.... In Cancer Alley, it is our children who are dying. In the Southwest and among farmworkers, it is our children who are dying. On Native American reservations, territories, and lands, it is our children who are dying. For Asian American sisters who labor in Silicon Valley, it is our children that are dying."[32] Chavis also highlighted the crisis of nuclear waste dumping on Native American land, which became another key concern of the Summit. Chavis contended that "there has been too much tolerance of the genocide against our Native American sisters and brothers. If we do not say anything else at this national Leadership Summit ... we must say that most of us have not taken up and embraced the suffering of our Native American sisters and brothers."[33]

During the October 25th Plenary Session entitled "Building a Multiracial and Multicultural Environmental Justice Movement," Native Americans, Black, Latinx, and Hawaiian delegates offered their perspectives on EJ via a series of panels called "Who We Are and Our Perspectives on Environmental Justice." Principal chief of the Cherokee Nation of Oklahoma Wilma P. Mankiller spoke first, outlining Native Americans' spiritual attachment to land and condemning the ways in which industry and government officials targeted Indian reservations for nuclear waste dumping, echoing Chavis' call to action. Mankiller noted that until she attended the Summit, she thought only Native people had to contend with environmental injustices: "We have always known that somehow or another that these very toxic facilities endued up in Native communities and poor communities." But she had not realized that environmental injustice was "a universal problem among poor people and particularly people of color. This is one reason I made such an extreme effort to get here, because this kind of networking and coalition building together is very, very important to us." Mankiller posed an important question that applied to all delegates: "People talk about progress and they talk about development when they want to develop our lands. Well, whose progress is it? And whose development is it? Is it our progress? Do we progress? Does that development help us in any way?"[34]

An important part of the proceedings was drafting the 17 Principles of EJ (see p. 11–12), one of the principal outcomes of the Summit and a foundational document that would animate the delegates, the organizations they represented, and thousands of others of activists who would later read them and connect them to their own struggles and strivings. Each ethnic group presented key issues during the drafting of the Principles, some of them dealing with the language used during the proceedings. According to Chavis, a Native American delegate requested that the phrase "people of color of the United States" be changed to "people of color in the United States" in the Principles to reflect ongoing colonialism affecting Native people.[35] There followed an interesting debate about using the term "people of color." Some delegates thought the term constrained them in a biological rather than a social context. A Native American woman argued that "people of color" is what connected all of the groups attending the Summit, that it was their common ground, a point seconded by several other delegates, including a Puerto Rican delegate from the University of Michigan. Another Native American delegate said that "people of color" was not a biological concept but "a political concept: It is a concept of people who were oppressed for many years—discriminated against politically, socially, economically, and culture.... Our voices have been denied for 500 years, and 'people of color' is a very appropriate concept at this time."[36]

Principles of Environmental Justice

Preamble

WE, THE PEOPLE OF COLOR, gathered together at this multinational People of Color Environmental Leadership Summit, to begin to build a national and international movement of all peoples of color to fight the destruction and taking of our lands and communities, do hereby re-establish our spiritual interdependence to the sacredness of our Mother Earth; to respect and celebrate each of our cultures, languages, and beliefs about the natural world and our roles in healing ourselves; to ensure EJ; to promote economic alternatives which would contribute to the development of environmentally safe livelihoods; and to secure our political, economic, and cultural liberation that has been denied for over 500 years of colonization and oppression, resulting in the poisoning of our communities and land and the genocide of our peoples, do affirm and adopt these Principles of EJ:

1 **Environmental Justice** affirms the sacredness of Mother Earth, ecological unity and the interdependence of all species, and the right to be free from ecological destruction.
2 **Environmental Justice** demands that public policy be based on mutual respect and justice for all peoples, free from any form of discrimination or bias.
3 **Environmental Justice** mandates the right to ethical, balanced, and responsible uses of land and renewable resources in the interest of a sustainable planet for humans and other living things.

4 Environmental Justice calls for universal protection from nuclear testing, extraction, production, and disposal of toxic/hazardous wastes and poisons and nuclear testing that threaten the fundamental right to clean air, land, water, and food.

5 Environmental Justice affirms the fundamental right to political, economic, cultural, and environmental self-determination of all peoples.

6 Environmental Justice demands the cessation of the production of all toxins, hazardous wastes, and radioactive materials, and that all past and current producers be held strictly accountable to the people for detoxification and the containment at the point of production.

7 Environmental Justice demands the right to participate as equal partners at every level of decision-making, including needs assessment, planning, implementation, enforcement, and evaluation.

8 Environmental Justice affirms the right of all workers to a safe and healthy work environment without being forced to choose between an unsafe livelihood and unemployment. It also affirms the right of those who work at home to be free from environmental hazards.

9 Environmental Justice protects the right of victims of environmental injustice to receive full compensation and reparations for damages as well as quality health care.

10 Environmental Justice considers governmental acts of environmental injustice a violation of international law, the Universal Declaration on Human Rights, and the United Nations Convention on Genocide.

11 Environmental Justice must recognize a special legal and natural relationship of Native Peoples to the U.S. government through treaties, agreements, compacts, and covenants affirming sovereignty and self-determination.

12 Environmental Justice affirms the need for urban and rural ecological policies to clean up and rebuild our cities and rural areas in balance with nature, honoring the cultural integrity of all our communities and providing fair access for all to the full range of resources.

13 Environmental Justice calls for the strict enforcement of principles of informed consent, and a halt to the testing of experimental reproductive and medical procedures and vaccinations on people of color.

14 Environmental Justice opposes the destructive operations of multinational corporations.

15 Environmental Justice opposes military occupation, repression, and exploitation of lands, peoples, and cultures, and other life forms.

16 Environmental Justice calls for the education of present and future generations which emphasizes social and environmental issues, based on our experience and an appreciation of our diverse cultural perspectives.

17 **Environmental Justice** requires that we, as individuals, make personal and consumer choices to consume as little of Mother Earth's resources and to produce as little waste as possible and make the conscious decision to challenge and reprioritize our lifestyles to ensure the health of the natural world for present and future generations.

The Summit closed with a ceremony organized by Lumbee leader Donna Chavis, who said that she hoped Summit attendees had embraced the opportunity to "share our cultures, and to learn more from each other," and a prayer offered by Tom Goldtooth.[37] For Chavis, Goldtooth, and other attendees, the Summit did allow everyone to share their cultures, learn from each other, and help them understand the depth and breadth of environmental injustices affecting not only their communities, but those across the nation and the globe. Goldtooth initially was not confident that the Summit would further his work on Native sovereignty. But after hearing testimony from other Native delegates and that of Black, Latinx, and Asian American delegates, he said, "it's a life-and-death situation that's going on. It's not just the communities here in Minnesota, it's not just Native people, but it's all people of color."[38] Summit co-organizer Dana Alston later wrote that "collectively, delegates surmounted the barriers that have historically divided us - regionalism, culture, gender, language, and class…. By the end of the summit, those gathered spoke with one voice as part of a movement to eradicate environmental racism and bring into being true social policy recommendations that would guide future justice and self-determination."[39]

The Summit facilitated the creation of regional EJ organizations such as the National Environmental Coalition of Native Americans, which formed in Las Vegas to protest nuclear waste disposal on tribal lands and to declare those lands nuclear-free zones, and the Asian Pacific Environmental Network, which Pamela Chiang helped organize in Oakland, California, to support the Laotian community facing oil refinery pollution. The Summit also inspired its participants to share their perspectives and EJ principles in other forums. In June 1992, Alston led a delegation of EJ leaders to the United Nations Conference on Environment and Development (UNCED) in Rio de Janeiro (the Rio Earth Summit) to present policy recommendations generated during the People of Color Leadership Summit.[40] An Iroquois delegation, including Oren Lyons and Leon Shenandoah, traveled to Rio de Janeiro to share Iroquois philosophies with the international delegates.[41] Native activists from the American Southwest also attended the 1992 World Uranium Hearings in Austria to not only protest the way in which Native people's land and health were sacrificed to create energy for non-Native people but also promote safe alternative energy.[42]

The Summit also had an impact on the national political level. In 1992, President George H.W. Bush established the first EPA Office of EJ. And in 1994, President William J. Clinton signed Executive Order 12898, which affirmed the federal government's commitment to incorporating EJ criteria in federal agencies' policies.[43] The EPA codified its definition of EJ as "the fair treatment and meaningful involvement of all people regardless of race, color, national origin, or income with respect to the development, implementation and enforcement of environmental laws, regulations and policies. Fair treatment means no group of people should bear a disproportionate share of the negative environmental consequences resulting from industrial, governmental and commercial operations or policies."[44] But little substantive action emerged from Washington, especially after the George W. Bush administration moved to limit EPA initiatives in the early 2000s.[45]

Given the weak response from American, Canadian, and Mexican government officials to ongoing environmental crises, EJ leaders across North America continued to stress the importance of grassroots work. As a result, they organized the Second National People of Color Environmental Leadership Summit ("Summit II"), held on October 23–27, 2002, in Washington, DC, which drew over 1,200 delegates from a variety of organizations, including academic institutions, civil rights groups, labor unions, and faith-based, youth, and community groups. Summit II delegates and attendees came from across North America, the Caribbean, South and Central America, and beyond; reflecting the international dimensions of the growing EJ movement, attendees also came from India, Nigeria, Peru, the Philippines, and the United Kingdom. According to co-organizer Devon Pena, "it was truly a global and multiethnic Summit."[46] During the summer of 2002, Summit II organizers issued a call for papers, which yielded papers on topics such as childhood asthma, energy, transportation, CJ, military toxics, brownfields redevelopment, sustainable agriculture, and occupational health and safety, especially of farmworkers, who continued to bear the brunt of pesticide exposure. These papers helped guide the workshops and hands-on training sessions.

Summit II demonstrated the incredible expansion of the EJ movement in North America since the first Summit in 1991, when the People of Color Environmental Groups Directory listed roughly 300 EJ groups in the United States. But in 2002, that list included over 1000 groups in the United States, Canada, and Mexico. The organizational infrastructure also expanded to include regional and international networks, EJ centers, and university-based legal clinics.[47] In addition, the number of books on EJ ballooned from just one in 1991 to over 100 in 2002, enlarging the study of EJ to include theory, policy, and legal practice. Robert D. Bullard, the author of that single 1991 book, *Dumping in Dixie*, said at Summit II:

"knowledge is power. It's important that we have researchers, writers, and academicians at the Summit. As people of color, we have to document our struggles and tell our stories."[48]

Robert Bullard's and Dana Alston's position paper from the 1991 People of Color Leadership Summit focused on the structural barriers EJ activists confront: "The concept 'environmental racism' emerged out of the struggle by people of color to dismantle exclusionary zoning, discriminatory land-use practices, industrial facility siting that target racial and ethnic communities, differential enforcement of regulations, and paternalism by national environmental groups.... Environmental racism is deeply imbedded in our laws, customs, and governmental practices. These practices systematically produce disparate environmental quality for people of color communities and white communities."[49] EJ activists and scholars, therefore, have called for structural change amounting to a transformation of modern political economy, to ensure not just compensatory or distributive justice but also participatory justice or procedural justice which enables people of color and marginalized community leaders to address the roots of environmental injustice rather than just the aftermath of it and to contribute ideas borne of their experiences with environmental crisis, including the crisis of climate change.

EJ activists in North America increasingly have formed coalitions with other groups equally affected by industrial and extractive projects to promote the value of their perspectives. "The Convening of Indigenous Peoples for the Healing of Mother Earth" conference attended by more than 200 leaders of Indigenous communities in Canada, the United States, and Mexico, which hosted the conference, reflected this effort by Indigenous peoples to unite across national boundaries to address the global crisis of climate change. Conference organizers released several statements that chronicled the extent to which Indigenous peoples bore the brunt of extractive industries that were altering the already fragile ecosystems of their territories. One statement declared that natural resource extraction "has left in its wake a legacy of contamination, waste and loss of life. Indigenous peoples are facing the negative impacts of pollution, mining, deforestation, logging, oil prospecting, dumping of toxic waste, genetic engineering, fertilizers and pesticides, and soil erosion, all of which contribute to a severe loss of biodiversity. All of these threaten food security, subsistence lifestyles, human health and our ability to sustain our peoples."[50]

Given this history, EJ activists have promoted an alternative vision of industrial society that advances sustainability for all people. The movement, according to EJ scholar David Pellow, "sought to openly integrate campaigns for justice on behalf of vulnerable human beings with the goal of environmental sustainability. From the movement's early days, activists sought environmental justice not only through shutting down polluting facilities, but also by demanding and creating access to parks and green space and affordable, healthy foods, safe neighborhoods, and for climate-related policies and practices that are socially just and ecologically sustainable." Pellow writes that EJ activists have thus contributed "a transformative vision of what an environmentally and socially just and sustainable future might look like, at the local, regional, national, and global scales."[51]

EJ activists have been on the front lines of several campaigns that both highlight the ways in which BIPOC communities suffer from energy development but also champion clean energy that can address a climate crisis that affects everyone. For example, the Idle No More movement organized by First Nations women in 2012 helped to link U.S. and Canadian Indigenous activists fighting fossil fuel development.[52] In turn, Canadian Indigenous groups crossed the border to join the resistance to the DAPL, which was initiated by women from the Standing Rock Sioux Tribe concerned about the pipeline's potential impact on the tribe's sacred sites and water supply from Lake Oahe (see Figure I.2). After the DAPL protests began in April 2016, Native and non-Native activists joined the resistance, speaking to the tremendous opposition to expanding fossil fuel infrastructure at a time of heightened awareness of climate change.[53]

Figure I.2 Hundreds of people marched on September 4, 2016, in protest of the Dakota Access Pipeline construction project and in support of Standing Rock Sioux sovereignty and sacred water and land resources. Photo by: Emma Cassidy, Survival Media Agency.

Although dominated by Indigenous activists, the Standing Rock protests demonstrated the ways in which EJ groups have supported each other. The Standing Rock Sioux activists were joined by Indigenous leaders from Central America, Palestinian youth, and Black activists. The Black Lives Matter organization issued this statement: "Environmental racism is not limited to pipelines on Indigenous land because we know that the chemicals used for fracking and the materials used to build pipelines are also used in water containment and sanitation plants in Black communities like Flint, Michigan. These same companies that build pipelines are the same companies that build factories that emit carcinogenic chemicals into Black communities."[54]

Standing Rock also became linked with the CJ movement that seeks to "keep" or "leave" fossil fuels in the ground, not only to prevent contamination of land, air, and water resources but also to address the climate crisis fossil fuels have caused. Dozens of EJ groups in Canada, Mexico, and the United States – including Indigenous Environmental Network, Freshwater Action Network Mexico, and Canadian Youth Climate Coalition – joined hundreds of other EJ groups around the world in the #Keep It in the Ground coalition that actors and activists formed to confront climate change.[55] And they sanctioned the notion of a "just transition" which speaks to the embrace of "a lower carbon economy that recognizes the trade-offs between ... competing needs and priorities (such as energy policy in the developing world) and seeks to address them in an equitable manner."[56]

In documenting the EJ struggles and stories of people throughout North America – working-class, poor, and BIPOC communities – the authors of the 11 chapters that follow showcase the varied approaches, methodologies, and subjects of EJ scholarship, which intersects with related fields such as gender and sexuality studies, food security and sovereignty, and reproductive justice; EJ scholars and activists increasingly focus on the impact of environmental injustices on human bodies, viewing them as "homes" or "environments" threatened by pollutants that impact reproduction and thus community survival.[57]

This theme of survival is reflected in the drafting of the 17 Principles of EJ at the 1991 People of Color Leadership Summit: "For people of color, environmental justice is not an abstract, impersonal concern; it is a here and now, life and death situation with a very human face. The degradation of the environment for people of color translates not merely into personal discomfort or aesthetic distortion but into diminished income, deterioration of health, mangled bodies, early death."[58] In short, the scholarship in this volume seeks to broaden our understanding of environment as a place, environmentalism as a social movement, and EJ as the foundation of a sustainable and just world.

Part I: Race, Place, and Environmental Justice in the United States

Chapter 1

Rob Gioielli, a professor of History at the University of Cincinnati-Blue Ash Honors College, specializes in urban environmental history. In his chapter, entitled "Urban Environmental Justice Movements in the United States," Dr. Gioielli begins the volume with an important overview of the social, political, and environmental factors shaping the evolution of the modern EJ movement. He contends that during the post-war period of the 1950s and 1960s, urban environmental activists addressed a series of crises triggered by urban renewal projects, "white flight," and highway construction, which exacerbated racial injustices and poverty conditions for marginalized urban constituencies across the country. Dr. Gioielli offers several short but instructive case studies of how activists in cities such as New York, St. Louis, and Seattle engaged federal anti-poverty initiatives to advance projects focused on eliminating environmental hazards such as lead paint poisoning, poor sanitation, and substandard housing. Dr. Gioielli also cites efforts by environmental activists to make these issues part of the broader national environmental movement that privileged non-human concerns. This chapter sets the stage for understanding the burgeoning EJ movement that emerged in the early 1980s.

Chapter 2

In the chapter entitled "Resilience at the Periphery: North America's Non-Urban Environmental Justice Movements," Dr. Elizabeth Grennan Browning provides a range of geographical, social, and environmental contexts for understanding the ways in which extractive and waste industries have targeted and exploited people of color and poor communities throughout rural North America. Dr. Browning, an assistant professor of History at the University of Oklahoma, first frames her subject with an overview of the social, political, and ecological challenges faced by people of color, working-class, and poor communities before exploring these challenges in a series of short case studies. She starts by examining the 1982 campaign against PCB dumping in rural Warren County, North Carolina, a seminal moment in EJ activism. She then explores the environmental and political challenges Latinx farmworkers have faced in the Central Valley of California, the world's most productive agricultural region, before shifting the story to Cancer Alley, Louisiana, where much of the nation's chemical production has long threatened the health and livelihoods of predominately Black communities and workers. She then documents the experiences of Native Americans and Canadian First

Nations' peoples contending with uranium, radium, and gold mining projects and oil and gas development before offering a short coda on the environmental dimensions of the rural carceral state, tracing the story from the internment of Japanese-Americans during World War II to the current prison-industrial complex.

Chapter 3

A professor of Geography and Native American & Indigenous Studies at Evergreen State College, Dr. Zoltán Grossman pioneered the study of interracial and interclass environmental coalitions in the United States. In his chapter, entitled "Intercultural Alliances," Dr. Grossman explores the environmental, social, cultural, and geographical issues that have brought previously hostile groups together to contest environmental crises stemming from dams, oil pipelines, gold mining, and other industrial projects that affected both Native and non-Native farmers, ranchers, and homeowners. Dr. Grossman's research draws on textual sources and regional histories as well as interviews with leaders of both Native tribal governments and EJ groups and representatives of non-Native sportfishing groups and environmental organizations which crossed cultural lines to protect common spaces. His four case studies highlight the extent to which EJ studies need to include a wide range of constituencies, from Indigenous peoples to white "settler communities" which have been used by corporate and political interests to sow racial discord and discourage collective action.

Part II: Indigenous Movements and Environmental Justice in the United States, Canada, Mexico, and the Caribbean

Chapter 4

Dr. Kyle Kajihiro offers a compelling account of important EJ issues in Hawai'i and U.S.-affiliated Pacific Islands such as Guam and the Marshall Islands. Kajihiro's chapter, entitled "Environmental Justice in Hawai'i and Oceania," focuses on the environmental degradation caused by industrial agriculture and the toxic legacy of the U.S. military's use of Hawaiian land and water for naval bases and operations, which have degraded Kānaka Maoli's (Native Hawaiians) cultural sites, fishing and farming, and ecosystems. Kajihiro, a professor of Ethnic Studies and Geography and the Environment at the University of Hawai'i at Mānoa, explores the rise of Hawaiian EJ groups such as Protect Kaho'olawe 'Ohana (PKO), which formed in 1976 to oppose U.S. naval exercises and to revitalize Kānaka Maoli's cultural practices. The PKO articulated the key ethical principle of "aloha 'āina" that animates Hawaiian EJ activism, as it reflects the Kānaka Maoli's love of Hawai'i's land, waters, and people and their obligation to care for them as family. Kajihiro also documents the efforts of Indigenous residents of the Marshall Islands to contend with the slow eco-violence caused by U.S. nuclear testing, military exercises, and rising sea levels triggered by climate change.

Chapter 5

In the chapter entitled "Alaska Native Environmental Activism," Dr. Holly Guise (Iñupiaq) draws on a range of primary sources to explore how Alaska Native elders, artists, activists, and journalists from the Indigenous-run *Tundra Times* newspaper generated an EJ movement to protect land rights, food sovereignty, and cultural traditions undermined by colonial policies and corporate resource extraction projects in the 20th and 21st centuries. An assistant professor of History at the University of New Mexico and a scholar of Alaska Natives' experiences during World War II, Dr. Guise uses several important case studies of Alaska Natives' resilience and resistance to dam construction, nuclear weapons testing and waste disposal, devastating oil spills such as the 1987 Exxon Valdez disaster, invasive oil pipelines such as the Pebble Mine project, as well as efforts by non-Native environmental organizations to erode their food sovereignty by restricting traditional hunting practices. She closes her chapter by assessing how climate change has brought new challenges to Alaska Native communities in the 21st century.

Chapter 6

Dr. Lydia Schoeppner begins her chapter, entitled "Indigenous Peoples in Canada and Beyond: The Inuit Circumpolar Council's climate change work," with a short survey of Canadian First Nations and Metis EJ issues before detailing the important work of the Inuit Circumpolar Council (ICC), which has been a leading Indigenous organization calling for national and international action on climate change since 1977. Dr. Schoeppner, a Faculty Fellow in Conflict Resolution Studies at Canadian Mennonite University in Winnipeg, documents the extraordinary efforts of Inuit people in Canada and Alaska to gain the attention of global leaders by chronicling the ways in which rising temperatures and persistent organic pollutants in the Arctic have eroded the ecosystem on which their culture and sustenance depend. The ICC sought to influence policy-making by attending global environmental forums such as the UN Conference on Environment and Development held in Rio De Janeiro in 1992 as well as organizing regional forums such as the Indigenous Peoples' Global Summit on Climate Change held in Alaska in April 2009; the ICC also staged dramatic events such as their Earth Day 2005 event during which actors Salma Hayek and Jake Gyllenhaal joined one thousand Inuit to form the message "Arctic Warning" in the snow to draw attention to both the crisis they faced in their homelands and its implications for the rest of the world. This account provides readers with essential information about the efforts of Indigenous peoples in Canada and Alaska to provide their perspectives and traditional ecological knowledge to help address the growing environmental injustice of climate change.

Chapter 7

In his chapter "Ecocide, Ethnic Rights, and Extractivism: Struggles for Environmental Justice in Mexico," Alessandro Morosin explores a series of case studies of Mexican EJ campaigns, drawing on and expanding his current study of EJ campaigns in Oaxaca, Mexico. Morosin, an assistant professor of Sociology and Criminology at the University of La Verne, broadens the chronological and geographical scope of this volume to include the oldest group of Indigenous peoples who have confronted the ecological and social violence of settler colonialism. Morosin examines EJ struggles in four disparate areas of Mexico using the analytical lenses of ecocide, ethnic rights, and extractivism, which has perpetuated natural resource exploitation that began during the colonial era. He traces the history of EJ groups that formed to counter not only foreign corporate extraction projects such as gold mining, but also corruption in the Mexican government and the violence of drug gangs which have invaded vulnerable rural communities. Morosin documents the incredible courage displayed by Mexican EJ activists in the face of cartel violence, government privatization of Indigenous lands, and transnational finance capital that in combination has threatened traditional cultural values, land-use practices, and political autonomy.

Chapter 8

In his chapter, entitled "Plundered Paradise: The Puerto Rican Struggle against Environmental Colonialism," A.J. Hudson shines a light on EJ campaigns in the Caribbean. A graduate of Yale University's School of the Environment, Hudson offers a historical overview of Puerto Rican environmental crises, starting with the creation of U.S.-owned sugar plantations in the early 1900s. He then examines the impact of post-WWII industrialization programs that generated a range of health problems, the ecological disasters and demographic displacement spawned by the U.S. Navy's use of the island of Vieques as a live firing range and munitions storage site, and the role of tourism in restricting Puerto Ricans' access to the island's recreational and economic resources. Hudson's previous work as a CJ Youth Organizer in Brooklyn and stint with the Climate Resilience Research project in Puerto Rico after Hurricane Maria devastated the island in 2018 provides context for Puerto Ricans' EJ struggles in both Puerto Rico and the United States.

Part III: Environmental Justice, Climate Justice, and Sustainability

Chapter 9

Kyle Whyte (Potawatomi), a member of the faculty of the School for Environment and Sustainability at the University of Michigan, draws on his extensive advocacy for Indigenous justice campaigns on campus and off, which includes his work on the Energy Equity Project and on the White House EJ Advisory Council. In a chapter entitled "Indigenous Environmental Justice, Renewable Energy Transition, and the Infrastructure of Sovereignty," Whyte documents how Indigenous leaders have worked to address both the historical legacies of environmental injustices of settler colonialism, especially land appropriation and invasive resource extraction, and their contemporary efforts to promote CJ in the critical areas of energy equity, renewable energy, and infrastructure development. Whyte frames the concept of *Infrastructure* as incorporating the multiple ways in which Native people confront the local, national, and global challenges of climate change to create sustainable communities that draw on Native traditions, practices, and institutional histories. Indigenous EJ necessitates government and corporate recognition of Indigenous peoples' self-determination, consent, and self-governance in determining how their resources will be utilized in the 21st century.

Chapter 10

Food justice and food sovereignty are especially relevant to farmworkers, restaurant workers, and BIPOC communities, many of which struggle to gain access to affordable and healthy food, especially for residents of rural Indian reservations and nearly 20 million people living in urban "food deserts" or "food apartheid." In his chapter, entitled "The Food Justice Movement," Dr. Justin Myers, an associate professor of Sociology at California State University-Fresno, adds an important dimension to the EJ story by examining the historical background of food justice movement activists connected by their common experience with racial and class inequities. Myers explores several examples of food justice activism, such as East New York Farms! (ENYF!), which has worked to purchase land for urban gardeners to provide affordable and culturally appropriate food for the community, and the Coalition of Immokalee Workers (CIW), which organized work stoppages, a hunger strike, and media awareness campaigns to address what its workers called "modern-day slavery" in Florida's huge agricultural operations. Myers also considers the story of fast-food workers who organized the Fight for $15 movement (FF15) to address structural poverty after toiling for decades on a minimum wage. The food justice movement fuses sustainability and social justice, working to create *just sustainabilities* for all people in the food system.

Chapter 11

The volume closes with the energetic and impassioned voices of young people in North America, who, because of their age, especially feel the urgency of the climate crisis. They have grown up in its shadows, protested political denial and inaction, and lived through one of its many manifestations such as Superstorm Sandy in New Jersey, a wildfire in California, or a heatwave in British Columbia. In a chapter entitled "'We Are Missing Our Lessons to Teach You One': Youth Activists on the Frontlines of Climate Justice," Dr. Jerusha Conner, professor of Education and Counseling at Villanova University, draws on her extensive research of youth climate activism to present a compelling history of young people from across North America fighting for CJ by forming environmental organizations in high schools and colleges such as the Sunrise Movement and 350.org, waging climate strikes and other protests, and offering their perspectives on the future in speeches, editorials, and social media. Identifying four main types and five central features of youth-led CJ activism, Dr. Conner traces its rise from the 1990s to its expansion in the 2010s and then its campaigns in the 2020s, which continued to grow despite the COVID-19 pandemic.

Notes

1 Andrew R. Highsmith, *Demolition Means Progress: Flint, Michigan, and the Fate of the American Metropolis* (Chicago, IL: University of Chicago press 2015), 1.
2 Julian Agyeman et al., eds., *Speaking for Ourselves: Environmental Justice in Canada* (Vancouver: UBC Press, 2009), 1–2.
3 For an excellent summary of Chester's environmental justice campaigns, see Luke W. Cole and Sheila R. Foster, *From the Ground Up: Environmental Racism and the Rise of the Environmental Justice Movement* (New York: New York University Press, 2000), Ch. 2.
4 *Laid to Waste: A Chester Neighborhood Fights for its Future,* https://www.youtube.com/watch?v=bfPQVpQ2kbg
5 "Learn About Environmental Justice," https://drrobertbullard.com/learn-about-environmental-justice/ Accessed August 22, 2022.

6 See Sonja Avlijaš, "Security for Whom? Inequality and Human Dignity in Times of the Pandemic," in *Pandemics, Politics, and Society: Critical Perspectives on the Covid-19 Crisis*, ed. Gerard Delanty (Berlin; Boston, MA: De Gruyter, 2021), 227–42.

7 Julie Sze, *Environmental Justice in a Moment of Danger* (Berkeley: University of California Press, 2020), 16–17.

8 I use the term Latinx throughout this essay, rather than Latino, because it is more inclusive of all genders. But the term is controversial; some activists prefer the term Latine. On the controversy, see Sarah Maslin Nir, "Some Republicans want to ban 'Latinx.' These Latino Democrats Agree." *The New York Times*, March 1, 2023. https://www.nytimes.com/2023/03/01/nyregion/connecticut-arkansas-latinx.html

9 "Environmental Racism: The Uneven Distribution of Risk," *PR Central*, May 2, 1998; http://www.hartford-hwp.com/archives/45/293.html see also https://www.appalachianhistory.net/2019/03/worst-industrial-tragedy-in-wv-history.html

10 Kathy Helms, "Victims of Nuclear Fallout Tell Their Stories," *Gallup Independent*, May 21, 2004.

11 Jace Weaver, ed., *Defending Mother Earth: Native American Perspectives on Environmental Justice* (Maryknoll, NY: Orbis Books, 1996), 47–48.

12 The people and the land of the Laguna Pueblo in New Mexico were left with the legacies of the Anaconda Copper Company's uranium mine, the largest in the world at one point. Nearly 100 million gallons of radioactive water flowed into both Navajo and non-Navajo environments from a 1979 accident near Church Rock, New Mexico. And 3.5 million pounds of radioactive tailings were found on the Lakota Reservation in South Dakota. See Donald Grinde and Bruce Johansen, *Ecocide of Native America: Environmental Destruction of Indian Lands and Peoples* (Santa Fe, NM: Clear Light Publishers, 1995), 211–13. On uranium mining in Native American communities, see Traci Brynne Voyles, *Wastelanding: Legacies of Uranium Mining in Navajo Country* (Minneapolis: University of Minnesota Press, 2015); and Peter Eichstaedt, *If You Poison Us: Uranium and Native Americans* (Sante Fe, NM: Red Crane Books, 1994). For interviews of Navajo miners, see "Memories Come to Us in the Rain and the Wind," http://www.inmotionmagazine.com/brugge.html

13 "'Black Survival in Our Polluted Cities,' 1970," in *Environmental Justice in Postwar America: A Documentary Reader*, ed. Christopher W. Wells (Seattle: University of Washington Press, 2018), 99.

14 On the case, see http://nationalhumanitiescenter.org/tserve/nattrans/ntuseland/essays/envjust.htm and https://www.sciencefriday.com/articles/robert-bullard-environmental-justice/

15 "Growers Spurn Negotiations on Poisons," *El Malcriado: The Voice of the Farm Worker*, January 1969. In *Environmental Justice in Postwar America*, 97–98.

16 On Love Canal, see Richard S Newman, *Love Canal: A Toxic History from Colonial Times to the Present* (New York: Oxford University Press, 2016); and Lois Marie Gibbs, *Love Canal and the Birth of the Environmental Health Movement* (Washington, DC: Island Press, 2010). On the neglect of Black residents' interests during the Love Canal crisis, see Elizabeth D. Blum, *Love Canal Revisited: Race, Class, and Gender in Environmental Activism* (Lawrence: University Press of Kansas, 2008).

17 Randolph Haluza-DeLay, Pat O'Riley, Peter cole, and Julian Agyeman, "Introduction: Speaking for Ourselves, Speaking Together: Environmental Justice in Canada," in *Speaking for Ourselves*, 15–21. Scholars of Canadian environmental justice campaigns have also recognized environmental injustices occurring in urban spaces in addition to rural landscapes and riverscapes occupied principally by Indigenous peoples. In Toronto, for example, people of color (especially immigrants from South Asia and China) and low-income residents face lead poisoning, substandard housing, and polluted soil and air from industrial plants and incinerator projects. Citizens for a safe environment formed in the South Riverdale area of Toronto after city officials proposed a waste incinerator. Roger Keil, Melissa Ollevier, and Erica Tsang, "Why Is There No Environmental Justice in Toronto? Or Is There?," in *Speaking for Ourselves*, 67–68.

18 Chief Bernard Ominayak, with Kevin Thomas, "These Are Lubicon Lands: A First Nation Forced to Step into the Regulatory Gap," in *Speaking for Ourselves*, 111, 116–17.

19 Josiah Rector, "The Spirit of Black Lake: Full Employment, Civil Rights, and the Forgotten Early History of Environmental Justice," *Modern American History* 1 no. 1 (2018): 45–66. For a history of EJ in Detroit, see Josiah Rector, *Toxic Debt: An Environmental Justice History of Detroit* (Chapel Hill: University of North Carolina Press, 2022).

20 "A Warren County PCB Protest Song" (1982), in *Environmental Justice in Postwar America*,132. Said one protestor, "It's one thing to be poor, it's another to be poor and poisoned." Quoted in *Environmental Justice in Postwar America*, 3.

21 Prominent EJ scholar Julie Sze argues that "at the core of the term *environmental justice* is a redefinition of 'the environment' to mean not only 'wild' places, but the environment of human bodies, especially in racialized communities." Sze, *Environmental Justice in a Moment of Danger*, 22.

22 "Political Difficulties Facing Waste-to-Energy Conversion Plant Siting, 1984, in *Environmental Justice in Postwar America*, 141.

23 Quoted in Cole and Foster, *From the Ground Up*, 3. Chemical Waste Management, Inc. (Chem Waste) had operated one of the largest toxic waste dumps in the country since the late 1970s and in 1988 proposed creating a new toxic waste incinerator at the Kettleman City dump, which would have led to thousands of trucks carrying toxic waste through the community.

24 On Buttonwillow, California, another predominately Latinx community targeted for toxic waste dumping, see *From the Ground Up*, Ch. 4.

25 U.S. General Accounting Office, *Siting of Hazardous Waste Landfills and Their Correlation with Racial and Economic Status of Surrounding Communities* (Washington, DC: Government Printing Office, 1983). https://www.gao.gov/assets/rced-83-168.pdf

26 United Church of Christ, *Toxic Wastes and Race in the United States: A National Report on the Racial and Socio-Economic Characteristics of Communities with Hazardous Waste Sites* (New York: Public Data Access, 1987), xiv; it is accessible at http://uccfiles.com/pdf/ToxicWastes&Race.pdf The United Church of Christ conducted a separate study 20 years later, reaching similar conclusions. See United Church of Christ, "Toxic Wastes and Race at Twenty, 1987–2007," https://www.ucc.org/what-we-do/justice-local-church-ministries/justice/faithful-action-ministries/environmental-justice/environmental-ministries_toxic-waste-20/; for 12 case studies of people of color and low-income "fenceline"

communities that are next to polluting companies and government facilities, see Steve Lerner, *Sacrifice Zones: The Front Lines of Toxic Chemical Exposure in the United States* (Cambridge, MA: The MIT Press, 2012).

27 Jeff Chang and Lucia Hwang, "It's a Survival Issue: The Environmental Justice Movement Faces the New Century," *ColorLines* 3, no. 2 (Summer 2000). http://www.hartford-hwp.com/archives/45/290.html; Moore served as executive director of Albuquerque's Southwest Network for Environmental and Economic Justice from 1993 to 2010.

28 Chang and Hwang, "It's a Survival Issue."

29 The representatives included 54 Native Americans, 158 African Americans, 62 Latinx, 24 Asian Americans and Pacific Islanders, 5 Hawaiians, 3 Alaska Natives, 154 men, and 144 women attended. Delegates included Dollie Burwell of Warren County (N.C.) Concerned Citizens, Peggy Shepard of West Harlem Environmental Action, Robert Bullard of the University of California at Riverside, Maurice Sampson of the Urban Recycling Institute of Philadelphia, Dolores Huerta, First Vice President of the United Farmworkers Union, and Pat Bryant, Executive Director of the Gulf Coast Tenants Union, which addressed the environmental injustices of cancer alley in Louisiana. Representatives of the Sierra Club, National Resources Defense Council, and Greenpeace also attended.

30 Southwest Organizing Group, "Letter to Big Ten Environmental Groups," in *Environmental Justice in Postwar America*, 137, 141. The recipients of the SWOP letter included the Sierra Club, the National Audubon Society, Friends of the Earth, EarthJustice, the National Wildlife Federation, the Wilderness Society, the National Parks Conservation Association, the Environmental Defense Fund, the Natural Resources Defense Council, and the Izaak Walton League. For an assessment of the letter's impact, see Marty Durlin, "The Shot Heard Round the West: What Resulted from Activists' 1990 Challenge to the Big Greens," *High Country News*, February 1, 2010. https://www.hcn.org/issues/42.2/the-shot-heard-round-the-west; and Marty Durlin, "The Group of Ten Respond: the Big Greens Grade Themselves," *High Country News*, February 1, 2010. https://www.hcn.org/issues/42.2/the-group-of-10-responds

31 "Proceedings: The First National People of Color Environmental Leadership Summit" (hereafter Proceedings), Charles Lee, ed. xvii. Center for Southwest Research University of New Mexico, Toney Anaya Papers 575 Box 35, Folder 22.

32 "Proceedings," 7, 8.

33 "Proceedings," 9.

34 "Proceedings," 26.

35 "Proceedings," 58.

36 "Proceedings," 62. Hawaiian delegates supported the change of wording from Native Americans to Native peoples to include all Indigenous peoples beyond those of Indian Country. Chavis explained that the four main ethnic groups consulted in advance of the summit – Black, Native, "Chicano," and "Asian-Pacific" – agreed to use "people of color" as a starting point.

37 Chang and Hwang, "It's a Survival Issue."

38 Chang and Hwang, "It's a Survival Issue."

39 Dana Alston, "The Summit: Transforming a Movement," *Race, Poverty & the Environment* 17, no. 1 (Spring 2010): 17.

40 Alston, "The Summit," 14–17.

41 Iroquois activists said attending the Rio conference, which included the World Conference of Indigenous Peoples, led to "the modern awakening of Iroquois

people to environmental consciousness" and to the creation of the Haudeno-saunee Environmental Task Force (HEFT), the goals of which were to encourage activities preventing local pollution and to improve the skills of Iroquois people in conducting scientific research and testing for toxicants. Barbara Graymont, *The Iroquois* (New York: Chelsea House, 2005), 120. Representatives of the Inuit Circumpolar Council (ICC), a leading Indigenous organization calling for national and international action on climate change since 1977, also sought to influence policy-making at the Rio conference.

42 Grace Thorpe, "No Nuclear Waste on Indian Land," 163; http://www.ratical.org/radiation/WorldUraniumHearing/LaurieGoodman.txt

43 https://www.archives.gov/files/federal-register/executive-orders/pdf/12898.pdf

44 United States Environmental Protection Agency, "'Learn About Environmental Justice,' Overviews and Factsheets," 2021, https://www.epa.gov/environmentaljustice/learn-about-environmental-justice

45 In 2004, the EPA acknowledged that it had failed to address environmental justice in its decision-making. https://www.epa.gov/sites/default/files/2015-12/documents/20040301-2004-p-00007.pdf; in 2009, President Barack Obama revived the office's original mission and put renewed focus on affected communities. Donald Trump tried to eliminate the office during his presidency.

46 "Environmental Justice for People of Color Summit draws 1,200 delegates to Washington," *The Black Commentator*, Issue # 16, November 2002. https://blackcommentator.com/16_re_print_pr.html; a prominent environmental justice scholar who attended the first Summit in 1991, Pena served as a professor at the University of Washington. For the outcome of the Second People of Color Environmental Leadership Summit, see the "Principles of Working Together," https://www.ejnet.org/ej/workingtogether.pdf; the website Environmental Justice/Environmental Racism is an excellent resource for primary documents. See https://www.ejnet.org/ej/

47 For example, the University of Michigan offered a master's and doctoral degree in environmental justice in 2002. "Environmental Justice for People of Color Summit draws 1,200 delegates to Washington."

48 Maria Sháa Tláa Williams, ed., "Environmental Justice for People of Color Summit draws 1,200 delegates to Washington."

49 "Proceedings." Appendix II: People of Color and the Struggle for Environmental Movement: "Position Paper by Robert Bullard and Dana Alston on behalf of the National Planning Committee for the First National People of Color Environmental Leadership Summit," 212.

50 "The Convening of Indigenous Peoples for the Healing of Mother Earth," March 2008, Palenque, Mexico. U.S. Committee on Senate Affairs, "Setting the Standard: Domestic Policy Implications of the UN Declaration on the Rights of Indigenous Peoples," June 9, 2011 (Washington, D.C.: Government Printing Office, 2011), 58.

51 David Naguib Pellow, *What is Critical Environmental Justice?* (Cambridge: Polity Press, 2018), 18.

52 On the Idle No More movement, see Febna Caven, *Cultural Survival Quarterly*, February 2013 https://www.culturalsurvival.org/publications/cultural-survival-quarterly/being-idle-no-more-women-behind-movement; for a study of both Canadian and American contexts, see *Environmental Racism in the United States and Canada: Seeking Justice and Sustainability*, ed. Bruce Johansen (Westport, CT: Praeger Publishing, 2020).

53 Although the pipeline would not traverse the Standing Rock Sioux Tribe's reservation, it would wind its way through land situated within the 1851 and 1868 treaties which the United States later abrogated. On Standing Rock and

DAPL, see especially Nick Estes and Jaskiran Dhillon, eds., *Standing with Standing Rock: Voices from the #NODAPL Movement* (Minneapolis: University of Minnesota Press, 2019); and Dina Gilio-Whitaker, *As Long as Grass Grows: The Indigenous Fight for Environmental Justice, from Colonization to Standing Rock* (Boston, MA: Beacon Press, 2019). For pre-DAPL campaigns to prevent the transfer of fossil fuels across Indian Country, see Zoltán Grossman, *Unlikely Alliances: Native Nations and White Communities Join to Defend Rural Lands* (Seattle: University of Washington Press, 2017), Ch. 5.

54 Quoted in Sze, *Environmental Justice in a Moment of Danger*, 43.
55 http://keepitintheground.org/
56 Quoted in Sze, *Environmental Justice in a Moment of Danger*, 45–46.
57 On Food sovereignty, see Kyle Powys Whyte, "Food Sovereignty, Justice, and Indigenous Peoples: An Essay on Settler Colonialism and Collective Continuance," in *Oxford Handbook of Food Ethics*, eds. Anne Barnhill et al. (New York: Oxford University Press, 2018), 345–66. See Rachel Stein, ed., *New Perspectives on Environmental Justice: Gender, Sexuality, and Activism* (New Brunswick, NJ: Rutgers University Press, 2004). For an excellent review of EJ Studies, see Pellow, *What is Critical Environmental Justice?* Ch. 1.
58 "Proceedings," 54.

Bibliography

Ageyman, Julian. *Speaking for Ourselves: Environmental Justice in Canada*. Vancouver: University of British Columbia Press, 2009.

Blum, Elizabeth D. *Love Canal Revisited: Race, Class, and Gender in Environmental Activism*. Lawrence: University Press of Kansas, 2008.

Bohme, Susanna Rankin. *Toxic Injustice: A Transnational History of Exposure and Struggle*. Berkeley: University of California Press, 2014.

Brown, Phil. *No Safe Place: Toxic Waste, Leukemia, and Community Action*. Berkeley: University of California Press, 1997.

Bullard, Robert D. *The Quest for Environmental Justice: Human Rights and the Politics of Pollution*. San Francisco: Sierra Club Books, 2005.

_____. *Dumping in Dixie: Race, Class, and Environmental Quality*. Milton Park: Routledge, 2018.

_____, and Beverly Wright. *Race, Place, and Environmental Justice after Hurricane Katrina*. Boulder, CO: Westview Press, 2009.

Cantzler, Julia Miller. *Environmental Justice as Decolonization: Political Contention, Innovation and Resistance over Indigenous Fishing Rights in Australia, New Zealand, and the United States*. Milton Park: Routledge, 2020.

Cole, Luke W., and Sheila R. Foster. *From the Ground Up: Environmental Racism and the Rise of the Environmental Justice Movement*. New York: New York University Press, 2000.

Daniel, Pete. *Toxic Drift: Pesticides and Health in the Post-World War II South*. Baton Rouge, LA: LSU Press, 2007.

Edenhofer, Ottmar, ed. *Climate Change, Justice, and Sustainability: Linking Climate and Development Policy*. New York: Springer, 2012.

Estes, Nick. *Our History Is the Future: Standing Rock Versus the Dakota Access Pipeline, and the Long Tradition of Indigenous Resistance*. New York: Verso, 2019.

Faber, Daniel J., ed. *The Struggle for Ecological Democracy: Environmental Justice Movements in the United States*. New York: Guilford Press, 1998.

Gilio-Whitaker, Dina. *As Long as Grass Grows: The Indigenous Fight for Environmental Justice, from Colonization to Standing Rock*. Boston, MA: Beacon Press, 2019.

Gioielli, Robert. *Environmental Activism and the Urban Crisis: Baltimore, St. Louis, Chicago*. Philadelphia, PA: Temple University Press, 2014.

Grijalva, James M. *Closing the Circle: Environmental Justice in Indian Country*. Durham: Carolina Press, 2008.

Grossman, Zoltán, and Alan Parker, eds. *Asserting Native Resilience: Pacific Rim Indigenous Nations Face the Climate Crisis*. Corvalis: Oregon State University Press, 2012.

Hurley, Andrew. *Environmental Inequalities: Class, Race, and Industrial Pollution in Gary, Indiana, 1945–1980*. Chapel Hill: The University of North Carolina Press, 1995.

Johansen, Bruce E. *Environmental Racism in the United States and Canada: Seeking Justice and Sustainability*. Westport, CT: Praeger, 2020.

LaDuke, Winona. *All Our Relations: Native Struggles for Land and Life*. Cambridge, MA: South End Press, 1999.

_____, and Sean Aaron Cruz. *The Winona LaDuke Chronicles: Stories from the Front Lines in the Battle for Environmental Justice*. Ponsford, MN: Spotted Horse Press, 2016.

Lerner, Steve. *Diamond: A Struggle for Environmental Justice in Louisiana's Chemical Corridor*. Cambridge, MA: MIT Press, 2006.

Maldonado, et al. *Climate Change and Indigenous Peoples in the United States: Impacts, Experiences and Actions*. New York: Springer Verlag, 2014.

Mascarenhas, Michael. *Where the Waters Divide: Neoliberalism, White Privilege, and Environmental Racism in Canada*. Lanham, MD: Lexington Books, 2012.

_____, ed. *Lessons in Environmental Justice: From Civil Rights to Black Lives Matter and Idle No More*. Thousand Oaks, CA: Sage Publishing, 2020.

McGurty, Eileen. *Transforming Environmentalism: Warren County, PCBs, and the Origins of Environmental Justice*. New Brunswick, NJ: Rutgers University Press, 2007.

Newman, Richard S. *Love Canal: A Toxic History from Colonial Times to the Present*. New York: Oxford University Press, 2016.

Pasternak, Judy. *Yellow Dirt: A Poisoned Land and the Betrayal of the Navajos*. New York: Free Press, 2011.

Pauli, Benjamin. *Flint Fights Back: Environmental Justice and Democracy in the Flint Water Crisis*. Cambridge, MA: MIT Press, 2019.

Pellow, David Naguib. *What Is Critical Environmental Justice?* Cambridge: Polity Press, 2017.

Rector, Josiah. "The Spirit of Black Lake: Full Employment, Civil Rights, and the Forgotten Early History of Environmental Justice." *Modern American History* 1, no. 1 (2018): 45–66.

_____. *-Toxic Debt: An Environmental Justice History of Detroit*. Chapel Hill: University of North Carolina Press, 2022.

Schlosberg, David. *Defining Environmental Justice: Theories, Movements, and Nature*. New York: Oxford University Press, 2020.

Spears, Ellen Griffith. *Baptized in PCBs: Race, Pollution, and Justice in an All-American Town*. Chapel Hill: The University of North Carolina Press, 2014.

Szasz, Andrew. *Ecopopulism: Toxic Waste and the Movement for Environmental Justice*. Vol. 1. Minneapolis: University of Minnesota Press, 1994.

Sze, Julie. *Sustainability: Approaches to Environmental Justice and Social Power*. New York: New York University Press, 2018.

_____. *Environmental Justice in a Moment of Danger*. Berkeley: University of California Press, 2020.

_____, Joanna Robinson, and Robert Gottlieb. *Noxious New York: The Racial Politics of Urban Health and Environmental Justice*. Cambridge, MA: MIT Press, 2006.

Taylor, Dorceta E. *Toxic Communities: Environmental Racism, Industrial Pollution, and Residential Mobility*. New York: New York University Press, 2014.

_____, and Ted I. K. Young. *Environment and Social Justice: An International Perspective*. Bingley: Emerald Publishing Limited, 2010.

Tompkins, Adam. *Ghostworkers and Greens: The Cooperative Campaigns of Farmworkers and Environmentalists for Pesticide Reform*. Ithaca, NY: Cornell University Press, 2016.

Voyles, Traci Brynne. *Wastelanding: Legacies of Uranium Mining in Navajo Country*. 1st ed. Minneapolis: University of Minnesota Press, 2015.

Waldron, Ingrid R. G. *There's Something in the Water: Environmental Racism in Indigenous & Black Communities*. Winnipeg: Fernwood Publishers, 2018.

Walker, Gordon. *The Routledge Handbook of Environmental Justice*. Milton Park: Routledge, 2017.

Washington, Sylvia, Heather Goodall, and Paul C. Rosier. *Echoes from the Poisoned Well: Global Memories of Environmental Injustice*. Lanham, MD: Lexington Books, 2006.

Wells, Christopher W. *Environmental Justice in Postwar America: A Documentary Reader*. Seattle: University of Washington Press, 2018.

Zimring, Carl A. *Clean and White: A History of Environmental Racism in the United States*. New York: New York University Press, 2015.

Part I

Race, Place and Environmental Justice in the United States

1 Urban Environmental Justice Movements in the United States

Rob Gioielli

In April of 1970 Nathan Hare declared that not only was the current trend for ecological concern relevant to the Black community, but that "the environmental crisis of whites (in both its physical and social aspects) already pales in comparison to that of blacks." This and other observations Hare makes in his seminal essay "Black Ecology" were primarily a rejoinder and challenge to the emerging environmental movement, which he correctly observed was "blatantly omitting" the voices of the Black community and their environmental interests. But what often goes unnoticed about this essay, published in the same month as the first Earth Day, was that Hare was primarily making an urban spatial critique. "Urban blacks have been increasingly imprisoned in the physical and social decay in the hearts of major central cities, an imprisonment which most emphatically seems doomed to continue. At the same time whites have fled to the suburbs and the exurbs, separating more and more the black and white worlds."[1]

"Black Ecology" is one of the most well-known of a wave of commentary, analysis and activism that emerged in the 1960s and 1970s in response to the sharp environmental inequalities of the post-World War Two American metropolis. During this period, the massive growth of the suburbs and concurrent disinvestment in older, industrial central cities both sharpened older forms of urban environmental injustice and created new ones. Most importantly, as Hare observed, the racialized nature of these structural shifts meant that both the acute toxic dangers and long-term burdens of post-industrial metropolis fell overwhelmingly on African Americans specifically, and people of color more broadly. Activists across the country worked to address the long-term structural factors that led to these shifts, but also to alleviate the day-to-day suffering of thousands of community residents. In the process, they forged a new urban environmental movement that also challenged the class and racial assumptions of the more mainstream versions of environmentalism. This activism would lay the foundation for the larger environmental justice movement that

DOI: 10.4324/9781003214380-3

emerged a decade later, providing models for organizing and activism that would last well into the next century.

This chapter will be focusing on this particular moment in postwar activism, showing its origins in both specific material conditions of the disinvested metropolis and the broader waves of social protest during the period. It will also show the threads and connections back to earlier forms of activism in the first decades of the century and make connections to the vibrant urban environmental justice movements of the late twentieth and early twenty first centuries.

In the process of this survey, a few key themes related to the history of urban environmental justice activism, and environmental justice more broadly, will begin to emerge. The first is that like other forms of environmentalism, this is a knowledge politics. The success or failure of a local movement or protest was often predicated on the ability of activists and residents to make specific and verifiable claims that not only had they, their family or community been harmed in some way, but that it had been from a specific source, actor or agency. The state, corporations and large institutions had set the terms of environmental health and politics around measurable scientific knowledge and data that was often difficult for poor, working-class communities of color to provide. More mainstream environmental groups were often made up of doctors, lawyers and university scientists who were either experts themselves, or were well versed in the worlds of medicine, law and science, and could access important information through professional networks. They also had the funds to pay for research and expertise when necessary.

These resources were scarce in many city neighborhoods, which leads to the second major theme, the importance of organizing and institutional infrastructure. Federal War on Poverty and Model Cities programs provided significant funding for community organizers, public health experts, research, surveys and youth programs that allowed urban communities to try and understand, quantify and attempt to ameliorate environmental issues. Although these and other federal initiatives were the most important and prominent during the period, there were hundreds of similar programs initiated by local churches and religious organizations, social service agencies, civil rights groups, trade unions, universities and hospitals. Many of them worked in collaboration with federal programs, because that's where the largest amount of funding came from.

This complex network of collaborations and coalitions had about a decade long-run from the mid-1960s to mid-1970s. The New Federalism reforms initiated under the Nixon administration did not end federal funding, but shifted it to block grants and other schemes that put a lot more of the dollars in the hands of mayors and others at the top of the metropolitan power structure. This shifted the structures and tone of urban environmental activism as organizations became more dependent on foundations and new forms of non-profit infrastructure that provided less flexibility and wanted to focus more on traditional social service provision. Many organizations also faced headwinds from traditional environmental funders, which saw concerns about toxic waste dumps as purely local or "Not in My Backyard Activism" that was not connected to the larger national movement. Nevertheless, many grassroots organizations were able to mount impressive movements to address local environmental justice concerns. Two decades into the twenty-first century, as urban climate justice and racial inequality in American cities move to the forefront of the environmental movement, these organizations are finally receiving the attention (and funding) they deserve.

One of the primary reasons urban environmental justice organizations had trouble getting funding in the 1980s and 1990s, and why these forms of activism in general have, until very recently, been generally ignored by scholars of environmental activism, is because they did not look like environmental organizations, which is the third key theme of this chapter. Depending on your perspective, the urban activism of the 1960s and 1970s either looked like civil rights or public health movements, straight-ahead community organizing or some combination of all of these. And that's because it was. Almost all of these activists did not define their work as environmental activism, and their movements as environmental. This is not only because the environmental movement itself was new and emerging during the period, but also because the structure, parameters and issues that it would address were quickly being defined and institutionalized by the white, middle-class and elite, and male-led organizations.[2] Although there was a fertile period in the 1970s where more heterodox activists had space to define issues and approaches, by the 1980s "environmentalism" became defined by large national mainstream groups that were focused on bureaucratic, legalistic and reformist approaches to issues around clean air and water, forests, wildlife and wilderness conservation.

This meant that who could be an environmentalist, and what an environmental movement was, did not contain room for urban activism by poor and working-class communities of color, setting the stage for the emergence of the environmental justice movement. This leads to the last important theme, which is the historical use of the term "environmental justice." Scholars generally agree that what we call the modern environmental justice movement emerged during the early 1980s in response to concerns over inequitable waste dumping in the American South.[3] Activists at the time also forthrightly claimed that their movement was new and unique, in order to separate themselves from a "mainstream" environmental movement that they saw as elitist and exclusionary. But then around the turn of the century, a number of scholars argued that although a self-identified environmental justice "movement" may be new, it was unfair to claim that environmentalism had never been concerned about urban or labor issues. In reality, America had a long history of activism around issues related to public health and toxics that shared similar characteristics to the newly emerging environmental justice movement.[4]

More recently, proponents of this long-history perspective have been successful, developing a broader narrative arc of activism and organizing from the early twentieth century to the present.[5] The term "environmental justice" is generally used for any sort of activism that is looking to address issues of environmental inequality, especially among historically marginalized racial and ethnic communities. The focus of this chapter, activism in the 1960s and 1970s, could be seen as a "precursor" to the environmental justice movement that emerged just a few years later. Yet I am hesitant to call it that for a number of reasons. First, it marginalizes the real connections that this activism had to the emerging environmental movement, and the possibilities, however tenuous, that that movement could have had a broader focus on urban and equity issues. Second, periodization and demarcations for social movements matter. They grow and evolve, shift and change. Sometimes they are successful and disband or are institutionalized by state actors. They not only can fall victim to infighting, burnout and sclerosis but also be torn asunder by state oppression. Acknowledging those shifts takes the experience of movement actors seriously, who oftentimes saw themselves much differently than historians generations later might categorize them.

Finally, calling the period of activism in the 1960s and 1970s environmentalism, instead of environmental justice, also accomplishes a self-consciously presentist goal: We need a better environmentalism, one that is not just more cognizant and inclusive of equity and justice issues, but one that puts them front and center. This is not to call for environmental justice organizations to be subsumed by the "mainstream" environmental organizations. Far from it. But we need a more holistic definition of environmentalism that sees the preservation of non-human nature and addressing climate justice as part of one, larger movement.

The Progressive Moment

"The third and most important obstacle to reform is the slum landlord," wrote Robert Hunter in the 1900 report, *Tenement Conditions in Chicago*. Through 200 pages he described, in scientific detail, how overcrowding, poor construction and, especially, inadequate plumbing had made thousands of Chicago apartments not only unsanitary, but deadly, with high rates of infectious disease outbreaks. The responsibility for these conditions, Hunter argued, lay squarely at the feet of the powerful, particularly landlords and property owners, and the local government. The goal of the report was to educate people, so they could lobby officials for change. "It should be understood that the Committee have no desire to present a harrowing picture of the misery of the tenement-house population simply to create a sensation," Hunter wrote. The goal was to put real pressure on city government officials to enact and enforce laws, a straightforward but challenging task, especially in turn-of-the-century Chicago.[6]

This report, which was written by Hunter but the result of a detailed study conducted by a special committee of Chicago's City Homes Association, which included noted reformer Jane Addams, was similar to dozens of studies conducted across the country during this period by Progressive reformers. But what was unique about *Tenement Conditions* was not only the level of scientific detail and sophistication of survey, but also what we might call its "power analysis." In other cities, reports would attribute horrible conditions to abstract factors like "market conditions" or the rapid growth of the city. Or they would even blame the victims, arguing it was the "poor housekeeping" of recent migrants from Eastern and Southern Europe, or rural parts of the American South, that caused such decrepit conditions and high rates of disease and mortality. By linking urban environmental conditions and social power, the Chicago report and the reform campaign it inspired are, on the other hand, one of the earliest examples of urban environmental justice.[7]

The Progressive era was an important but challenging time for urban environmental activism. The thousands of reformers who made up the Progressive movement were committed to urban life, and to understanding and working to address the many problems that had been created by the rapid growth and industrialization of American cities in the late nineteenth and early twentieth centuries. But because their movement was almost completely middle class in orientation, they carried all the positives and negatives of bourgeois reformism. On the plus side, this meant a commitment to adequately researching social problems, and understanding them from a theoretically neutral perspective. They had access to wealth and resources, and the social power connected to a city's major institutions. On the negative, they brought the prejudices and predispositions of their class to their analysis and activism, which could run the gamut from soft paternalism to harsh forms of racism.

These patterns and tensions would play out during the Progressive era and would also shape policy and the arc of social movements in the decades after World War Two. One of the most important aspects of Progressivism that shaped all types of environmentalism was a commitment to scientific and technical understanding of public health and environmental hazards. Progressives had a fundamental faith in scientific knowledge, and a belief that not only could most problems be solved through a rigorous technical analysis, but that privileging expertise and experts in politics and policy was the best way to address the host of social and cultural issues that were caused by rapid industrialization and urbanization. When this focus on research and expertise was harnessed to address social inequalities, as it was by Jane Addams and others at Hull House, the results were impressive and effective.[8] But a focus on expertise also helped reinforce existing class, race and gender power structures, as white men maintained almost exclusive access to higher education in America, and shifted environmental reform to a technocratic domain. The urban smoke movement was a case in point. In the early twentieth century, dozens of grassroots groups in major industrial cities like Cincinnati and St. Louis formed to try and address the constant clouds of particulate matter from coal-fired railroad engines, factories and home furnaces. Middle-class women were the engine behind these organizations, but their voices faded into the background as the issue became the domain of a new class of almost exclusively male engineers and bureaucrats.[9]

One of the key contradictions of many of these policies is that Progressive reformers were almost always city residents themselves and were in general committed to trying to make cities vibrant, healthy and successful places for all residents. Thus, as severely flawed as some of their analysis and practices might have been, especially in terms of reinforcing and perpetuating forms of racial inequality, they did have a broadly communitarian philosophy, which is something that would gradually erode from much of the metropolitan white middle class in the decades after World War Two.

This positivist faith in expertise laid the groundwork for conflicts in the postwar era, when urban community members and grassroots organizations openly challenged engineers, doctors and bureaucrats about the harm being done to their communities by various urban renewal and highway projects, or the scale of various public health issues. But Progressivism also built the foundation for postwar urban environmental issues through the ways that urban experts conceptualized the city. Policies and practices like industrial use zoning, real estate investment mapping (which resulted in redlining), public housing and urban renewal had devastating consequences in many American cities in the 1950s and 1960s. All of these had their origins in the Progressive era push to try and understand, quantify and manage the industrial metropolis.

Urban Destruction

Hundreds of people crammed into Baltimore's War Memorial Hall in August 1969 and spent hours tearing into local bureaucrats, engineers and officials on the front dais. On paper, the hearings were over a new route for the Rosemont Expressway, a highway planned for the city's west side. But they quickly morphed into something else. As local journalist James Dilts noted, the night became "a hearing on the road itself," on the entire interstate highway complex planned for metropolitan Baltimore. Local Black activist and community leader Hezekiah Morris best summed up the attitude toward all of the planned highways. "We believe that a road is an anathema, an eating disease penetrating through the city, taking neighborhood properties, graveyards, anything else in its way.... It leaves behind destruction, rats and everything, and the world is turned upside down, people fighting against each other, separate."[10]

At the local level, the Rosemont Hearings were the culmination of more than two decades of fighting over the routing for Baltimore's highway system. But they were also the product of a series of broader, national policies, programs and structural forces that fundamentally remade American cities in the decades after World War Two. These policies said that not only was it a good idea to build a six-lane highway in the middle of a vibrant, urban neighborhood, but that this highway was necessary and vital to preserving the city. They were programs that told urban residents, especially the most marginalized, that the solution to their run-down apartments and tenements was the destruction of their neighborhood and movement into new, modern high-rise housing. And they were structural forces that actively drained cities of residents and investment, leaving city leaders and residents with fewer and fewer resources to clean-up, repair or replace potholed streets, decrepit apartments and rat-infested empty lots.[11]

Much of the historical literature uses the term "urban decline" as a primary descriptor for what happened to central cities in the decades after World War Two. I have long bristled against that construction because it naturalizes decisions made by powerful actors for a variety of political, social and economic reasons, reinforcing the arguments of generations of urban commentators that what happened to American cities was a product of forces outside human control and not the result of specific choices made with full knowledge of their impact. Urban destruction is a better term because it highlights the intentionality of action and responsibility of policy makers and everyday citizens and emphasizes the material, environmental and ecological reality of what happened to American cities.

Although there are dozens of stories of how this played out on the ground, St. Louis is an especially stark and instructive example. As one of the great river cities of the nineteenth century, St. Louis continued to be an industrial powerhouse through the early twentieth century, even though the railroad made Chicago America's new great emporium of the West. Through the 1940s, city leaders were planning for a city they hoped would soon surpass 1 million residents. The city was dense and lively, home to immigrant communities from Germany and Italy, and a significant African American population. The manufacturing economy was diverse, and not dependent on one major industry like Detroit or Pittsburgh. Its location at the confluence of the Mississippi and Missouri Rivers was still of vital importance, connecting the farm economies of the Western and Southern states.

Despite these strengths, St. Louis faced a number of challenges. Perhaps most importantly, it was a deeply segregated city. There was a large historic Black population dating to the founding of the city that boomed with the Great Migration of the 1920s. But similar to other industrial cities like Chicago or Cleveland, Black residents were confined to living in two tightly controlled belts of housing: one in the Mill Creek Valley, in the central part of the city, and another on the North Side. It was virtually impossible for any Black family, regardless of their income level, to rent or purchase a home outside of these two areas. This had a significant impact not only on housing choice, but also on living conditions. Landlords and property owners took advantage of this lack of choice by making minimal repairs, subdividing apartments, and ignoring many housing and building codes.[12] City officials attempted to solve some of these problems in the 1950s with the Mill Creek Valley Urban Renewal project. Many of these homes in this region were some of the most degraded in the country, with the majority not even having indoor plumbing or running water. Officials claimed that residents would be provided with replacement dwellings in new, modern public housing projects like the Pruitt-Igoe Homes. But there was never enough capacity, and the renewal project just put more pressure on the existing north side Black communities, as did the continued migrations of African Americans from the rural South to St. Louis in the decades after World War Two.[13]

By the 1960s, population pressure and segregation were part of a number of factors that would create an environmental and public health crisis in St. Louis. The neighborhoods of north St. Louis had some of the city's oldest housing stock and had been subject to "redlining" by local banks and investors for more than 20 years, which identified them as poor long-term investments, which meant that it was virtually impossible for anyone to get a conventional mortgage to purchase or fix up the property. This did not mean that there was not money to be made by white property owners. As scholars like Nathan Connolly and Keeanga-Yamahtta Taylor have shown, exploiting racialized housing markets has always been profitable. Landlords knew that Black poor and working-class St. Louisans faced few options. The rents they charged were not overwhelmingly high because they knew their tenants couldn't afford much. But they rationalized this supposed "service" to the community by conducting the bare minimum of maintenance. The buildings were already old and suffered from decades of neglect that was compounded by every small issue. Basements flooded, plaster ceilings caved in, appliances were old and rusty.[14]

For decades, the solution for many landlords was to just slap a new coat of paint on the walls, covering up stains and cracks, and making a new tenant feel like they were moving into a decent apartment. But more often than not that was one particular type of paint: white lead. Lead that is oxidized with an acid forms a white powder. When mixed with oil it forms a bright, durable paint. White lead was the base coat of choice for oil painters for centuries and is in thousands of European masterworks in museums around the world. But in the early twentieth century, corporations began mass marketing it to consumers, and it was used in millions of homes across the United States beginning in the 1920s. Lead paint was being phased out in many newer homes by the 1950s but still existed as a base coat in older buildings, and because it was inexpensive and brightened up a worn out home or apartment, white lead was the paint of choice for many inner-city landlords, particularly in older, industrial cities like St. Louis.[15]

Ivory Perry did not know much about lead paint in 1965 when he began working for the St. Louis Human Development Corporation (HDC), a local social service agency funded by federal War on Poverty programs. A veteran activist in the local Civil Rights Movement, by 1969 he was a housing coordinator at the Union-Sarah Gateway Center, which served a number of communities on the north side. He spent his days helping people find spots in public housing, negotiate leases and forcing recalcitrant landlords to conduct basic maintenance. In 1969, he got to know Wilbur Thomas, a young scientist working at Washington University. Perry mentioned that many of the children of the families he worked with suffered from an abnormal amount of cold and flu-like symptoms. Thomas, knowing the prevalence of lead paint in many of the community's buildings, gave Perry a special solution that, if sprayed on paint, would turn a bright red. Over the next few weeks, he sprayed it in hundreds of buildings and saw an ocean of red. Almost all had lead paint.[16]

Armed with this sudden knowledge that poisonous lead permeated homes with thousands of children in St. Louis, Perry embarked on a campaign that would see some immediate short-term successes but in the long term would fail to reckon with powerful, intersecting, destructive forces that were poisoning thousands of children in St. Louis. A few months after Perry began spraying homes, he worked with members of the St. Louis Board of Alderman to pass a lead ordinance. Modeled after similar laws in cities like Chicago, it empowered city officials to force a landlord to do repairs if a child was poisoned. This gave Perry and other activists a powerful new tool to rally around and try and force landlords to make repairs. But they quickly realized that this form of "secondary prevention," as it was known in the public health field, made local children living test subjects, having to undergo a serious poisoning incident before any action could be taken. And new research was showing that especially for young children, the impact of a single lead poisoning incident could lead to irreversible, long-term neurological damage.[17]

The story of lead poisoning in St. Louis, and in dozens of American cities during this period, is an example of what scholar Rob Nixon has called the "slow violence" of environmental injustice.[18] Normally when we think of racial violence, our minds conjure images of violence that is quick, sharp, destructive and obvious: lynching, the mass shootings of the Holocaust, or the police brutality that is still all too prevalent in many communities of color. But the violence of lead poisoning was incremental. Children breathed lead dust day after day, and it was not until they were noticeably sick that their parents took them to a doctor and possibly had them tested for lead poisoning. The neurological damage caused by lead took even longer to surface, as children struggled in school, or with their emotions or impulse control. This form of violence can be just as destructive as more immediate forms, but also harder to respond to or organize against.[19]

One of the primary reasons activists and victim's families were able to see the impact of this violence and try to prevent or ameliorate it was because of the infrastructure of community organizing and activism that had developed in the city. Perry and other activists were full-time social workers with the Human Development Corporation, the local branch of the federally financed War on Poverty. During this period many War on Poverty agencies supported and even encouraged community organizing, the thinking being that having poor people advocate for themselves was more effective (and cheaper) than just providing expanded social welfare benefits. But the most important support for lead activism came through Wilbur Thomas, who was actually employed as part of a specific project at Washington University to do community-based research and outreach on environmental issues. The project was based at the Center for the Biology of Natural Systems (CBNS), a groundbreaking, interdisciplinary environmental studies center founded by Barry Commoner. During the 1960s and 1970s, Commoner was one of the country's most well-known environmental voices and one of the few that merged social justice concerns with ecological ones.[20] Although he was not involved directly with lead poisoning activism in St. Louis, by providing access to funding and scientific expertise, Commoner helped local activists and residents make key connections between the environmental and socioeconomic conditions that Black families faced.

Thomas would be the most important voice for this vision of environmentalism in St. Louis, albeit for a short period of time. In 1970 he wrote a series of essays for *Proud*, a local magazine, entitled "Black Ecology," in which he laid out the specific environmental issues that urban Black residents faced, and how these were the result of poverty and racial discrimination. In April of that year, he worked with Freddie Mae Brown, another social worker and colleague of Perry's to write "Black Survival: A Collage of Skits" which they performed with a group of local teenagers for a number of audiences during the first Earth Day protests. In the skits Thomas plays a Black professor, who helps convince a group of college students that "ecological issues" do matter to the African American community, just in ways that were profoundly different than how environmentalism was being discussed in the mainstream media.[21]

Because of the presence of Commoner and Thomas, the lead poisoning movement in St. Louis was somewhat unique with how residents and activists were able to articulate that the discrimination and disinvestment that they faced produced its own set of environmental issues. But urban environmental movements existed across the country during this period and were most likely to emerge where residents had the same type of institutional support for organizing and scientific expertise to identify environmental challenges and address them.

Seattle: A Drive on Rats

Although not nearly as old as St. Louis, Seattle had exploded in population in the middle decades of the twentieth century, and especially during World War Two. By the 1960s, the neighborhood known as the "Central District" contained the majority of the metropolitan region's Black population but was also subject to a significant amount of disinvestment, especially in older and increasingly dilapidated housing stock. This, combined with a rapid increase in the number of empty lots, illegal dumping and the region's humid, temperate climate, led to an explosion in the Central District's rat population. By the end of the 1960s, this was one of the major concerns of district residents and became the original target of the Seattle Model Cities Environmental Health Project's rat eradication programs. During a series of annual "Fall Drive on Rats" events at the end of the 1960s and beginning of the 1970s, volunteers, community activists and public health workers cleared vacant lots, inspected manholes, surveyed hundreds of properties and used thousands of pounds of poison bait, all with the goal of significantly lowering the district's rat population.[22]

In a vacuum the rat project looked similar to other public health efforts across the country, especially in neighborhoods where neglect led to explosions of the rat population. But this was different. By 1965, it was widely acknowledged that far from saving cities, federally sponsored urban renewal programs had actually led to more community destruction and resident dislocation, and that federal anti-poverty efforts would ultimately be futile in cities with failing housing and increasing loss of jobs. Passed by

Congress in 1966, the federal Model Cities program was an attempt to address both issues by pairing physical reconstruction and economic development funds together in a series of demonstration projects or "model cities" across the country. One of the goals behind the initiative was to make the funding flexible, so community members could design what was needed in their own neighborhoods.[23]

In Seattle, this led to the Model Cities Environmental Health Project. Headed by Isaac Banks, who was hired away from the city's public health service, the project looked to address environmental issues that connected to larger social justice concerns. Residents were surveyed about what specific issues were most urgent, and then community members were hired and trained to execute various projects. To Banks and others, economic opportunity and environmental improvement were not just related but intimately tied together. After working on general cleanup and rat control, the program expanded to help residents control cockroach populations in their homes and address lead poisoning. Although the Seattle Model Cities program only lasted a few years (subject to Nixon administration policy changes discussed below), it was a robust example of urban environmentalism that looked to tackle the immediate environmental harms as well as the broader socioeconomic inequalities that undergirded those issues.[24]

New York: The Garbage Offensive

New York faced urban environmental issues that were similar to many major American cities, but because of its size, scale and diversity, they varied across the five boroughs, with different types of organizing and responses. Over the course of the 1960s and 1970s, there were a variety of protests over garbage pickups and sanitation, lead poisoning, rats and overall public health and environmental concerns. One of the groups to try and address multiple issues was the Young Lords. Originally formed in Chicago, the Young Lords were a radical leftist organization in the vein of the Black Panthers that looked to connect Third World liberation struggles (in this case the continued colonial occupation of Puerto Rico) to activism and organizing in the United States. Similar to the Panthers, the Young Lords also understood that to build a constituency, a group needed to work to address the real-world problems of community residents, one of which was garbage. Apartments in East Harlem, where the majority of Puerto Rican migrants lived in Manhattan, were tremendously overcrowded, and shifts in postwar consumption and the explosion of "disposable" packaging meant that the average American, even in poorer neighborhoods, produced a lot more trash than they had before World War Two. The community also had a significant amount of abandoned buildings and empty lots, which were magnets for all sorts of trash and illegal dumping. Although it was one of the city's largest agencies, the sanitation department did not have the infrastructure or staff to address New York's exploding growing trash problem, and poor and marginalized neighborhoods like East Harlem ended up getting the least amount of attention.[25]

Young Lords' activists started by just sweeping the streets and collecting trash, then tried to put pressure on the city for more frequent pickups and attention to larger neighborhood problems. For example, the city had no policy for abandoned cars, and East Harlem became a communal dumping ground for hundreds of rusted out hulks that lined the streets. Protests escalated over the course of 1969 with a series of garbage blockades and other forms of civil disobedience in the city streets. This "Garbage Offensive" was tremendously successful, forcing city and even state leaders to address the issues with increased funding and full restructuring of the sanitation department's operations and procedures. The Young Lords continued this urban environmental activism with a focus on childhood lead poisoning. Families in East Harlem faced issues similar to those in St. Louis: older, dilapidated apartments and landlords who ranged from reluctant to negligent in addressing maintenance issues, which increased children's exposure to lead paint. During the winter of 1969–1970, the Young Lords engaged in a "Lead Offensive" which combined door-to-door testing of thousands of children, surveys of living conditions and, eventually, protests and sit-ins at city office and hospitals. The result of this activism was a sea-change in the city's housing and public health infrastructure to address childhood lead poisoning.[26]

Scholars such as Johanna Fernandez have shown that for the Young Lords this community environmental activism was not an end in itself, but part of a larger vision for radical political organizing. Nevertheless, the Young Lords' work on garbage and lead poisoning show similar themes to other urban environmental movements during this period. The founders of the group's New York chapter cut their teeth in War on Poverty agencies earlier in the 1960s, where they learned the basics of community organizing and an understanding of the complex web of political, social and economic forces that shaped people's lives and living conditions in the city. The lead campaign was also built on partnerships and collaborations with doctors, nurses and non-medical staff in the Metropolitan Hospital, the large public hospital in the southern part of East Harlem. The hospital was significantly overburdened and underfunded but was the main source of medical care for many community residents of limited means. Hospital staff provided expertise for understanding the causes and impact of lead poisoning. They would also work with the Young Lords on a later series of protests to address the larger issues of structural inequality at the hospital.

These movements in St. Louis, Seattle and New York are just three examples of the dozens of urban environmental movements that occurred across the country during the late 1960s and early 1970s. In addition to lead poisoning, rats and trash, they also looked to address destructive highway construction and urban renewal projects, air pollution and lack of access to parks and greenspace. Many of these movements are well documented, either in local histories, or in broader narratives of civil rights activism and community organizing during this period. But up until recently, most have not been considered part of the environmental movement. At the time, many of these activists would not have said they were environmentalists, either because they saw themselves as community activists first, or because they did not see a direct connection between their work and the emerging "mainstream" environmental movement. But they are nevertheless an important part of the history of environmental activism in the United States, and a fuller understanding of them upends, in many ways, our scholarly and popular understandings of all the social movements from the period.[27]

Shifting Currents in the 1970s

In much of our popular memory, the protests and activism of "The Sixties" are discussed and portrayed as something that just emerged, as if by magic, in a cloud of raised fists, marijuana smoke and fringed jackets. The emphasis is too often on the specific issues or problems that activists and protestors were responding to, as if injustice, war and ecological destruction occurred for the first time in 1967 or will always provoke a tumult of rebellion. The intersecting social conditions that helped people understand the issues, build solidarity with each other, form movements and try and chart a path for change are often ignored or misunderstood. Vietnam was a war fought by draftees aged 18–25 when this age group was going through a demographic boom, with the largest percentage of young Americans ever having access to higher education. The Civil Rights Movement was built on the groundwork of decades of quiet organizing by activists in the South and the North and occurred in a specific Cold War context in which powerful, majority white institutions believed it was necessary to accommodate social change because of geopolitical global priorities.[28]

The environmental movement was no different. As Rachel Carson warned in her 1962 book *Silent Spring*, modern technologies were being developed and deployed by corporations and governments at a scale and pace that made it almost impossible to understand their short and long-term impacts. But those same corporations, universities and government agencies also housed a rapidly growing cohort of biologists, chemists and other scientists and scholars. They had the expertise, or had access to it through colleagues, friends and family members, to see the harm being caused by new chemicals or nuclear technology and took action to try and understand it and inform the broader public.[29] We have often taken that key infrastructure of the environmental movement for granted because many activists and historians of environmental activism come from middle class or elite backgrounds, and we have naturalized access to this knowledge and expertise.

But for millions of Americans, this access was not automatic. They did not work as professors or researchers or have family members that did. They did not live down the street from journalists or lawyers who might help them understand how to publicize or litigate an issue. Most importantly, they usually did not have the time or resources to do the research or organize themselves. That this why this decade, from roughly 1965 to 1975, is so important for urban environmentalism. Through federal War on Poverty or Model Cities programs, university outreach efforts and federal grants, there was a critical mass of infrastructural support for urban environmental organizing. It's important to note that almost none of these institutions or funding was intended to provide support for environmental activism specifically. But activists and communities took advantage of the capacity available to shape the work around their concerns and issues.

Understanding the importance of this type of infrastructure to urban environmentalism is vital if we are going to make sense of what happened over the course of the 1970s, and the role of these types of urban activists within the larger environmental movement. The election of Richard Nixon is the most important turning point here. In 1969, he embraced the environment as a (relatively) non-controversial set of issues that could shore up his support among centrist voters across the country. This included bureaucratic reorganizations that led to the creation of the EPA, but also support for landmark legislation such as the Clean Air Act Amendments and the National Environmental Policy Act. Under Nixon and especially the centrist Republican environmental bureaucrats like William Ruckelshaus and Russell Train, the federal government developed a robust regulatory infrastructure and funded and supported similar bureaucracies and policy-making at the state level. This almost instant institutionalization of a set of major environmental concerns at the federal level (NEPA was enacted almost four months before Earth Day) quickly centralized not only America's environmental regulatory state, but also its environmental nongovernmental organizations, which almost immediately became focused on shaping federal policy and/or litigating important cases in the courts.[30]

At the same time however, the Nixon administration was dismantling the very infrastructure that had supported the emergence of urban environmentalism. Within the context of the politics of the early 1970s, Nixon could not try and outright destroy various federal social welfare programs, as Reagan would attempt to do a decade later. Instead, the money was redirected as a part of the "New Federalism": Funding levels would not decrease, but instead monies would go directly to the state and municipal governments in the form of block grants and other schemes. This was sold as a more efficient and democratic way of providing federal assistance, the argument went, because local governments could make the best decision about what programs or projects needed funding. The reality, however, was that funding was being shifted from the neighborhood level to city hall, which big city mayors and other political leaders were more than happy to accept, because it provided them with two significant benefits. They now had bigger checks for pet projects that provided more direct funding for their friends and benefactors in big institutions like local banks, real estate firms, labor unions and construction companies, while simultaneously removing significant support for community organizing and activism that was a constant thorn in their side.[31]

These two policy shifts put environmentalism down a path that would eventually see a full split between more middle-class, "mainstream" environmentalism and what would become known as the environmental justice movement. But this would not happen until the end of the 1980s. Despite the challenges urban environmental activists faced because of the erosion of infrastructural support at the grassroots level, at the national level there were real efforts to make connections between urban issues and the broader environmental movement. Although these were ultimately unsuccessful, they provide a glimpse of an alternative path, of what a more diverse, grassroots and equity-focused environmentalism could have looked like in the United States.

Missed Opportunities

A survey of the work of various American environmental groups in the 1970s shows that many of them were trying to contend with what it meant to engage directly with urban communities and the environmental issues residents of these places faced. This work started with Environmental Action (EA), which grew out of Environmental Teach-In, the group that organized the initial Earth Day protests. Based in Washington, DC, EA was staffed by younger activists and focused on political work at the national level. Based on this profile, it might be easy to dismiss EA as a conservative, reformist organization, but its young staff had experience in the anti-war and civil rights movements and understood that urban populations faced different and substantively more severe environmental challenges. They communicated this through their research and publication program, which included their journal *Environmental Action*. "Until now the ecology movement has operated on the assumption that everyone breathes the same air and drinks the same water; therefore, it is in everyone's interest to take care of the air and water," EA staff member Sam Love wrote in a

1972 essay. "The harsh reality that our movement must accept is that the rich have a cleaner environment than the poor and middle class."[32] Two years earlier staffer Dennis Clark had made a similar argument. Urban Blacks needed to do a better job of paying attention to environmental issues, but environmentalists also needed to be more inclusive. "The prospects for this contribution, however, will not be realized unless the organizations interested in environmental problems include blacks in their efforts to broaden public support for their work. Considering the past conditions, this means an effort to ensure black participation at every level of policy-making and activity," Clark wrote.[33]

As a new organization staffed by experienced activists, EA's more radical and equity-focused perspective is perhaps not surprising. More interesting is support that came from The Conservation Foundation. Founded by Fairfield Osborn Jr., president of the New York Zoological Society in 1948, the foundation was a research and public policy think tank whose work was originally based around more traditional conservationist issues, but it became more political under the leadership of Russell Train in the late 1960s, who moved the organization to Washington, DC and began more direct lobbying of the politicians and federal agencies. After Train moved into federal government service in 1969, director Sydney Howe brought a larger focus on civil rights and equity, funding research on urban environmental issues and sponsoring a 1973 conference on environmental quality and social justice.[34]

Howe's pivot to social justice would prove controversial at the foundation, whose board wanted the organization to focus on traditional conservation issues. He was forced out in 1973, and after a couple of years doing research at Washington, DC think tanks, he was hired as the director of the Urban Environment Conference, or UEC. Created by Michigan Senator Philip Hart in 1972, the UEC was designed as a vehicle to help center the concerns of urban and working-class constituencies in environmental issues. The UEC was involved in a number of environmental educational and coalition building efforts during the 1970s, including the groundbreaking 1976 conference "Working for Environmental and Economic Justice and Jobs: A National Conference." Held at the United Auto Workers' Black Lake retreat center, the Black Lake Conference, as it became known, was an effort to address what was emerging as a major friction point among labor, environmental and urban social justice activists: jobs and disinvestment. As the pace of urban plant closures and shutdowns increased, corporate interests pointed to environmental regulations as a source of rising costs that undercut profitability. This conference and other efforts, like the group Environmentalists for Full Employment, were attempts to try and strengthen these coalitions before they were cleaved apart by economic issues.[35]

Despite the strong economic headwinds of the 1970s, environmentalists greeted the election of Democrat Jimmy Carter in 1976 with significant optimism. Not only was this the first Democrat in the White House in almost a decade, but during that time the political and social ground had shifted significantly. A whole cohort of young people had invaded Washington as activists turned lobbyists and now governmental officials, many of them from environmental organizations. But the ground for the environmental movement had shifted as well. By the time of Carter's inauguration in the winter of 1977, the country was deep into a period of economic stagnation, combined with rising inflation. The new EPA leaders understood that nature protection, especially if it impacted the economy, was a tough sell. Environmental health, protecting people against disease and cancer, was a much more politically defensible niche than "birds and bunnies," EPA Director Costle argued in an internal memo. Prominent Democrats agreed. "I'm not sure there ought to be an environmental movement in the terms we had in the late Sixties. There ought to be a public health movement," Sen. Gary Hart, D-Colorado, told *Forbes* magazine.[36]

The EPA would shift a lot of its focus toward public health and regulating toxic substances, which it had significantly increased power to do because of the passage of the Toxic Substances Control Act in 1976. But EPA and other federal agencies would also look to expand the constituency for environmental issues, especially to the urban African American community, but also build connections to traditional environmental organizations. For example, EPA Deputy Director Barbara Blum gave numerous versions of a speech entitled "Cities: An Environmental Wilderness" to the Sierra Club and other groups where she called on them to pay more attention to urban issues. "Because the environmental movement has focused so much of its attention in the past on the wilderness, it has been charged as elitist. We had better face the fact that there is more than a grain of truth to this charge. No person is an island, and yet we all know that in our midst there are islands everywhere, vast, ugly, dilapidated, teeming islands filled with poverty, unemployment, chronic sickness, depression and sometimes despair and violence."[37] The EPA also published numerous brochures and pamphlets on urban environmental issues and had a series of research and educational pieces on the urban environment in the *EPA Journal*, its primary internal and external communications organ. Outside of the EPA, the Department of Health, Education and Welfare sponsored the Community Urban Environment (CUE) program at the National Urban League. This included a widely distributed slideshow that showed a fictional famous pool player "Chi Slim" educating friends about the specific environmental issues urban Black residents face, and how these are rooted in racism and inequality.[38]

All these educational and bridge building efforts culminated in 1979 with the "City Care" conference in Detroit. Co-sponsored by the EPA,

National Urban League and the Sierra Club, the conference was part of the latter group's efforts under President J. William Futrell to recruit more African American members and to expand its base beyond its traditional upper middle-class white constituency that was primarily interested in wilderness preservation and conservation issues. The conference focused on jobs and economic development as well as urban environmental issues as an attempt to overcome deepening concerns over disinvestment in America's older industrial cities. By all accounts the event was a success, and the EPA and Sierra Club followed up with a series of regional conferences over the next few years.

It would be easy to argue that top-down efforts by federal agencies and major national organizations were at best ineffectual and at worst window dressing for larger regulatory and policy regimes that enforced deep structural and environmental inequalities. But their scale and sincerity were real, and education has always been a key part of the environmental movement, especially in urban areas. The "Chi Slim" slideshow and various other workshops and conferences were in many ways designed to help urban constituencies understand how issues they were already experiencing were *environmental* issues, and to make connections between broader forms of inequality and injustice. They were based around an understanding of environmentalism as a knowledge politics but were also optimistic that if urban communities gained that knowledge, they would be able to take action. Although from a twenty-first-century vantage point it is easy to see the 1970s as one moment in the long decline and disinvestment in America's older, industrial cities, for many at the time this was a period of optimism. Urban African Americans and other racial and ethnic groups were benefitting from a significant expansion of post-civil rights political power at the local and national level. There was a belief that with expanded voting rights and access to the regulatory state, historically marginalized groups had the ability to voice their concerns and have them addressed.

Conclusion: The Birth of Environmental Justice

It is important to note the optimism of the late 1970s, of the possibilities of creating a truly diverse and inclusive environmental movement that took the concerns of marginalized urban residents seriously, because of what happened just a few years later. The material reality of these issues would not change. In fact, they got demonstrably worse. But they would find voice outside of traditional environmental institutions and even the federal government in a new movement: environmental justice.

During the 1970s, it appeared that environmental reform had become a new, relatively uncontroversial pillar of American governance. The passage of major federal environmental regulations had been largely uncontroversial, and politicians on both the left and right espoused strong environmental positions. But this surface level bi-partisanship masked a growing resentment among many corporate leaders, large landowners, mining and forestry interests, especially in the American West, that federal regulation had gone too far. They provided significant support for Ronald Reagan's presidential campaign in 1980, and once elected, the anti-regulatory president rewarded them with two key executive appointments: James Watt as the new Secretary of Interior and Anne Gorsuch as the EPA Administrator. While Watt was the more flamboyantly anti-regulation, Gorsuch also worked to minimize the federal role in environmental control, devolving a significant amount of enforcement to the state level.[39]

The work of both Gorsuch and Watt, and the Reagan administration's general anti-regulatory stance, engendered a fierce backlash from environmental organizations, which spent the first part of the 1980s fighting to save core environmental protections. They were largely successful, but in the process, efforts to expand their constituency and speak to urban issues and address concerns over environmental inequality fell by the wayside. To be fair, the core white, middle-class constituencies of groups like the Sierra Club had never fully supported the broader coalition building efforts.[40] Once core environmental protections were under attack, it was easy to throw what were considered marginal concerns to the wayside.

At the same time, however, new forms of environmental inequality were emerging all across the United States, especially in American cities. Urban disinvestment continued and inequality deepened during the 1980s, as deindustrialization continued to lay waste to older industrial cities, and the Reagan Administration worked to dismantle the remaining pieces of the social welfare state. Concern over toxic substances led to the passage of the TSCA and other regulatory measures. But the flip side of this meant that historically poor and marginalized communities were targeted for new dumps or cleanup operations, because it was believed they didn't have the power to oppose them, or they were desperate enough for jobs to accept anything.[41]

The reality for many of these communities was far different. Over the course of the 1980s and 1990s, residents and activists realized that whatever jobs factories, dumps, power plants and incinerator facilities brought to their communities was offset by an inordinate amount of noxious and disease-causing pollutants in their communities. They conducted community research to show higher levels of cancer and other long-term health impacts and worked to build coalitions and mobilize with other civil rights and labor organizations with the goal of opposing new facilities and shutting down existing ones. Although they were definitely engaging in environmental activism, they did not see themselves as traditional or "mainstream" environmentalists. They were part of a new movement, environmental justice, which consciously worked to show how larger so-cial inequalities, especially forms of structural racism, created severe and striking forms of immediate environmental harm for certain communities. Many of the early urban environmental justice movements were focused on known or planned sites of toxic pollution such as incinerators or waste dumps. This allowed activists to focus attention and organizing on these sites, and they were often successful in preventing their construction or shutting them down. In movements in New York and Los Angeles, for example, activists were also able to develop a larger, intersectional critique that built upon the particular experience of women and working-class people of color in dealing with long-term toxic hazards.[42]

As they educated themselves about environmental, sustainability and public health issues, community organizations expanded their work to fo-cus on food justice and insecurity, transportation and mobility, and parks and greenspace. For example, the Little Village Environmental Justice Organization in Chicago, which serves a majority Latinx neighborhood, began when local parents were concerned about toxic exposure to their children during a major local school renovation. They then broadened their work to protest local toxic sites, and to create local parks, gardens and greenspaces, and improve community connectivity and transit.[43] Com-munity gardening and urban farming have been some of the most robust forms of recent urban environmental justice organizing, as communities look to turn vacant lots into productive properties and also counteract the lack of access to healthy food options.[44] Much of this activism comes not only from an attempt to address immediate needs within a community, but also through an understanding of structural inequalities within regional, national and global food systems. Significant amounts of transit activism and justice work come from a similar perspective. For example, lobbying for a new bus line helps solve short-term needs but does not address the long-term overinvestment in automobility versus public transit.[45]

Of course, as urban environmental justice movements expand in so-phistication, expertise and funding, the landscape of issues and problems is also changing rapidly. Over the past decade, many activists and com-munity members have begun to focus on climate justice. As the planet warms, sea levels will rise, heat waves will be longer and more severe, as will rainstorms, hurricanes and floods. These impacts will be felt most acutely in cities, and poor and marginalized communities will generally be the most vulnerable and currently have the fewest resources to deal with the problems. Thus much of newer organizing work is about mobilizing community members around imminent climate threats and working with local and national governments for long-term infrastructural solutions.[46]

One of the arguments that climate justice activists make is that middle-class and wealthy city residents will be less vulnerable to climate change because of their ability to mobilize resources or relocate. And this is true. But as threats worsen, there is the question of whether or not anyone will truly be able to "escape" climate impacts. If this is the case, perhaps it is time to reconsider the direction of the environmental movement broadly, and urban activism specifically. Environmental organizations and activists need to move beyond just becoming more equity minded (which many are finally making progress on) but take cues from urban environmental justice organizations and think broadly about larger structural inequalities and injustices and what kind of societies we want to live in moving forward. It would be helpful to look back to the 1970s and consider the possibilities and opportunities of a movement that merges the concerns of different groups for a broader vision of a more just and sustainable society.

Notes

1 Nathan Hare, "Black Ecology," *The Black Scholar* 1, no. 6 (1970): 2.
2 Samuel P. Hays and Barbara D. Hays, *Beauty, Health, and Permanence: Environmental Politics in the United States, 1955-1985*, Studies in Environment and History (New York: Cambridge University Press, 1987); Adam Ward Rome, *The Bulldozer in the Countryside: Suburban Sprawl and the Rise of American Environmentalism*, Studies in Environment and History (New York: Cambridge University Press, 2001).
3 Robert D. Bullard, *Dumping in Dixie: Race, Class, and Environmental Quality*, Third edition (Boulder, CO: Westview Press, 2000); Eileen McGurty, *Transforming Environmentalism: Warren County, PCBs, and the Origins of Environmental Justice* (New Brunswick, NJ: Rutgers University Press, 2009).
4 Robert Gottlieb, *Forcing the Spring: The Transformation of the American Environmental Movement*, Rev. and updated ed. (Washington, DC: Island Press, 2005); Martin V. Melosi, "Environmental Justice, Political Agenda Setting, and the Myths of History," *Journal of Policy History* 12, no. 1 (January 1, 2000): 43–71.
5 Ellen Griffith Spears, *Baptized in PCBs: Race, Pollution, and Justice in an All-American Town* (Chapel Hill: The University of North Carolina Press, 2014).
6 Robert Hunter, *Tenement Conditions in Chicago* (Chicago, IL: City Homes Association, 1901).
7 Maureen A. Flanagan, "The City Profitable, the City Livable: Environmental Policy, Gender, and Power in Chicago in the 1910s," *Journal of Urban History* 22, no. 2 (1996): 163–90.
8 Harold Platt, "Jane Addams and the Ward Boss Revisited: Class, Politics, and Public Health in Chicago, 1890-1930," *Environmental History* 5, no. 2 (2000): 194–222.
9 David Stradling, *Smokestacks and Progressives: Environmentalists, Engineers and Air Quality in America, 1881-1951* (Baltimore: Johns Hopkins University Press, 1999).
10 Interstate Division for Baltimore City, *Hearings on the Rosemont Bypass*, August 9, p. 196, 68–69.
11 Francesca Russello Ammon, *Bulldozer. Demolition and Clearance of the Postwar Landscape* (New Haven, CT: Yale University Press, 2016); Andrew R.

Highsmith, *Demolition Means Progress: Flint, Michigan, and the Fate of the American Metropolis* (Chicago: University of Chicago Press, 2015); Christopher Klemek, *The Transatlantic Collapse of Urban Renewal: Postwar Urbanism from New York to Berlin* (Chicago, IL; London: The University of Chicago Press, 2012); Raymond A. Mohl, "Stop the Road: Freeway Revolts in American Cities," *Journal of Urban History* 30, no. 5 (July 1, 2004): 674–706.

12 Colin Gordon, *Mapping Decline: St. Louis and the Fate of the American City* (Philadelphia: University of Pennsylvania Press, 2008).

13 Joseph Heathcott and Maire Agnes Murphy, "Corridors of Flight, Zones of Renewal: Industry, Planning and Policy in the Making of Metropolitan St. Louis, 1940-1980." *Journal of Urban History* 31, no. 2 (2005): 151–89; Clarence Lang, *Grassroots at the Gateway: Class Politics and Black Freedom Struggle in St. Louis, 1936-75* (Ann Arbor: University of Michigan Press, 2009).

14 Robert Eugene Quinn and Michael A. Mendelson, *The Decline of an Urban Housing Entrepreneur: Congratulations or Condolences?* (Edwardsville: Center for Urban and Environmental Research and Services, Southern Illinois University at Edwardsville, 1977); N. D. B. Connolly, *A World More Concrete: Real Estate and the Remaking of Jim Crow South Florida*, Historical Studies of Urban America (Chicago, IL; London: The University of Chicago Press, 2014); Keeanga-Yamahtta Taylor, *Race for Profit: How Banks and the Real Estate Industry Undermined Black Homeownership* (Chapel Hill: University of North Carolina Press, 2019).

15 Christian Warren, *Brush with Death: A Social History of Lead Poisoning* (Baltimore: Johns Hopkins University Press, 2000). See also Elizabeth Fee, "Public Health in Practice: An Early Confrontation with the "Silent Epidemic" of Childhood Lead Paint Poisoning," *Journal of the History of Medicine and Allied Sciences* 45, no. 4 (1990): 570–606; Jane Lin-Fu, "Modern History of Lead Poisoning: A Century of Discovery and Rediscovery," in *Human Lead Exposure* ed. Herbert Needleman (London: CRC Press, 1992).

16 George Lipsitz, *A Life in the Struggle: Ivory Perry and the Culture of Opposition* (Philadelphia: Temple University Press, 1995) Ch. 7; William Paul Locke, "A History and Analysis of the Origin and Development of the Human Development Corporation of Metropolitan St. Louis, Missouri, 1962-1970" (Ph.D: St. Louis University, 1974).

17 "Sit-in Against Lead Poisoning," *St. Louis Globe and Democrat*, November 12, 1970; "Citizens Group Urges Lead Poison Warning," *St. Louis Globe-Democrat*, November 20, 1970; People's Coalition on Lead Poisoning, "Chronology of Events Between the Coalition Against Lead Poisoning and the Court System of St. Louis and the State of Missouri," 1971, Folder 348, Freedom of Residence Committee Addenda, State Historical Society of Missouri, St. Louis Research Center.

18 Rob Nixon, *Slow Violence and the Environmentalism of the Poor* (Cambridge, MA: Harvard University Press, 2011).

19 Gerald Markowitz and David Rosner, *Lead Wars: The Politics of Science and the Fate of America's Children* (Berkeley: University of California Press, 2014).

20 Wilbur Thomas, "Environmental Field Program: Second Annual Report to the W.K. Kellog Foundation," January – December 1970, Box 423, Folder Research Proposals of Center Senior Fellows 71–72, Commoner Papers; Barry Commoner, "By Using Nature as a Lab" *SR* (St. Louis Post-Dispatch Sunday Magazine), May 7, 1966; Kelly Moore, *Disrupting Science: Social Movements, American Scientists and the Politics of the Military, 1945-1975* (Princeton, NJ:

Princeton University Press, 2008) Ch. 4; Michael Egan, *Barry Commoner and the Science of Survival: The Remaking of American Environmentalism* (Cambridge, MA: MIT, 2007).

21 Freddie Mae Brown and the St. Louis Metropolitan Black Survival Committee, "Black Survival: A Collage of Skits," in *Earth Day – The Beginning: A Guide for Survival*, Environmental Action (New York: Bantam Books, 1970); Wilbur Thomas, "Black Survival in Our Polluted Cities," *Proud*, April 1970.

22 Jeffrey C. Sanders, *Seattle and the Roots of Urban Sustainability: Inventing Ecotopia*, History of the Urban Environment (Pittsburgh, PA: University of Pittsburgh Press, 2010) Ch. 2.

23 Bernard J. Frieden and Marshall Kaplan, *The Politics of Neglect: Urban Aid from Model Cities to Revenue Sharing* (Cambridge, MA: MIT Press, 1975), 14–67.

24 Sanders, *Seattle and the Roots of Urban Sustainability*.

25 Johanna Fernández, *The Young Lords: A Radical History* (Chapel Hill: The University of North Carolina Press, 2020); Matthew Gandy, "Between Borinquen and the Barrio: Environmental Justice and New York City's Puerto Rican Community, 1969–1972," *Antipode* 34, no. 4 (September 1, 2002): 730–61.

26 Fernández, *The Young Lords*.

27 For a broader discussion of this point, see Robert Gioielli, *Environmental Activism and the Urban Crisis: Baltimore, St. Louis, Chicago* (Philadelphia, PA: Temple, 2014).

28 Charles M. Payne, *I've Got the Light of Freedom: The Organizing Tradition and the Mississippi Freedom Struggle* (Berkeley: University of California Press, 2007); Melvin Small, *Antiwarriors: The Vietnam War and the Battle for America's Hearts and Minds*, Vietnam–America in the War Years, vol. 1 (Wilmington, DE: Scholarly Resources, 2002).

29 Rome, *The Bulldozer in the Countryside*; Christopher C. Sellers, *Crabgrass Crucible: Suburban Nature and the Rise of Environmentalism in Twentieth-Century America* (Chapel Hill: University of North Carolina Press, 2012).

30 Brooks Flippen, *Nixon and the Environment* (Albuquerque: University of New Mexico Press, 2000).

31 Timothy Conlan, *New Federalism: Intergovernmental Reform from Nixon to Reagan* (Washington, DC: Brookings Institution Press, 1988).

32 Sam Love, "Ecology and Social Justice: Is There a Conflict?" *Environmental Action*, August 5, 1972.

33 Dennis J. Clark, "Urban Blacks and Environmental Activism," *Environmental Action*, November 28, 1970.

34 The Conservation Foundation, *1971 Annual Report* (New York: The Conservation Foundation, 1971); William Steif, "Conservation Committee Wants Director Fired," *Albuquerque Tribune*, February 28, 1973.

35 For full discussion of the Black Lake Conference and its place within labor, urban and environmental politics, see Josiah Rector, *Toxic Debt: An Environmental Justice History of Detroit* (Chapel Hill: University of North Carolia Press, 2022) Ch. 6.

36 Jean Briggs, "The Price of Environmentalism – The Backlash Begins," *Forbes*, June 15, 1977; Douglas Costle, "Health and the Environment," *EPA Journal*, July/August 1978; Jeffrey K. Stine, "Environmental Policy During the Carter Presidency," in *The Carter Presidency: Policy Choices in the Post-New Deal Era*, ed. Gary M. Fink and Hugh Davis Graham(Lawrence: University Press of Kansas, 1998); Rafe Pomerance, to Tom Bennett, December 26, 1972, Box 31,

Folder 9, Urban Environment Conference Papers, Walter J. Reuther Library, Wayne State University (UEC Papers).

37 Barbara Blum, "Cities: An Environmental Wilderness," speech to Sierra Club, San Francisco 1978, Box 27, Folder 23, UEC Papers.

38 National Urban League Housing Division, *Project Cue: Pollution Slide Show*, Box 25, Folder 12, UEC Papers.

39 James Morton Turner and Andrew C. Isenberg, *The Republican Reversal: Conservatives and the Environment from Nixon to Trump* (Cambridge, MA: Harvard University Press, 2018).

40 Neil Goldstein, to Gene Perrin, Oct. 25, 1978, Box 242, Folder 3, Sierra Club Records, University of California Bancroft Library; Mark Stapke, to Judy K., October 31, 1978, Box 242, Folder 3, Sierra Club Records.

41 Müller, Simone M. "Hidden Externalities: The Globalization of Hazardous Waste," *Business History Review* 93, no. 1 (2019): 51–74.

42 Laura Pulido, "Rethinking Environmental Racism: White Privilege and Urban Development in Southern California," *Annals of the American Association of Geographers* 90, no. 1 (2000): 12–40; Julue Sze, *Noxious New York: The Racial Politics of Urban Health and Environmental Justice* (Cambridge, MA: MIT Press, 2007).

43 Leslie Kern and Caroline Kovesi, "Environmental Justice Meets the Right to Stay Put: Mobilising Against Environmental Racism, Gentrification, and Xenophobia in Chicago's Little Village," *Local Environment* 24, no. 9 (2018): 952–66.

44 Kristin Reynolds and Nevin Cohen *Beyond the Kale: Urban Agriculture and Social Justice Activism in New York City* (Athens: University of Georgia Press, 2016).

45 Robert Bullard, Glenn S Johnson and Angel O Torres, *Highway Robbery: Transportation Racism & New Routes to Equity* (Cambridge, MA: South End Press, 2004).

46 Ashley Dawson, *Extreme Cities: The Peril and Promise of Urban Life in the Age of Climate Change* (London: Verso, 2017).

Selected Bibliography

Fernández, Johanna. *The Young Lords: A Radical History*. Chapel Hill: The University of North Carolina Press, 2020.

Flanagan, Maureen A. "The City Profitable, the City Livable: Environmental Policy, Gender, and Power in Chicago in the 1910s." *Journal of Urban History* 22, no. 2 (1996): 163–90.

Gioielli, Robert. *Environmental Activism and the Urban Crisis: Baltimore, St. Louis, Chicago*. Philadelphia, PA: Temple, 2014.

Gottlieb, Robert. *Forcing the Spring: The Transformation of the American Environmental Movement*. Washington, DC: Island Press, 2005.

Hurley, Andrew. *Environmental Inequalities: Class, Race and Industrial Pollution in Gary, Indiana, 1945 to 1980*. Chapel Hill: University of North Carolina Press, 1995.

Lipsitz, George. *A Life in the Struggle: Ivory Perry and the Culture of Opposition*. Philadelphia, PA: Temple University Press, 1995.

Rector, Josiah. *Toxic Debt: An Environmental Justice History of Detroit*. Chapel Hill: University of North Carolina Press, 2022.

Sanders, Jeffrey C. *Seattle and the Roots of Urban Sustainability: Inventing Ecotopia*, History of the Urban Environment. Pittsburgh, PA: University of Pittsburgh Press, 2010.

Sze, Julie. *Noxious New York: The Racial Politics of Urban Health and Environmental Justice*. Cambridge, MA: MIT Press, 2007.

Taylor, Keeanga-Yamahtta. *Race for Profit: How Banks and the Real Estate Industry Undermined Black Homeownership*. Chapel Hill: University of North Carolina Press, 2019.

2 Resilience at the Periphery

North America's Nonurban Environmental Justice Movements

Elizabeth Grennan Browning

Within debates about the historical origins and enduring legacies of environmental racism, how might considerations of space and race transcend multiple geographic scales, from local to regional, national, and global? At the same time, what might we learn about environmental justice by focusing on nonurban areas? As Nicholas Low and Brendan Gleeson have argued, "there is an urgent need for international environmental and ecological justice movements to transcend the 'politics of place' in order that the nature of industrial commodity production may itself be problematized."[1] Environmental justice scholars have recognized since the mid-1990s that studies of environmental inequality must examine more than just quantitative, statistics-based approaches focused on distributional inequality, to consider ideological or qualitative aspects of environmental racism across geographical scales.[2]

This ideological problematization of environmental racism requires us to reexamine the dynamic relationship between rural and urban places and to look anew at the liminal and contested space between what have traditionally been perceived as sharp (but in actually are quite fluid) spatial boundaries. At the rural-urban interface, it often appears as though urban life takes primacy over rural life as the infrastructure of metropolitan areas expand and encroach into the countryside. However, the cultural imaginary of rural life continues to shape urban residents' perceptions of the good life, from leisure to epistemologies of civic virtue and right relationship with, or proper stewardship of, the environment. Despite this notion of rural communities representing a repository of virtue and wholesome environmental goods, poor rural communities have increasingly borne the brunt of environmental health hazards. Over the past number of decades, rural areas have more frequently become dumping grounds for locally undesirable land uses, including feedlots, slaughterhouses, landfills, incinerators, and toxic waste sites.[3] Due to limited economic resources, rural areas often come to rely financially on polluting industries and are thus less likely to mount a coordinated resistance against environmental injustices

DOI: 10.4324/9781003214380-4

stemming from such industries.[4] Rural traditions and cultural practices such as gardening and hunting can further exacerbate the likelihood of residents' toxic exposures, especially for Native communities whose traditional subsistence practices lead to the consumption of contaminated produce and wild-caught fish and meat.[5] When it comes to environmental contamination, rural areas also frequently receive less intense bureaucratic scrutiny compared to urban areas from institutions like the Environmental Protection Agency because such administrative bodies often must prioritize remediation of urban toxic sites due to population density.[6]

At the same time that we should account for the unique environmental harms encountered by nonurban communities, we must also acknowledge that drawing hard sociological and geographical distinctions between rural and urban spaces has become increasingly obsolete. The percentage of Americans living in urban areas has far outpaced those living in rural areas over the past century: in 1900, more than 60 percent of Americans resided in rural areas (defined as small towns with a population less than 2,500; farms; and open countryside), while today, more than 80 percent of Americans reside in urban areas. However, the population of rural areas has remained stable at approximately 50 million people. Over the past half century as rural and small-town America have become less isolated from mainstream economic and cultural influences, the spatial and social boundaries demarcating rural and urban America have increasingly blurred through mutual interdependencies and the flow of information, people, and goods.[7] Still another set of significant shifts in the makeup of nonurban America over the past century is the rise of corporate, industrialized agriculture. Global markets and federal policies have contributed to farm consolidation where large landholders have absorbed small, diversified family farms and now rely on agrichemicals and other artificial inputs to propagate a monocrop agricultural system—mainly, the production of corn and soy. Over the course of the twentieth century, the share of the national workforce employed in the agricultural sector declined from 41 percent to 2 percent.[8] The evolving features of nonurban landscapes and communities require that we look at how these places are bound up in spatial networks of socioeconomic and cultural production.

Understanding the evolution of one particular place's interlocking sets of social-ecological relations requires looking beyond its immediate boundaries to consider what geographer Doreen Massey has called "stretched out" social relations—that is, how "places are produced by other places."[9] Global capitalism's "treadmill of production" (the ever-increasing ratcheting up of natural resource extraction to power energy-intensive technological systems) has created social ecologies of privilege and "wastelanding"—the socioecological processes of externalizing the negative effects of environmental degradation on communities with relatively limited political power—that span from the most densely populated areas to the most desolate regions of North America. Nonurban places are in part defined by their diametric opposition to cities, but we cannot understand one without the other, and, in fact, they are inextricably intertwined in shaping hierarchies of power regarding which communities become vulnerable to pollution and degradation. Although urban space often takes precedence in scholarly and media representations of environmental justice issues, rural, suburban, and exurban places offer new insights into how environmental injustices emerge, progress, and inspire reaction.

Settler colonialism provides a crucial theoretical perspective for interrogating the intersection of racial and spatial marginalization within nonurban manifestations of environmental racism.[10] As an ongoing process, settler colonialism invokes white, Anglo-American dominance through territorial control and racialized labor exploitation.[11] Through genocidal policies, forced dispossession, and cultural assimilation, white settlers diminished Native sovereignty, occupied Native land, and appropriated Native resources. Settler colonial logics of white supremacy underlay the broader categorization of racial hierarchy throughout US history, with Native, African American, Asian, and Latinx groups facing distinct forms of discrimination and oppression within the settler-colonial power structure. Foregrounding race and space as key, intertwined categories of analysis in longstanding settler colonial practices allows for a deeper study of nonurban environmental injustice.

Through this settler-colonial lens, Rob Nixon's explanation of environmental inequity as "slow violence" becomes clear: "a violence of delayed destruction that is dispersed across time and space, an attritional violence that is typically not viewed as violence at all." The kinds of "long dyings"—both environmental and human—at work in the gradual unfolding of environmental crises beg for a better representation in strategic planning and social memory.[12] This chapter considers the conceptual framing of nonurban environmental injustices as operating in places that have historically been deemed "closer to nature" than the city (i.e., agrarian farmland, bucolic suburbs and exurbs, and rugged wilderness) and thus conventionally presumed less prone to environmental degradation. It is in these rural, suburban, and remote spaces that the environmental injustices appear imperceptible, or the pace of these forms of violence seem to move more slowly than urban places because the long-perceived "corrupting" force of the city is not accelerating such problems. By drawing on key case studies—including stories of polychlorinated biphenyl (PCB) dumping in Warren County, North Carolina, a farm-worker community's toxic exposures in California's Central Valley, cancer clusters in Louisiana's Cancer Alley, Native dispossession from nuclear colonialism, and the intersecting histories of the carceral state and rural landscapes—this chapter will examine to what extent nonurban environmental movements take shape from frontline communities' access—or lack thereof—to financial resources, political power, and media attention. By integrating environmental history and political ecology, this chapter situates North American environmental justice crises as coupled-system issues where human and environmental processes interact to (re)produce degradation and its associated risks. Studying the history of nonurban manifestations of environmental inequities will show that resilience to these crises requires a polycentric governance model rooted in feminist political ecology, and an accounting for the deep histories of environmental injustices across time and space.

Defining a Movement in Warren County, North Carolina

Over the course of two weeks in the summer of 1978, the owner of a waste-hauling company enlisted his two sons to help him illegally dump 30,000 gallons of oil contaminated with PCBs along the shoulders of 210 miles of state roads in rural North Carolina. Hired by Robert Ward, the owner of a transformer manufacturing company, the truckers drove under the cover of night since the dumping was in violation of the Toxic Substances Control Act. In light of new government regulations regarding PCBs, Ward pursued these illegal measures in an effort to sidestep escalating disposal costs. Due to the chemical substance's toxicity, Congress banned the production of PCBs the following year in 1979.[13]

To facilitate the remediation of the 40,000 cubic yards of soil along the contaminated roadsides (which marked the largest PCB spill in US history), the state of North Carolina planned to build a landfill in the small town of Afton, in rural Warren County 60 miles north of Chapel Hill in the northeast corner of the state. Approximately 62.5 percent African American and predominantly poor, Warren County's proportion of Black residents was higher than any other of the 100 counties in the state, and only several other counties had higher poverty rates. Like other socioeconomically vulnerable counties in North Carolina, Warren County became the home of many former slaves during Reconstruction, and it continued to have a higher percentage of Black residents compared to other counties in the state through the twentieth century. Local residents viewed the dump siting controversy as a manifestation of racism, akin to their continued struggles for school desegregation and voter registration.

Concerned about groundwater contamination, property values, and the loss of economic development due to the stigma of hosting a hazardous waste facility, Warren County residents tried to stop the landfill through a series of unsuccessful legal battles waged over three years against the state and the US Environmental Protection Agency. Undeterred, the state began constructing the landfill in the summer of 1982, even though officials had to receive a special waiver from EPA because the site's water table was higher than legally allowed. Having exhausted all their legal resources, residents organized a collective action campaign at the landfill site, leading daily protests over the course of six weeks as the contaminated soil arrived at the landfill. Protestors sacrificed their bodies by laying down on the highway to prevent dump trucks from accessing the landfill. On the first day of protests, 55 individuals were arrested, and by the conclusion of the campaign when 7,223 truckloads of waste had been dumped at the site, police had made 523 arrests. Organizers from the United Church of Christ (UCC) and the Southern Christian Leadership Conference, among other civil rights organizations, consistently participated in the protests. When a highway trooper arrested and jailed UCC Reverend Benjamin Chavis, a leader in the protests, for allegedly driving too slowly, Chavis declared, "This is racism. This is environmental racism"—a term which came to define the larger movement.[14] Despite the remarkable commitment of community activists, Governor Jim Hunt refused to meet with protesters and directed extensive police resources to ensure the operation of the dump, as he sought to attract chemical and microchip companies to boost the state's economic development.

Warren County was not the first time that minority communities protested environmental harms within their communities in the United States. Several other high-profile campaigns helped set the stage, including Dr. Martin Luther King, Jr.'s support of striking Black sanitation workers in Memphis in 1968, and in 1979 protests by middle-class African American homeowners in Houston who fought to prevent the siting of a sanitary landfill in their suburban neighborhood by filing the first class-action lawsuit to use civil rights law to challenge the siting of a waste facility (*Bean v. Southwestern Waste Management, Inc.*).[15]

Even with these earlier manifestations of EJ activism, Warren County's media coverage and widespread organizing introduced a new sense of environmentalism, interpreting environmental problems as civil rights issues that required attention to distributive and procedural justice.[16] However, environmental justice activists soon recognized the distributive justice model's limitations and focused on a wholesale elimination of risk rather than ending racial and socioeconomic inequalities in the distribution of environmental harms. Two landmark studies inspired by Warren County crystallized the environmental justice activists' framing of environmental racism and the movement's core strategies: the 1983 General Accounting Office (GAO) report *Siting of Hazardous Waste Landfills and Their Correlation with Racial and Economic Status of Surrounding Communities* and the 1987 UCC report *Toxic Wastes and Race in the United States*, which established that toxic waste sites were more likely to be sited in communities with large BIPOC populations. GAO reported that in eight Southern states 75 percent of commercial hazardous waste landfills were located in predominantly Black communities, while African Americans made up only 20 percent of the region's population. UCC estimated that 60 percent of Black and Hispanic Americans nationwide lived near an "uncontrolled toxic waste site," an abandoned or closed site that posed an endangerment to human health.[17] These authoritative studies became the foundation for future EJ research and provided important data that bolstered activists' positions.

Environmental justice organizers in Warren County continued their fight over the landfill, providing a model for community control over remediation. As voter registration became part of the anti-dump campaign, Warren County political representation was transformed in November 1982, with African Americans elected to county-wide offices in significant numbers for the first time. In becoming more politically active, Black voters elected Black State Representative Frank Balance, who introduced legislation that prohibited the siting of another hazardous-waste facility within 25 miles of the Warren County site and pushed forward a plan to detoxify the site. The battle over environmental injustices in Warren County spurred a broader politicization of Black activists around securing political and economic equality.

Concerned about their community's environmental health, African American residents asserted control over the remediation process. In 1994, water leaking into the landfill threatened to rupture the protective liner. The state planned to pump the water out and take it to another hazardous waste dump in Alabama or Utah. However, local advocates refused to offload the problem onto another overburdened community. Demanding a say in decision-making about the crisis, residents succeeded in pressuring the governor to establish a citizens' advisory panel on the cleanup process. The panel prioritized transparency in selecting detoxification technology options and secured other community demands, including establishing cleanup targets, implementing a quota of local contractors to complete the work, and ensuring that new infrastructure such as streetlights would remain after the remediation process was complete.[18]

The legacies of the Warren County battle are manifold. It helped to ignite a national environmental justice movement, inspired scientific studies that spurred national consciousness about environmental inequities, and modeled community engagement and oversight on remediation projects. Perhaps one of the most significant legacies was the political empowerment of African American residents, and the community's enduring consciousness of environmental racism, particularly with respect to nonurban spaces populated predominantly by people of color.[19]

Environmental Health Disparities within California's San Joaquin Valley Farm-Worker Communities

Located in Kings County in California's San Joaquin Valley, just off California's main north-south corridor Interstate 5, Kettleman City is a small, rural farm-worker community of approximately 1,245 residents, as of 2020. The entire population is Latinx; all residents reported speaking Spanish at home, and nearly half of the town's children live below the federal poverty line. Residents have suffered high levels of asthma, birth defects, cancer, and miscarriages. Between 2007 and 2010, 11 children born to Kettleman City residents had serious birth defects such as brain damage, cleft lips and palates, and deformities. Three of these infants died.[20]

In 2010, the California Department of Public Health determined in an investigation of health records from surrounding areas that since 1987 there appeared to be no "common underlying cause" for the defects, and that the rate of birth defects was only slightly higher than surrounding areas. Advocates critiqued the study, pointing out that the Kettleman City cluster primarily occurred over a 15-month period, and that bringing in data from the previous two decades was tantamount to comparing apples and oranges.[21] Although it was notoriously difficult to conclusively link these health concerns to environmental exposures within legal frameworks, residents' lived experiences made clear the detrimental effects that the surrounding environmental hazards presented to their health.

California's Central Valley is the world's most agriculturally productive area, but this productivity does not come without its costs: Kettleman City is bordered on all sides by agricultural fields and has suffered from pesticide drift and polluted runoff. Making up just 2 percent of the US agricultural land, the Central Valley receives 25 percent of the nation's applied pesticides.[22] Sociologist Jill Harrison has found that 90 percent of aerially applied pesticides have the potential to drift from application sites to unintended areas, such as residential locations.[23] Air pollution in Kettleman City is further exacerbated by diesel emissions from passing trucks, as well as benzene and old oilfield emissions. In 2006, residents learned that their groundwater was also contaminated due to the municipal wells' elevated levels of natural arsenic and benzene. Residents were forced to rely on limited supplies of bottled water from the state and traveled half an hour to purchase additional bottled water sufficient for their daily needs. Environmental justice activists succeeded in getting the state to construct a new water quality treatment plant which opened in 2020.[24]

In what has become a foundational case for the American environmental justice movement, Kettleman City also drew the attention of environmental justice advocates through concern over the health effects of the 1,600-acre Kettleman Hills Hazardous Waste Facility, which opened three-and-a-half miles away from the community in 1980, without the community's knowledge or consent. The largest hazardous waste site west of the Mississippi, the Kettleman facility is owned by Houston-based Chemical Waste Management, Inc., (Chem Waste), a subsidiary of the largest waste-handling company and recycler in North America, the multinational corporation Waste Management Inc. Each year the facility processes over 400 tons of hazardous waste, asbestos, pesticides, petroleum, and PCBs. Due to mishandling waste and neglect of quality-control measures, over the past three decades, the company has been fined over $2 million by state and federal agencies.[25]

In 1988, Chem Waste proposed developing a toxic waste incinerator at the dump site, which would have brought an additional 5,000 truck-loads of toxic waste through Kettleman annually. Municipal officials and industry executives did not inform residents about the first community hearing about the incinerator, held in January 1988, but residents caught wind of the incinerator plans through communication with Greenpeace organizers. Approximately 200 Kettleman City residents traveled 40 miles to the county seat of Hanford to attend the hearing. The Planning Commission chair tried to seat the group 300 feet from the commission with a Spanish translator, but the group insisted on being at the front of the room and subsequently testified to the commission that this treatment reminded them of decades prior when African Americans were forced to sit at the back of the segregated bus. Resident Maricela Alatorre later reflected, "The incident summed up what the County felt for the people out here in Kettleman City. Our rights were second to this huge corporation."[26]

After the Commission approved the incinerator, residents formed the grassroots organization El Pueblo Para el Aire y Agua Limpio (People for Clean Air and Water). El Pueblo advocates researched the area's pollution and discovered that the San Joaquin Valley recorded the second-worst pollution for a US air basin, after Los Angeles. El Pueblo drew on the region's long history of labor activism and farmworkers' rights, including the 1960s work of the United Farm Workers Union, led by Larry Itliong, Cesar Chavez, and Dolores Huerta.[27] The California Rural Legal Assistance program, under the direction of its general counsel Ralph Abascal, also worked to protect farmworkers from environmental harms such as pesticide exposure, organizing a lawsuit against the Environmental Protection Agency that resulted in a ban on approximately 85 percent of pesticides available at the time.[28]

El Pueblo also discovered a concerning California Waste Management Board report (known as the Cerrell Report, see Source 1) from 1984 funded by California taxpayers. The report indicated to government officials and corporate leaders that it would be relatively easy to site garbage incinerators within communities that had less than 25,000 residents, who were predominantly poor, rural, Catholic, with limited education, and working in resource-extractive industries.[29] That these communities would not have the political power to resist such unwanted land uses disturbed the leaders of El Pueblo who recognized that this community profile exactly matched Kettleman City. Furthermore, they found a similar pattern in Chem Waste's toxic dumps within the small California towns of Buttonwillow (85 percent Latinx) and Westmorland (87 percent Latinx and 3 percent Black). Chem Waste's toxic dump in Emelle, Alabama, is the largest facility of its kind in the United States, and likely the world, and the surrounding community is 80 percent African American. At the time of the proposed Kettleman incinerator, Chem Waste's incinerators were located in Chicago's South Side, Port Arthur, Texas, Sauget, Illinois, and East St. Louis—all communities with predominantly Black and Brown populations.[30]

El Pueblo appealed the Planning Commission's approval of the incinerator plan to the County Supervisors but received the same decision. Local government leaders in Kings County, none of whom hailed from Kettleman City, supported the construction of the incinerator because California's compensated siting model meant that Kings County, where every town except Kettleman City was majority white, would receive over 16 percent of the County's annual budget from tax revenue received from Chem Waste, doubling the revenue provided by the toxic waste dump alone. Under the leadership of El Pueblo, Kettleman City residents filed a lawsuit. The judge ruled in the residents' favor, finding that the permitting process had not meaningfully included the residents. Despite residents' continuous requests to participate in the California Environmental Quality Act (CEQA) review process, officials never provided them with a Spanish translation, which effectively shut residents out of negotiations because the community was almost 40 percent monolingual in Spanish.[31]

Environmental justice activist groups El Pueblo, Greenpeace, and Citizen Action continued their campaigns and Kettleman City gained national media attention as Chem Waste appealed the judgment. Ultimately, Chem Waste announced in September 1993 that it would no longer pursue the incinerator plan, citing shifting economic conditions and public policies around incineration. Despite these stated claims regarding the motives behind their decision, it was clear that corporate leaders had been deeply affected by the dedicated activism of Kettleman City residents. The Chem Waste Kettleman Hills Facility manager even hand-delivered the decision to one of El Pueblo's leaders, noting how grassroots organizing had informed their deliberations.[32]

As El Pueblo anticipated and fought the slow violence wrought not only by the proposed incinerator, but also by water contamination and pesticide drift, the organization emphasized that factors behind the disproportionate pollution exposure stemmed from structural racism that also manifested in higher poverty rates, heightened mortgage foreclosure rates, low educational attainment, and high rates of incarceration. Drawing on the area's remarkable tradition of labor organizing among farmworkers, El Pueblo identified connections among exploited labor, environmental threats, and vulnerable bodies. Kettleman City's history reveals that environmental justice historical analysis must consider the embodied experiences of environmental toxicity and center the intersectional nature of frontline communities' political organizing—that is, the ways in which environmental consciousness is informed by anti-racist practices.[33]

Grassroots Organizing in Louisiana's Chemical Corridor

Located on the banks of the Mississippi, 25 miles west of New Orleans, Diamond, a subdivision of Norco, Louisiana, is an African American community with many residents who trace their history to ancestors who worked the sugarcane fields as slaves on the area's Trépagnier Plantation (which later became the Diamond Plantation) along the Mississippi Delta's German Coast. In 1811, planter Jean-Francois Trépagnier was killed during one of the largest slave rebellions in American history.[34] Today Diamond is situated in the middle of the "chemical corridor," an 85-mile-long region which stretches from Baton Rouge along the Mississippi down to southeastern New Orleans, and long dubbed "cancer alley" by local residents. Home to over 150 petroleum refineries and petrochemical plants, the region accounts for a quarter of petrochemical production in the United States. Poor air quality and other environmental harms within the chemical corridor particularly burden low-income minority communities because of their limited access to public health resources. Since the 1980s residents of parishes within this region have documented high rates of cancer, miscarriages, and respiratory illnesses such as asthma and bronchitis. Frontline communities such as Diamond face disproportionate risks from industrial facilities' toxic air emissions, including known and likely carcinogens such as ethylene oxide and chloroprene. Environmental justice scholars have noted how frontline communities' status as sacrifice zones fits within the capitalist imperative and legacy of neoliberal governance regarding environmental affairs: communities' disproportionate exposure to toxic emissions results from both lax environmental regulations and corporations' focus on generating goods with artificially low prices, regardless of the health impacts on surrounding communities.[35]

New Orleans Refining Company (subsequently acquired by Shell Petroleum Company) began building the area's first oil refinery near Diamond in 1916, and in the early 1950s Royal Dutch/Shell began building the chemical plant after purchasing land where freed slaves had formerly resided, tearing down houses of Diamond Plantation slaves' descendants, and forcing them to rebuild at the fence line of the refinery and chemical plant. Black residents' frustration grew as the plants continued to expand near their homes, and in the mid-1970s a group of women in Diamond organized protests demanding that Shell buy their properties and relocate them.[36]

Local schoolteacher Margie Eugene-Richard spearheaded the grassroots movement, the Concerned Citizens of Norco, in the late 1980s, garnering media attention and the support of environmental justice activists (see Source 2). Eugene-Richard lived just across the fence from Shell Chemical and had also worked on a temporary basis cleaning the plant's facilities during periods of maintenance and repair. Eugene-Richard recalled that one time when she was cleaning, she was startled to see that the collar of her sweater turned green. Concerned about toxic substances coming home to her children, she looked up the chemicals that were in her midst at the plant and discovered that they were known carcinogens.[37] She was certain that government environmental agencies were not adequately monitoring the full range of chemicals that residents were exposed to, so she and her allies began a "bucket brigade" in 1996, using bucket air samplers to provide a better sense of neighboring plants' toxic pollution. This data collected through citizen science measures lent credibility to advocates' demands.[38]

Despite activism by Eugene-Richard and others over the course of several decades, Shell ignored requests for relocation. The company quietly and slowly purchased neighboring properties at rock-bottom prices as they came on the market. It was not until 1999 that Shell officials met face-to-face with Diamond residents on the community's home turf. About a year later, Shell offered to buy out 75 Diamond lots located closest to the plants to create a "buffer zone" around the plant. Shell claimed this buffer zone was not necessary on account of safety concerns but rather was important on the grounds of preventing noise and visual nuisances for the community. This voluntary buyback program essentially cut the community in half, leaving 129 families without the option, and splitting a close-knit community where low-income neighbors relied on each other to survive. One resident who sat outside the relocation zone questioned the ethics of Shell's decision to only pursue limited relocations: "I am smelling the same thing as the people on those two streets [that are being relocated].... Where is Shell's moral sense? Why are they doing this? The Shell company is a multi-billion-dollar company and I'm quite sure that buying these few little properties is not going to break them. They can afford it."[39]

Diamond residents feared not only the health implications of toxic air emissions, but also periodic toxic spills and explosions, as well as ground contamination from decades of blowing dust. Diamond's four roads were sandwiched between a Shell Chemical plant and the sprawling Shell/

Motiva oil refinery. Residents suffered a constant sensory barrage of acrid industrial smells akin to bleach and ammonia, deafening noises from the factory and railroads, and stacks topped by bright flares burning off excess gas. Residents noted that these nuisances from the facility caused nausea, headaches, sore throats, dizziness, and difficulty breathing. In 1973, an explosion killed two Diamond residents. When an explosion killed seven Shell workers in 1988, it blew out windows and doors in Diamond homes, and rained down sheetrock ceilings, prompting middle-of-the-night evacuations across the community. In addition to chronic concerns over emergencies, Diamond residents complained of being deprived of their families' deep roots in agriculture. Recalling their parents' and grandparents' vegetable gardens that were vital to their families' subsistence over the years, residents lamented that they could not even grow flowers in their backyards because of Shell's toxic emissions. The hazards of Diamond were multitudinous, but most residents could not afford to leave because their homes' proximity to Shell plants lowered their property values.[40]

Even though Diamond was a small town, it experienced similar patterns of environmental racism as large, segregated cities such as Chicago and Baltimore because of the racialized history of land-use planning. While redlining and zoning have protected white residential areas from industrial development, communities of color face more mixed-use zoning, including industrial and commercial zones.[41] As Diamond's Black residents organized to demand that Shell relocate them, some of their neighbors in the adjacent white community of Norco—separated by railroad tracks and woodland—claimed that the Black residents were lying or exaggerating about the problem in an effort to push a settlement that would award them financial compensation from Shell. As sociologist Arlie Hochschild has documented, many rural white Louisianans claimed that it was worth lifting environmental regulations and risking environmental degradation in order to protect jobs in the state. However, many rural Black Louisianans supported more rigorous regulation because they faced significant toxic exposures and did not benefit from employment opportunities at nearby industrial plants.[42] Through citizen science and grassroots organizing, Diamond residents eventually succeeded in negotiating a comprehensive settlement from Shell, which expanded the buyout program from the two closest streets adjacent to the Shell plant to encompass the town's entire four streets, while also offering a home improvement loan option for those who decided to remain in the neighborhood. This widely publicized negotiation provided an empowering model for other small communities that struggled against major corporations to establish safer residential environments. In addition to the homeowners' improvement and buyout programs, the settlement included $200 million from Shell for capital improvements in upgrading equipment to reduce emissions over seven years, as well as $5 million devoted to the creation of a community development district in Norco, supervised not by Shell employees, but by local residents focused on developing the town's infrastructure, small businesses, job training programs, support programs for the elderly, and other community wellbeing initiatives.[43]

Extractive Capitalism and Indigenous Dispossession from #NoDAPL to Nuclear Colonialism

The interwoven North American histories of settler colonialism and environmental racism come into focus when examining the environmental injustices experienced by Indigenous peoples who exercise treaty rights to tribal resources and sacred sites that have become affected by oil and gas pipeline projects as well as other resource extraction projects such as nuclear ore mining.[44] In April 2016 LaDonna Brave Bull Allard of the Standing Rock Sioux Tribe and her grandchildren founded the Sacred Stone Camp to protest the construction of the Dakota Access Pipeline (DAPL), stretching from North Dakota to Illinois, and part of the Bakken Oil Pipeline—the largest crude oil pipeline in the United States—out of concern for protecting their ancestral heritage and preventing future water pollution in the upper Missouri River and Lake Oahe, the Standing Rock Reservation's main water source. At the height of the #NoDAPL protests in September 2016, tribal citizens from nearly 300 federally recognized Native American tribes and their allies (3,000–4,000 demonstrators in total) gathered at the camp and engaged in direct action, including chaining their bodies to construction vehicles. Security guards responded to the peaceful protest by attacking protestors with rubber bullets, pepper-spray, water cannons, and guard dogs.[45] Although the protests succeeded in temporarily blocking the pipeline, President Donald Trump's January 2017 executive order accelerated the Dakota as well as the Keystone XL pipelines' construction, and in June crude oil began pumping through the Dakota pipeline. The DAPL collective action continues to inspire Indigenous and environmental justice protests against pipelines and fracking across the United States.[46]

Standing Rock signifies a landmark battle for environmental justice vis-à-vis Indigenous resistance against land-based violence, and it marked an important moment of Native and non-Native solidarity in challenging the unjust histories of Native dispossession and violent extractive economic production enforced through state-sanctioned violence.[47] Commenting on the #NoDAPL campaign, Dian Million has noted:

> Indigenous places are often imagined as isolated empty places, disposable, or usable places subordinate to national need. Indigenous peoples are not isolated, in a past, outside of capital, or without

capitalist relations: we are central to them. We get past some kinds of 'geographical' differences where we foreground other relations: the relations revealed, for instance, between the necessity and desire for life and clean water in African American communities in Flint, Michigan, juxtaposed with these needs in Standing Rock.[48]

Recognizing the distinct natures of environmental injustices in metropolitan and nonurban places, as well as African American and Indigenous communities, Million drew connections across these divides in the universal desire for basic dignities and tools for survival secured through access to environmental goods.

Another history of Indigenous dispossession spurred by the capitalist logic of extractive economies that helps us understand the connectivity of nonurban and urban places is Native communities' experiences with industrial mining in Canada's territorial north. Beginning in the 1930s, the traditional land of the Sahtu Dene, an Indigenous group from the Great Bear Lake region of the Northwest Territories, became the site of radium and uranium mining and related infrastructure at the service town of Port Radium. In 1944, the Canadian government nationalized the mine as part of the Allied atomic bomb program.[49] Local Dene people joined the ranks of unskilled labor at the mine, loading and unloading ore in burlap sacks that traveled by barge to an Alberta railhead. In addition to the occupational health concerns from Native workers' exposures to radioactive material, the radium and uranium production introduced new environmental hazards as workers dumped radioactive tailings near Port Radium, and into Great Slave Lake, among other local pothole lakes (shallow bodies of water formed in outwash plains by retreating glaciers).[50]

In 1962 uranium and radium mining ended at Port Radium, although it was repurposed as a silver mine through 1982. Over the ensuing decades, high rates of cancer among Native people who were former ore carriers prompted an Indigenous collective action campaign demanding remediation. In response, a collaboration formed between the national government and community, resulting in a series of environmental and health studies. The studies did not determine that uranium exposure was directly responsible for the high cancer rates and concluded that the radioactive tailings did not harm the environment.[51] However, the Sahtu Dene have disagreed with the studies' conclusions and pushed for further examination of the mine's environmental health impacts. Beyond concerns over physical health effects, the Sahtu Dene have explained the detrimental mental health effects experienced by those who witnessed their community's and surrounding landscape's transformation via industrial development.[52]

The Sahtu Dene's situation is not unique across North America and fits within a broader history of "nuclear colonialism" where geopolitical battles over resources have turned Indigenous territories into "sacrifice zones" through uranium mining, nuclear testing, and nuclear waste siting.[53] This chain of nuclear material production has disproportionately burdened Indigenous communities with environmental harms while benefiting states and corporations.[54]

Other forms of industrial mining besides uranium demonstrate similar patterns indicating extractive capitalism's harmful effects on Indigenous communities in the territorial North. From the 1930s through 2005, gold mining at Yellowknife on the north shore of Great Slave Lake in Canada's Northwest Territories has negatively affected the nearby Dene communities of Dettah and Ndilo. The ore smelting process produced toxic emissions of the arsenic trioxide, a known carcinogen, and studies from the 1970s revealed that 90 percent of Dene children had arsenic levels above the safety threshold of one part per million.[55] At the same time that these Dene communities encountered a multitude of health challenges stemming from the nearby mines, they also did not receive significant economic benefits due to discriminatory hiring patterns and minimal training opportunities, among other factors. For example, in 1968, five mines in Yukon Territory had a workforce that was only 3.4 percent Indigenous, and six mines in the Northwest Territories had a workforce that was 5.4 percent Indigenous.[56] As companies have abandoned mines over the decades, remediation has become a key concern for Indigenous communities. A study from 2002 found that 36 percent of First Nations communities were located within 31 miles of a mine.[57]

Such nonurban areas that are not devoted to agricultural production have long been susceptible to the Euro-American discourse of wasteland—framing the land as an empty space, fit for the siting of toxic waste, military arms testing, and other unwanted land usages. Traci Brynne Voyles has explained the process of "wastelanding" as "a fully colonial project of rendering resources extractable and lands and bodies pollutable rather than merely a problem of the distribution of environmental 'bads.'"[58] Valerie Kuletz has identified the Euroamerican cultural perception of the desert as a barren wilderness as justifying desert lands' domination by the American military-industrial complex, and the exploitation of the Indigenous spaces within the American West as sites of colonial nuclearism and "national sacrifice zones."[59] Militarization, nuclearism, and the concentration of scientific power is especially apparent at major nuclear weapons lab sites, including Los Alamos and Sandia in New Mexico, and the Naval Air Weapons Station China Lake in the Mojave Desert's Indian Wells Valley.

In examining Indigenous homelands within the framework of nuclear colonialism, environmental justice scholars have explored the context of settler colonialism at the intimate level of the body. This operates at both material and ontological levels. As "wasteland" areas became environmentally degraded, pollution affects both the bodily health and the political power of the colonized. In examining the history of uranium mining on Diné (Navajo) lands, Traci Brynne Voyles explains that wasteland "becomes a place where pollution and environmental degradation collect, settle, and form sediment that makes a lasting impact on human and nonhuman bodies," and, likewise, "wasteland discourses collect and sediment to give shape to power relations between peoples and geographies, creating a highly spatialized set of power relations that invoke place as well as race."[60]

Gregory Hooks and Chad L. Smith have proposed the idea of a "treadmill of destruction," emerging from geopolitics rooted in settler colonialism, to help explain why Native American nonurban lands are often located nearby closed military bases that are hazardous on account of their unexploded ordnance.[61] Throughout the nineteenth century, the US federal government and American settlers violently dispossessed Native Americans through the so-called Indian Wars, and the removal of Native Americans to reservations west of the Mississippi. As the federal government mobilized for two world wars and the Cold War, land acquired for military operations shared similar characteristics with land on which Native American reservations were concentrated: it was primarily located in Western states, remote, arid, and undesirable from the perspective of white settlers in terms of agricultural and industrial productivity. The military's newly acquired lands were often adjacent to Indian reservations. For example, in 1940, President Franklin Delano Roosevelt established Southern Nevada's 3.5-million-acre Nellis Range Complex, the world's largest gunnery range at the time. During the Cold War, the US military incorporated the bombing range into the nuclear weapons complex—the largest "peacetime militarized zone on earth"—at the Nevada Test Site, renamed the Nevada National Security Site in 2010.[62] This area encompasses the traditional lands of the Western Shoshone and Southern Paiute people. And 50 years after the Wounded Knee Massacre of 1890, the US government seized approximately 342,000 acres of the Pine Ridge Reservation in South Dakota to establish a bombing range for pilot training, requiring 125 Oglala Lakota to sell their land for three cents an acre.[63]

Due to the dynamic and interconnected natures of coupled human-natural systems, the militarization of Western landscapes has had extensive environmental and public health consequences reaching far beyond local communities. As detailed by Voyles, during the ramp-up of military-industrial partnerships during World War II, the federal Defense Plant Corporation enlisted Basic Magnesium, Incorporated (BMI), an Anaconda Copper Mining company subsidiary, to establish a plant in Henderson, Nevada, in 1941, that produced magnesium—the "miracle metal" that combined with aluminum to create strong but lightweight alloy parts for airplanes, helicopters, and rockets. During the Cold War, a new company, the Pacific

Engineering Production Company of Nevada converted the facility to join a nearby Navy plant in producing the highly combustible inorganic compound ammonium perchlorate, which was used in the US military's rocket fuel. Easily dissolved, perchlorate contaminated local groundwater, and flowed downhill via the Lower Colorado River basin, creating a multitude of health problems for people who relied on the river water for bathing, drinking, and farming. One important health problem resulting from the water contamination was hypothyroidism, which is associated with adults' metabolic complications and infants' developmental problems.

This toxic contaminant had a particularly pernicious afterlife for the local water sources of the Torres Martinez and Agua Caliente bands of the Cahuilla Nation in the Coachella Valley of Southern California. In response to water shortages and groundwater depletion, the Coachella Valley Water District pumped in untreated Colorado River water laden with dissolved perchlorate, pesticides, and nitrates. From 1996 to 2015, the Water District ignored demands from the Agua Caliente Cahuillas to stop pumping, or at least treat the Colorado River intake to meet EPA contamination standards. A landmark federal lawsuit in 2017 acknowledged the Agua Caliente Cahuillas as possessing a "sovereign right to drinkable groundwater underneath their reservations." At the southern end of the Coachella aquifer sits the Torres Martinez Reservation where the tribal EPA discovered that perchlorate had contaminated tap water at dangerous levels. However, without a federal maximum contaminant level designation by the US EPA, no regulatory measure exists to monitor and prevent elevated rates of perchlorate.[64] As Indigenous communities continue to suffer enduring legacies from "nuclear colonialism," their resistance and related claims to sovereignty and the right to a healthy environment underscore the importance of examining the linked histories of toxic heritage stemming from the industrial-military complex's appropriation of Indigenous land.

Environmental Justice and the Carceral State in Rural America

Over the course of the twentieth century, rural environments in North America served as key sites of imprisonment within the carceral state and dynamically shaped the experiences of incarcerated individuals. As Connie Chiang has argued, during World War II, the US War Relocation Authority (WRA) and War Department exiled Japanese Americans to incarceration camps that were located far from urban areas and strategic military sites, due to security considerations and anti-Japanese sentiment. Eight camps were located in the interior West, and two were located in Arkansas. Seven of the ten camps were located on federal land, including land administered by the Bureau of Reclamation, Bureau of Indian Affairs, and Farm Security Administration. Officials claimed to select sites such as Manzanar and Tule

Lake in California based on their potential for cultivation and agricultural productivity so that the camps would be self-sufficient and contribute to the war effort, but financial cost often trumped these considerations. The camps' living quarters lacked insulation and the walls cracked in extreme desert heat and winter cold, leading to unsanitary conditions and the intrusion of flies, mosquitoes, scorpions, and other unwanted creatures. With construction materials and labor in demand during the war, inexperienced construction crews used subpar lumber and pipes. Most of the detainees were from the Pacific Coast and were forced to endure unfamiliar and harsh rural environments, including arid and windy intermountain Western landscapes prone to dust storms.[65]

As WRA officials turned toward isolated, rural environments as a means of exerting social control over Japanese Americans and facilitating the exploitation of their labor for the war effort, detainees also capitalized on their environments' complexities as an instrument of resistance. Officials deemed incarcerated Japanese Americans' labor in nature as a means of reform and assimilation. Through gardening and recreation in nature (including hiking, fishing, and camping within the camps' grounds), Japanese Americans asserted some degree of control over their detainment and developed their own sense of American identity that pushed against the government's and American public's exclusionary measures.

Still another form of resistance came from incarcerated Japanese Americans' efforts to control their own labor within their harsh prison environments. In the Sevier Desert at the Central Utah (Topaz) Relocation Center, alkaline soil corroded the water pipelines, leading to leaks, standing water, and mosquito infestations. Imprisoned Japanese Americans detested their assigned labor repairing the pipelines due to the strenuous and unpleasant tasks of digging ditches and laying pipes, and on account of their belief that it was the responsibility of the federal government to fix the camp's infrastructure. Detainees protested the inadequate pay that they received from the WRA. Ultimately the WRA enlisted enough detained men by threatening to fire them and terminate their paychecks and allowances for family clothing—leading many within the camp to equate their work with forced labor.[66] While the arid, rural environment of the WRA camps created difficult living conditions for Japanese Americans, incarcerated individuals also turned toward the natural environment as an instrument that allowed them to exert some degree of power through supporting their survival and resistance.

In the postwar era, rural America witnessed the rise of prisons as a "growth industry," especially since the 1980s at which point most new prisons came to be built in non-metropolitan areas. Although most prisoners in America were originally residents of urban areas, most prisons are located in rural areas. As globalization has restructured socioeconomic networks over the past decades, rural areas have suffered from the attrition of farming, manufacturing, and mining industries, as evidenced by farm consolidation, corporate downsizing, and factory closings. Although rural communities have embraced the prison industry as a means of creating jobs and stimulating economic growth, increasing evidence indicates that prisons have negative effects on the social and natural environments of rural areas.

Many prison towns have not witnessed substantive job growth and have faced unintended negative consequences, including pollution and political disenfranchisement. Ruth Wilson Gilmore found that on average fewer than 20 percent of prison jobs go to the current residents of a town with a new state prison. Higher-paying correctional officer and management jobs require years of education and training, leaving many rural residents ineligible for such work. Communities with prisons tend to attract big box-store chains that displace locally owned businesses, leading to limited increases in tax revenues and a loss of reinvestment from local companies. Rural prison towns also tend to face difficulty bringing in other new business investments due to several factors, including the notion of a prison as an "undesirable neighbor," and on account of prisons' relative lack of regulatory environmental controls. California prison towns like Avenal and Tehachapi have faced significant water quality issues, on account of higher rates of nitrates and other pollutants. Finally, the disproportionate number of prisons in rural areas has resulted in a loss of political power and representation for urban areas, as political redistricting has led to increased rural census figures from incarcerated populations, even as most of these prisoners are disenfranchised. The voting power of predominantly Black and Hispanic urban prison populations has been diluted due to this redistricting process, which has benefitted predominantly white, rural areas where new prisons have cropped up over the past several decades.[67]

Conclusion

The overlapping intersectionalities of race, class, and place have shaped patterns of political power in nonurban North America, bringing disproportionate environmental hazards to Black, Brown, and Indigenous communities. As land development in North America has catalyzed urban sprawl and the growth of exurbs at the rural-urban fringe over the past half century, unintended environmental problems arose—including pollution, wildfires, limited water supplies, and public health threats from animal-to-human disease transmission—that further exacerbate the distribution of environmental harms.[68]

In order to promote the equitable development of nonurban areas alongside responsible conservation initiatives in the face of a rapidly changing climate, it is imperative to democratize governance and management of natural resources and political decision-making. Political ecologists have shown that ownership and control over property produces not only economic power, but also political power. Globalization, industrialization, and consolidation have pushed out smaller farmers and nonurban landowners within the forestry and agricultural sectors, prompted the dissipation of solid blue-collar jobs in the wake of corporate relocations, and left a legacy of economic-political marginalization among rural populations as well as underinvestment in nonurban infrastructure.[69]

Political ecologists have also recently revisited the process of making and remaking the commons—or "communing"—in an effort to bring a feminist perspective to providing an alternative to neoliberal governance. Key within this feminist political ecology approach is economist Elinor Ostrom's idea of polycentricity, a complex governance system that facilitates collective action in response to disturbance and change by bringing together a diversity of governing bodies and stakeholders in semi-autonomous and overlapping decision-making centers.[70] Polycentric climate governance for climate justice is especially important for enhancing the resilience of nonurban areas that do not have as much economic and political capital as metropolitan centers.[71]

Environmental history is particularly well-poised to contribute to a polycentric model of climate justice by drawing out the complex roots of historic environmental injustices, identifying stakeholders whose voices have long been neglected in public debates, and uncovering interwoven layers of causality across social, political, economic, and environmental spectrums. Environmental historians' craft of storytelling can also offer powerful narratives that contribute to public deliberations in meaningful ways and break through politicized media circuits. Prioritizing the inclusion of historically disadvantaged groups within political decision-making and environmental planning offers an important pathway forward toward restorative justice and a more equitable future.

Notes

1 Nicholas Low and Brendan Gleeson, *Justice, Society and Nature: An Exploration of Political Ecology* (London: Routledge, 1998), 131.
2 Laura Pulido, "A Critical Review of the Methodology of Environmental Racism Research," *Antipode* 28, no. 2 (1996): 142–59; Michael K. Hieman, "Race, Waste, and Class: New Perspectives on Environmental Justice," *Antipode* 28, no. 2 (1996): 111–21; David N. Pellow, "Environmental Inequality Formation: Toward a Theory of Environmental Injustice," *American Behavioral Scientist* 43, no. 4 (2000): 581–601; David R. Simon, "Corporate Environmental Crimes and Social Inequality: New Directions for Environmental Justice Research," *American Behavioral Scientist* 43, no. 4 (2000): 633–45; Arn Keeling and John Sandlos, "Environmental Justice Goes Underground?

Historical Notes from Canada's Northern Mining Frontier," *Environmental Justice* 2, no. 3 (2009): 121.

3 Hilda E. Kurtz, "Scale Frames and Counter-Scale Frames: Constructing the Problem of Environmental Injustice," *Political Geography* 22, no. 8 (2003): 887–916; Thomas E. Shriver and Gary R. Webb, "Rethinking the Scope of Environmental Injustice: Perceptions of Health Hazards in a Rural Native American Community Exposed to Carbon Black," *Rural Sociology* 74, no. 2 (2009): 270–92.

4 Stephanie A. Malin and Kathryn Teigen DeMaster, "A Devil's Bargain: Rural Environmental Injustices and Hydraulic Fracturing on Pennsylvania's Farms," *Journal of Rural Studies* 47, part A (2016): 278–90.

5 Elizabeth Hoover, *The River Is in Us: Fighting Toxics in a Mohawk Community* (Minneapolis: University of Minnesota Press, 2017).

6 Laura A. Bray, "Settler Colonialism and Rural Environmental Injustice: Water Inequality on the Navajo Nation," *Rural Sociology* 86, no. 3 (2021): 591; Laura McKinney, "Reinventing Rural Environmental Justice," in *Reinventing Rural: New Realities in an Urbanizing World*, ed. Gregory M. Fulkerson and Alexander R. Thomas (Lanham, MD: Lexington Books, 2016), 57–76.

7 U.S. Census Bureau, *Table 1. Urban and Rural Population: 1900 to 1990* (Washington, DC: U.S. Census Bureau, October 1995), https://www2.census.gov/programs-surveys/decennial/tables/1990/1990-urban-pop/urpop0090.txt; U.S. Census Bureau, GCT-P1. Urban/Rural and Metropolitan/Nonmetropolitan Population: 2000, in *Census 2000 Summary File 1* (Washington, DC: U.S. Census Bureau, 2010), Bur. https://www.census.gov/programs-surveys/geography/guidance/geo-areas/urban-rural/2000-urban-rural.html; see also Daniel T. Lichter and David L. Brown, "Rural America in an Urban Society: Changing Spatial and Social Boundaries," *Annual Review of Sociology* 37 (2011): 565–92.

8 Douglas B. Jackson-Smith and Eric Jensen, "Finding Farms: Comparing Indicators of Farming Dependence and Agricultural Importance in the United States," *Rural Sociology* 74, no. 1 (2009): 375–55.

9 Doreen Massey, *Space, Place, and Gender* (Minneapolis: University of Minnesota Press, 1994), 22.

10 Bray, "Settler Colonialism."

11 Evelyn Nakano Glenn, "Settler Colonialism as Structure: A Framework for Comparative Studies of US Race and Gender Formation," *Sociology of Race and Ethnicity* 1, no. 1 (2015): 52–72; Patrick Wolfe, "Land, Labor, and Difference: Elementary Structures of Race," *The American Historical Review* 106, no. 3 (2001): 866–905.

12 Rob Nixon, *Slow Violence and the Environmentalism of the Poor* (Cambridge, MA: Harvard University Press, 2011), 2

13 Eileen Maura McGurty, *Transforming Environmentalism: Warren County, PCBs, and the Origins of Environmental Justice* (New Brunswick, NJ: Rutgers University Press, 2007), 1; Dollie Burwell and Luke W. Cole, "Environmental Justice Comes Full Circle: Warren County Before and After," *Golden Gate University Environmental Law Journal* 1 (2007): 11; Robert D. Bullard, *Dumping in Dixie: Race, Class, and Environmental Quality* (Boulder, CO: Westview Press, 1990). For a related history of PCB contamination in Anniston, Alabama, see Ellen Griffith Spears, *Baptized in PCBs: Race, Pollution, and Justice in an All-American Town* (Chapel Hill: University of North Carolina Press, 2014).

14 Burwell and Cole, "Environmental Justice Comes Full Circle," 23–27. Darryl Fears and Brady Dennis, "'This Is Environmental Racism': How a

Protest in a North Carolina Farming Town Sparked a National Movement," *The Washington Post*, April 6, 2021, https://www.washingtonpost.com/climate-environment/interactive/2021/environmental-justice-race/

15 Bean v. Southwestern Waste Management Corp., 482 F. Supp. 673 (S.D. Tex. 1979).

16 Procedural justice refers to the due process of law—the fairness of the application of the law as opposed to the spirit of the law itself. Distributive justice denotes the distribution of burdens and benefits across society on account of policies, laws, social norms, and institutional practices.

17 US General Accounting Office Resources Community and Economic Development Division, *Siting of Hazardous Waste Landfills and Their Correlation with Racial and Economic Status of Surrounding Communities* (Washington, DC: Government Printing Office, 1983); Commission for Racial Justice of the United Church of Christ, *Toxic Wastes and Race in the United States: A National Report on the Racial and Socio-Economic Characteristics of Communities with Hazardous Waste Sites* (New York: United Church of Christ, 1987).

18 Burwell and Cole, "Environmental Justice Comes Full Circle," 31–34.

19 Burwell and Cole, "Environmental Justice Comes Full Circle," 36–40.

20 Land, Chemicals, and Redevelopment Division of the US EPA Region 9, "Draft Environmental Justice Analysis for the Kettleman Hills Facility Proposed TSCA Permit," CAT000646117 (San Francisco: EPA, 2019), 17, https://www.epa.gov/sites/default/files/2019-08/documents/final_draft_kettleman_ej_analysis.pdf

21 Jesse McKinley, "No Cause Found for Cluster of Birth Defects," *New York Times*, February 9, 2010, A18.

22 "We Speak for Ourselves: The Struggle of Kettleman City," in *From the Ground Up: Environmental Racism and the Rise of the Environmental Justice Movement*, ed. Luke W. Cole and Sheila R Foster (New York: New York University Press, 2000); Julie Sze, "Denormalizing Embodied Toxicity: The Case of Kettleman City," in *Racial Ecologies*, ed. Leilani Nishime and Kim D. Hester Williams (Seattle: University of Washington Press, 2018), 109–10.

23 Jill Lindsey Harrison, *Pesticide Drift and the Pursuit of Environmental Justice* (Cambridge, MA: MIT Press, 2011).

24 See also Carolnia Balazs, Rachel Morello-Frosch, Alan Hubbard, and Isha Ray, "Social Disparities in Nitrate-Contaminated Drinking Water in California's San Joaquin Valley," *Environmental Health Perspectives* 119, no. 9 (2011): 1272–78.

25 Sze, "Denormalizing Toxicity," 110.

26 Alatorre, qtd. in Cole and Foster, *From the Ground Up*, 7.

27 Laura Pulido, *Environmentalism and Economic Justice: Two Chicano Struggles in the Southwest* (Tucson: University of Arizona Press, 1996); Laura Pulido and Devon Peña, "Environmentalism and Positionality: The Early Pesticide Campaign of the United Farm Workers' Organizing Committee, 1965–71," *Race, Gender & Class* 6, no. 1 (1998): 33–50.

28 Sze, "Denormalizing Toxicity," 111.

29 Cole and Foster, *From the Ground Up*, 1–3.

30 Cole and Foster, *From the Ground Up*, 3–4.

31 Cole and Foster, *From the Ground Up*, 8–9.

32 Cole and Foster, *From the Ground Up*, 9.

33 Sze, "Denormalizing Embodied Toxicity."

34 Daniel Rasmussen, *American Uprising: The Untold Story of America's Largest Slave Revolt* (New York: Harper, 2011), 108–10; Adam Rothman, *Slave*

Country: American Expansion and the Origins of the Deep South (Cambridge, MA: Harvard University Press, 2005), 108, 111.

35 Steve Lerner, *Diamond: A Struggle for Environmental Justice in Louisiana's Chemical Corridor* (Cambridge, MA: MIT Press, 2004), 1-3; Idna G. Castellón, "Cancer Alley and the Fight Against Environmental Racism," *Villanova Environmental Law Journal* 32, no. 1 (2021), https://digitalcommons.law.villanova.edu/elj/vol32/iss1/2; Wesley James, Chunrong Jia, and Satish Kedia, "Uneven Magnitude of Disparities in Cancer Risks from Air Toxics," *International Journal of Environmental Research and Public Health* 9, no. 12 (2012): 4365–85.

36 Lerner, *Diamond*, 11–12.

37 Lerner, *Diamond*, 122–23.

38 Dara O'Rourke and Gregg P. Macey, "Community Environmental Policing: Assessing New Strategies of Public Participation in Environmental Regulation," *Journal of Policy Analysis and Management* 22, no. 3 (2003): 383–414.

39 Lerner, *Diamond*, 127–37.

40 Lerner, *Diamond*, 9–10, 64–65.

41 Lerner, *Diamond*, 3.

42 Arlie Russell Hochschild, *Strangers in Their Own Land: Anger and Mourning on the American Right* (New York: New Press, 2016).

43 Lerner, *Diamond*, 250–51.

44 Jeanette Wolfley, "Embracing Engagement: The Challenges and Opportunities for the Energy Industry and Tribal Nations on Projects Affecting Tribal Rights and Off-Reservation Lands," *Vermont Journal of Environmental Law* 19, no. 2 (2018): 115–63.

45 Jack Healy, "From 280 Tribes, a Protest on the Plains," *New York Times*, September 11, 2016.

46 Julie Sze, *Environmental Justice in a Moment of Danger* (Berkeley: University of California Press, 2020), chap 1.

47 Sze, *Environmental Justice in a Moment of Danger*, 28.

48 Dian Million, "'We are the Land, and the Land is Us': Indigenous Land, Lives, and Embodied Ecologies in the Twenty-First Century," in *Racial Ecologies*, 25.

49 Robert Bothwell, *Eldorado: Canada's National Uranium Company* (Toronto: University of Toronto Press, 1984).

50 Peter C. van Wyck, "The Highway of the Atom: Recollections Along a Route," *Topia* 7 (2002): 99–115; Keeling and Sandlos, "Environmental Justice," 117.

51 Déline First Nation and Indian and Northern Affairs Canada, "*Canada-Déline Uranium Table, Final Report Concerning Health and Environmental Issues Related to the Port Radium Mine*," (Ottawa: Indian and Northern Affairs Canada, 2005).

52 Déline First Nation, *If Only We Had Known: The History of Port Radium as Told by the Sahtuot'ine* (Déline: Déline First Nation, 2005).

53 Valerie L. Kuletz: *The Tainted Desert: Environmental Ruin in the American West* (New York: Routledge, 1998); Winona LaDuke and Ward Churchill, "Native America: The Political Economy of Radioactive Colonialism," *Journal of Ethnic Studies* 13, no. 3 (1985): 107–32; Traci Brynne Voyles, *Wastelanding: Legacies of Uranium Mining in Navajo Country* (Minneapolis: University of Minnesota Press, 2015).

54 Keeling and Sandlos, "Environmental Justice," 118.

55 Lisa Sumi and Sandra Thomsen, *Mining in Remote Areas: Issues and Impacts* (Ottawa: MiningWatch Canada, 2001), 19; Keeling and Sandlos, "Environmental Justice," 119–20.

56   Department of Indian Affairs and Northern Development, "Native Labour in the Northern Mining Industry," qtd. in Keeling and Sandlos, "Environmental Justice," 120.
57   William Hipwell, Katy Mamen, Viviane Weitzner, and Gail Whiteman, *Aboriginal People and Mining in Canada: Consultation, Participation and Prospects for Change: Working Discussion Paper* (Ottawa: North-South Institute, 2002), 4.
58   Voyles, *Wastelanding*, 10–11, 24. See also Elizabeth Grennan Browning, "Wastelanding and Racialized Reproductive Labor: 'Long Dyings' in East Chicago from Urban Renewal to Superfund Remediation," *Environmental History* 26, no. 4 (2021): 749–75.
59   Kuletz, *The Tainted Desert.*
60   Voyles, *Wastelanding*, 23. See also Tao Leigh Goffe, "'Guano in Their Destiny': Race, Geology, and a Philosophy of Indenture," *Amerasia Journal* 45, no. 1 (2019): 27–49; Melvin Oliver and Thomas Shapiro, *Black Wealth, White Wealth: A New Perspective on Racial Inequality* (New York: Routledge, 1995), 5; Laura Pulido, "Rethinking Environmental Racism: White Privilege and Urban Development in Southern California," *Annals of the Association of American Geographers* 90, no. 1 (2000): 16; George Lipsitz, "The Racialization of Space and the Spatialization of Race: Theorizing the Hidden Architecture of Landscape," *Landscape Journal* 26, no. 1 (2007): 10–23; Sherene Razack, *Race, Space, and the Law: Unmapping a White Settler Society* (Toronto: Between the Lines, 2002); Sarah Alisabeth Fox, *Downwind: A People's History of the Nuclear West* (Lincoln: University of Nebraska Press, 2014).
61   Gregory Hooks and Chad L. Smith, "The Treadmill of Destruction: National Sacrifice Areas and Native Americans," *American Sociological Review* 69, no. 4 (2004): 558–75.
62   Kuletz, *Tainted Desert*, 38–39.
63   Hooks and Smith, "The Treadmill of Destruction," 565.
64   Traci Brynne Voyles, *The Settler Sea: California's Salton Sea and the Consequences of Colonialism* (Lincoln: University of Nebraska Press, 2021), 190–96.
65   Connie Y. Chiang, *Nature Behind Barbed Wire: An Environmental History of the Japanese American Incarceration* (New York: Oxford, 2018), chap. 2.
66   Connie Y. Chiang. "Imprisoned Nature: Toward an Environmental History of the World War II Japanese American Incarceration." *Environmental History* 15 (2010): 236–67.
67   Tracy Huling, "Building a Prison Economy in Rural America," in *From Invisible Punishment: The Collateral Consequences of Mass Imprisonment*, ed. Marc Mauer and Media Chesney-Lind (New York: The New Press, 2002).
68   For communities developed by socioeconomically privileged Americans beyond the suburban fringe on former wildlands, see Lincoln Bramwell, *Wilderburbs: Communities on Nature's Edge* (Seattle: University of Washington Press, 2014).
69   Zachary D. Swick, "Adaptive Policy and Governance: Natural Resources, Ownership, and Community Development in Appalachia," *Appalachian Journal* 42, no. 1/2 (2014/2015): 38–62.
70   For polycentricity, see Elinor Ostrom, *Understanding Institutional Diversity* (Princeton, NJ: Princeton University Press, 2005); Vincent Ostrom, Charles M. Tiebout, and Robert Warren, "The Organization of Government in Metropolitan Areas: A Theoretical Inquiry," *American Political Science Review* 55, no. 4 (1961): 831–42. For feminist political ecologies of polycentrism, see Floriane Clement, Wendy Jane Harcourt, Deepa Joshi, and Chizu Sato, "Feminist

Political Ecologies of the Commons and Commoning." *International Journal of the Commons* 13, no. 1 (2019): 1–15; Zofia Lapniewska, "Reading Elinor Ostrom through a Gender Perspective," *Feminist Economics* 22, no. 4 (2016): 129–51.
71 Chukwumerije Okereke, "Equity and Justice in Polycentric Climate Governance," in *Governing Climate Change: Polycentricity in Action?*, ed. Andrew Jordan, Dave Huitema, Harro van Asselt, and Johanna Forster (New York: Cambridge University Press, 2018), 320–37; Fernando Tormos-Aponte, and Gustavo A. García-López, "Polycentric Struggles: The Experience of the Global Climate Justice Movement," *Environmental Policy and Governance* 28, no. 4 (2018): 284–94.

Bibliography

Balazs, Carolnia, Rachel Morello-Frosch, Alan Hubbard, and Isha Ray. "Social Disparities in Nitrate-Contaminated Drinking Water in California's San Joaquin Valley." *Environmental Health Perspectives* 119, no. 9 (2011): 1272–78.

Bothwell, Robert. *Eldorado: Canada's National Uranium Company*. Toronto: University of Toronto Press, 1984.

Bramwell, Lincoln. *Wilderburbs: Communities on Nature's Edge*. Seattle: University of Washington Press, 2014.

Bray, Laura A. "Settler Colonialism and Rural Environmental Injustice: Water Inequality on the Navajo Nation." *Rural Sociology* 86, no. 3 (2021): 586–610.

Browning, Elizabeth Grennan. "Wastelanding and Racialized Reproductive Labor: 'Long Dyings' in East Chicago from Urban Renewal to Superfund Remediation." *Environmental History* 26, no. 4 (2021): 749–75.

Bullard, Robert D. *Dumping in Dixie: Race, Class, and Environmental Quality*. Boulder, CO: Westview Press, 1990.

Burwell, Dollie, and Luke W. Cole. "Environmental Justice Comes Full Circle: Warren County Before and After." *Golden Gate University Environmental Law Journal* 1 (2007): 9–40.

Castellón, Idna G. "Cancer Alley and the Fight against Environmental Racism." *Villanova Environmental Law Journal* 32, no. 1 (2021). https://digitalcommons.law.villanova.edu/elj/vol32/iss1/2

Chiang, Connie Y. "Imprisoned Nature: Toward an Environmental History of the World War II Japanese American Incarceration." *Environmental History* 15 (2010): 236–67.

———. *Nature Behind Barbed Wire: An Environmental History of the Japanese American Incarceration*. New York: Oxford University Press, 2018.

Clement, Floriane, Wendy Jane Harcourt, Deepa Joshi, and Chizu Sato. "Feminist Political Ecologies of the Commons and Commoning." *International Journal of the Commons* 13, no. 1 (2019): 1–15.

Cole, Luke W., and Sheila R Foster. *From the Ground Up: Environmental Racism and the Rise of the Environmental Justice Movement*. New York: New York University Press, 2000.

Commission for Racial Justice of the United Church of Christ. *Toxic Wastes and Race in the United States: A National Report on the Racial and Socio-Economic*

Characteristics of Communities with Hazardous Waste Sites. New York: United Church of Christ, 1987.

Déline First Nation. *If Only We Had Known: The History of Port Radium as Told by the Sahtuot'ine*. Déline: Déline First Nation, 2005.

Fox, Sarah Alisabeth. *Downwind: A People's History of the Nuclear West*. Lincoln: University of Nebraska Press, 2014.

Glenn, Evelyn Nakano. "Settler Colonialism as Structure: A Framework for Comparative Studies of US Race and Gender Formation." *Sociology of Race and Ethnicity* 1, no. 1 (2015): 52–72.

Goffe, Tao Leigh. "Guano in Their Destiny': Race, Geology, and a Philosophy of Indenture." *Amerasia Journal* 45, no. 1 (2019): 27–49.

Harrison, Jill Lindsey. *Pesticide Drift and the Pursuit of Environmental Justice*. Cambridge, MA: MIT Press, 2011.

Hieman, Michael K. "Race, Waste, and Class: New Perspectives on Environmental Justice." *Antipode* 28, no. 2 (1996): 111–21.

Hipwell, William, Katy Mamen, Viviane Weitzner, and Gail Whiteman. *Aboriginal People and Mining in Canada: Consultation, Participation and Prospects for Change: Working Discussion Paper*. Ottawa: North-South Institute, 2002.

Hochschild, Arlie Russell. *Strangers in Their Own Land: Anger and Mourning on the American Right*. New York: New Press, 2016.

Hooks, Gregory, and Chad L. Smith. "The Treadmill of Destruction: National Sacrifice Areas and Native Americans." *American Sociological Review* 69, no. 4 (2004): 558–75.

Hoover, Elizabeth. *The River Is in Us: Fighting Toxics in a Mohawk Community*. Minneapolis: University of Minnesota Press, 2017.

Huling, Tracy. "Building a Prison Economy in Rural America." In *From Invisible Punishment: The Collateral Consequences of Mass Imprisonment*, edited by Marc Mauer and Media Chesney-Lind, 197–213. New York: The New Press, 2002.

Jackson-Smith, Douglas B., and Eric Jensen. "Finding Farms: Comparing Indicators of Farming Dependence and Agricultural Importance in the United States." *Rural Sociology* 74, no. 1 (2009): 375–55.

James, Wesley, Chunrong Jia, and Satish Kedia. "Uneven Magnitude of Disparities in Cancer Risks from Air Toxics." *International Journal of Environmental Research and Public Health* 9, no. 12 (2012): 4365–85.

Keeling, Arn, and John Sandlos. "Environmental Justice Goes Underground? Historical Notes from Canada's Northern Mining Frontier." *Environmental Justice* 2, no. 3 (2009): 117–25.

Kuletz, Valerie L. *The Tainted Desert: Environmental Ruin in the American West*. New York: Routledge, 1998.

Kurtz, Hilda E. "Scale Frames and Counter-Scale Frames: Constructing the Problem of Environmental Injustice." *Political Geography* 22, no. 8 (2003): 887–916.

LaDuke, Winona, and Ward Churchill. "Native America: The Political Economy of Radioactive Colonialism." *Journal of Ethnic Studies* 13, no. 3 (1985): 107–32.

Lapniewska, Zofia. "Reading Elinor Ostrom through a Gender Perspective." *Feminist Economics* 22, no. 4 (2016): 129–51.

Lerner, Steve. *Diamond: A Struggle for Environmental Justice in Louisiana's Chemical Corridor*. Cambridge, MA: MIT Press, 2004.

Lichter, Daniel T., and David L. Brown. "Rural America in an Urban Society: Changing Spatial and Social Boundaries." *Annual Review of Sociology* 37 (2011): 565–92.

Lipsitz, George. "The Racialization of Space and the Spatialization of Race: Theorizing the Hidden Architecture of Landscape." *Landscape Journal* 26, no. 1 (2007): 10–23.

Low, Nicholas, and Brendan Gleeson. *Justice, Society and Nature: An Exploration of Political Ecology.* London: Routledge, 1998.

Malin, Stephanie A., and Kathryn Teigen DeMaster. "A Devil's Bargain: Rural Environmental Injustices and Hydraulic Fracturing on Pennsylvania's Farms." *Journal of Rural Studies* 47, part A (2016): 278–90.

Massey, Doreen. *Space, Place, and Gender.* Minneapolis: University of Minnesota Press, 1994.

McGurty, Eileen Maura. *Transforming Environmentalism: Warren County, PCBs, and the Origins of Environmental Justice.* New Brunswick: Rutgers University Press, 2007.

McKinney, Laura. "Reinventing Rural Environmental Justice." In *Reinventing Rural: New Realities in an Urbanizing World*, edited by Gregory M. Fulkerson and Alexander R. Thomas, 57–76. Lanham, MD: Lexington Books, 2016.

Million, Dian. "'We Are the Land, and the Land Is Us': Indigenous Land, Lives, and Embodied Ecologies in the Twenty-First Century." In *Racial Ecologies*, edited by Leilani Nishime and Kim D. Hester Williams, 19–33. Seattle: University of Washington Press, 2018.

Nixon, Rob. *Slow Violence and the Environmentalism of the Poor.* Cambridge, MA: Harvard University Press, 2011.

Okereke, Chukwumerije. "Equity and Justice in Polycentric Climate Governance." In *Governing Climate Change: Polycentricity in Action?*, edited by Andrew Jordan, Dave Huitema, Harro van Asselt, and Johanna Forster, 320–37. New York: Cambridge University Press, 2018.

Oliver, Melvin, and Thomas Shapiro. *Black Wealth, White Wealth: A New Perspective on Racial Inequality.* New York: Routledge, 1995.

O'Rourke, Dara, and Gregg P. Macey. "Community Environmental Policing: Assessing New Strategies of Public Participation in Environmental Regulation." *Journal of Policy Analysis and Management* 22, no. 3 (2003): 383–414.

Ostrom, Elinor. *Understanding Institutional Diversity.* Princeton, NJ: Princeton University Press, 2005.

Ostrom, Vincent, Charles M. Tiebout, and Robert Warren. "The Organization of Government in Metropolitan Areas: A Theoretical Inquiry." *American Political Science Review* 55, no. 4 (1961): 831–42.

Pellow, David N. "Environmental Inequality Formation: Toward a Theory of Environmental Injustice." *American Behavioral Scientist* 43, no. 4 (2000): 581–601.

Pulido, Laura. "A Critical Review of the Methodology of Environmental Racism Research." *Antipode* 28, no. 2 (1996): 142–59.

———. *Environmentalism and Economic Justice: Two Chicano Struggles in the Southwest.* Tucson: University of Arizona Press, 1996.

Pulido, Laura. "Rethinking Environmental Racism: White Privilege and Urban Development in Southern California." *Annals of the Association of American Geographers* 90, no. 1 (2000): 12–40.

_____, and Devon Peña. "Environmentalism and Positionality: The Early Pesticide Campaign of the United Farm Workers' Organizing Committee, 1965–71." *Race, Gender & Class* 6, no. 1 (1998): 33–50.

Rasmussen, Daniel. *American Uprising: The Untold Story of America's Largest Slave Revolt.* New York: Harper, 2011.

Razack, Sherene. *Race, Space, and the Law: Unmapping a White Settler Society.* Toronto: Between the Lines, 2002.

Rothman, Adam. *Slave Country: American Expansion and the Origins of the Deep South.* Cambridge, MA: Harvard University Press, 2005.

Shriver, Thomas E., and Gary R. Webb. "Rethinking the Scope of Environmental Injustice: Perceptions of Health Hazards in a Rural Native American Community Exposed to Carbon Black." *Rural Sociology* 74, no. 2 (2009): 270–92.

Simon, David R. "Corporate Environmental Crimes and Social Inequality: New Directions for Environmental Justice Research." *American Behavioral Scientist* 43, no. 4 (2000): 633–45.

Spears, Ellen Griffith. *Baptized in PCBs: Race, Pollution, and Justice in an All-American Town.* Chapel Hill: University of North Carolina Press, 2014.

Swick, Zachary D. "Adaptive Policy and Governance: Natural Resources, Ownership, and Community Development in Appalachia." *Appalachian Journal* 42, no. 1/2 (2014/2015): 38–62.

Sze, Julie. "Denormalizing Embodied Toxicity: The Case of Kettleman City." In *Racial Ecologies*, edited by Leilani Nishime and Kim D. Hester Williams, 107–122. Seattle: University of Washington Press, 2018.

_____. *Environmental Justice in a Moment of Danger.* Berkeley: University of California Press, 2020.

Tormos-Aponte, Fernando, and Gustavo A. García-López. "Polycentric Struggles: The Experience of the Global Climate Justice Movement." *Environmental Policy and Governance* 28, no. 4 (2018): 284–94.

US General Accounting Office Resources Community and Economic Development Division. *Siting of Hazardous Waste Landfills and Their Correlation with Racial and Economic Status of Surrounding Communities.* Washington, DC: Government Printing Office, 1983.

van Wyck, Peter C. "The Highway of the Atom: Recollections Along a Route." *Topia* 7, (2002): 99–115.

Voyles, Traci Brynne. *Wastelanding: Legacies of Uranium Mining in Navajo Country.* Minneapolis: University of Minnesota Press, 2015.

_____. *The Settler Sea: California's Salton Sea and the Consequences of Colonialism.* Lincoln: University of Nebraska Press, 2021.

Wolfe, Patrick. "Land, Labor, and Difference: Elementary Structures of Race." *The American Historical Review* 106, no. 3 (2001): 866–905.

Wolfley, Jeanette. "Embracing Engagement: The Challenges and Opportunities for the Energy Industry and Tribal Nations on Projects Affecting Tribal Rights and Off-Reservation Lands." *Vermont Journal of Environmental Law* 19, no. 2 (2018): 115–63.

3 Intercultural Alliances

Zoltán Grossman

*Our number-one objective in this life must be to find common ground....
It does us no good to forge forward in the struggle to survive if we forget
that we must all fit in the same canoe. We share this land. We share these
resources. We share a common future.... Our customs and traditions may
not fit into the same molds that Western society embraces, but that doesn't
make them wrong. It makes them different. If we are to paddle the river of
life together, we must all learn to understand, appreciate, and, yes, celebrate
these differences.*"

(Billy Frank Jr. (Nisqually))

Environmental justice movements may succeed in their goals by working separately, keeping to their own racial or ethnic community, and not
engaging with other communities. But experiences and studies show that
different communities are far more powerful if they combine forces across
lines of race, ethnicity, and class, to form coalitions or alliances that are
less vulnerable to corporate "divide-and-conquer" tactics.

Environmental justice alliances offer an opportunity to go beyond
treatment of intercultural conflict as a natural condition. Social scientists commonly examine racial or ethnic conflict, but few have studied
examples of cooperation based on common interest against an outside
threat.[1] Fewer have studied the complex relationships between intercultural conflict and cooperation.[2] Even fewer have looked at mutual
community interests based on a common territorial identity or "sense
of place."[3]

Strategic alliances have been formed around the world in settings stratified by race, class, and ethnicity.[4] They involve formal coalitions of organizations, or looser alliances of social sectors or constituencies. In the
U.S. and Canada, environmental justice alliances that cross racial or ethnic
cultural lines need to consider centuries of white supremacy and settler
colonialism.

DOI: 10.4324/9781003214380-5

Any intercultural or cross-cultural alliance must wrestle with tensions arising from differences in positionality of race, gender, and class, and values attached to natural and human places.[5] For example, pesticide use may be opposed by an environmental group because of the effects on wildlife or consumers, and by a social justice group because of the poisoning of farmworkers, so an alliance would have to incorporate both perspectives.[6]

Wealthy white communities are at the top of the North American social hierarchy, and so are more able to protect their communities from environmental harm. Although white working-class communities are affected by pollution, they have also a relative advantage over communities of color in using mobility to avoid contamination.[7]

Some of the most powerful environmental justice alliances bring together different communities of color with diverse interests of their own, whether African American, Latinx, Asian, or Indigenous, as "people of color have always been and continue to be leaders in the fight for a more equitable and ecologically just world."[8] Other alliances attempt to bridge the even wider historical gap between white communities and particular communities of color. Environmental justice alliances of any variety face multiple challenges in "uniting around shared principles while engaging difference." Processes necessary for "cultivating solidarity across difference and inequality ... include uniting around shared principles while engaging difference; acknowledging and managing inequalities; making space for each other; attention to managing conflicts; and actions that confirm the shared commitments and negotiated identity."[9]

Developing intercultural alliances follow similar strategies as developing interclass coalitions. Interclass organizing is not just a matter of bringing together middle-class and working-class constituencies, but recognizing their different interests, languages, and organizing styles. In Pacific Northwest alliances during the 1990s "spotted owl wars," for example, miscommunication between mainstream environmentalists and timber workers came not just from different goals or priorities, but from different cultures and "languages" of organizing.[10]

Mainstream environmental activists are mainly "values-based" and joined the movement because of *what they believe* and their moral concerns. Their emphasis is on raising consciousness around an issue, so they may move on to another issue. Workers (much like communities of color) are more "interest-based" and join a social justice movement because of *who they are* and their direct experiences. Their emphasis is on defending their communities and committed to an issue until it is resolved.

Mainstream "values-based" environmental organizations are relatively better resourced, white, and upper middle-class than more grassroots "interest-based" groups on the frontlines of environmental harm. Grassroots relationships between land-based communities appear to make for more effective alliances than campaigns centered on mainstream "Big Green" or "grasstops" organizations.[11] Grassroots environmental movements must navigate "an array of shifting alliances" and rarely get the same media or public attention as mainstream groups.[12]

In rural North America, environmental justice alliances face an uphill climb. Even if they are directly affected by environmental harm, many rural whites view environmentalists as privileged urban outsiders.[13] In the American West, in particular, the predominant discourse from the private "property rights" movement (from the 1980s "sagebrush rebels" to today's Bundyites) reinforces settler privilege, white supremacy, and freedom from environmental regulations.[14] But the stories of collaboration between white ranchers and farmers, environmentalists, and Native American tribal nations, in efforts such as Oregon's Rural Organizing Project or Montana's Northern Plains Resource Council, have been largely unheard.[15]

Collaborative governance in the American West "where patchworks of public, private, and tribal interests characterize the region's resources" is facilitated by "unlikely alliances, or partnerships among diverse actors who have historically been at odds." These "unlikely alliances" are "likely to arise in the presence of a crisis, when appropriate leadership is present, when some of the actors have interacted effectively in the past, and when actors need to pool resources."[16] Perhaps the most telling example of unlikely alliances across cultural lines is between Native American nations and their rural white neighbors.

Native/Non-Native Unlikely Alliances

Indigenous nations and their settler neighbors have long been archetypal enemies in conflicts over natural resources. In the late 20th century U.S., tribal nations fighting for their treaty rights dealt with local white farmers, ranchers, commercial fishers, or sport fishers as the main obstacle to securing treaty-guaranteed access to fish, game, or water. As the tribes secured these rights, many rural whites in certain regions joined an anti-Indian movement to oppose tribal sovereignty and treaty rights. Yet in some of these same areas beginning in the late 1970s, environmental justice alliances unexpectedly brought together Native Americans and rural white resource users in areas of the country where no one would have predicted or even imagined them.[17]

The evolution went through four general and often overlapping stages. First, Native nations asserted their cultural autonomy and tribal sovereignty. Second, a backlash from some rural whites created a conflict around the use of natural wealth (such as minerals, water, or fish). Third, the two groups initiated dialogue around common outside threats. Finally, the communities increased collaboration around the protection of their livelihoods. The neighboring groups felt that if they continued to fight over common resources, there may not be any left to fight over. The stages of this evolution were complicated by divisions within both Native nations and white communities.

Indigenous-settler environmental justice alliances have included communities confronting mines, hydroelectric dams, logging, nuclear waste, military projects, and other environmental threats (see Figure 3.1). Natives and rural whites in each area took different paths from conflict to cooperation and experienced varied levels of success in improving relations between them. In certain instances, a significant number of rural whites came to see Native sovereignty and treaty rights as a legal tool to protect their common space from an "outsider" common enemy and redefined their common community of interest as including their Native neighbors.

While my research used many textual sources and regional histories, the bulk of the sources were about 80 interviews with people on their experiences in crossing cultural lines to protect a common place. They include sportfishing group leaders and fishing guides, farmer and rancher group leaders, tribal government leadership, Native environmental organizers, and rural white environmental organizers, schoolteachers, small business owners, and others.

This chapter examines four primary sets of case studies: centered on fish, water and dams in the Pacific Northwest, military projects in Nevada

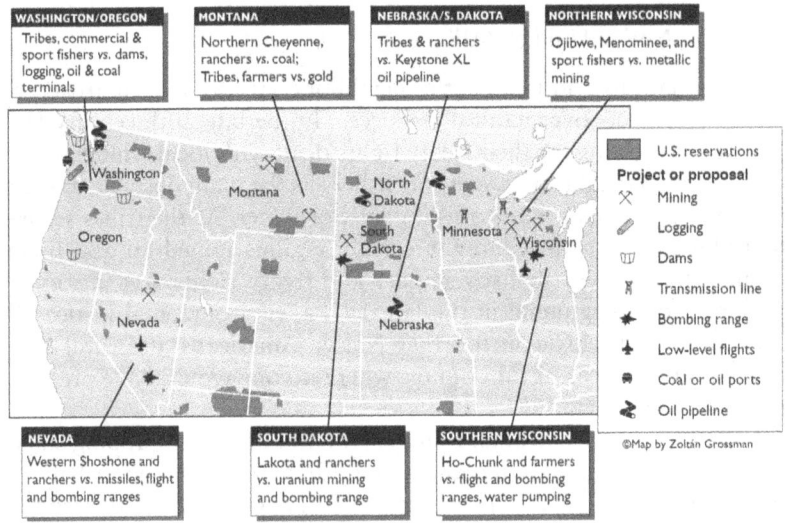

Figure 3.1 Examples of Native/non-Native environmental alliances. Map by Zoltán Grossman.

and southern Wisconsin, on the mining of sacred sites in the Northern Plains, and on fishing and mining in northern Wisconsin. Each represents different paths that any intercultural alliance might take, reflecting larger questions in intercultural relations around the world. The study focuses narrowly on the relationship between rural Native and settler communities and does not closely examine relations between Native peoples and urban-based organizations, which have their own sets of dynamics.[18]

Implications of the Alliances

Intercultural environmental alliances could be portrayed as quaint or irrelevant examples of common ground between two allegedly "disappearing" U.S. populations: reservation Indians, and rural whites who still value the land and its natural wealth. The popularized image of "Cowboys and Indians" has been ingrained in the national (and global) consciousness as a cultural template of irreconcilable enemies.

Yet despite Hollywood's clichéd stereotype of cowboys eternally fighting Indians, many white ranchers and farmers see their lifestyle as endangered by corporate economic trends, much as tribal members have seen their land-based cultures under siege. Native and white rural communities are confronted to different degrees by environmentally damaging projects that would not be tolerated in more populated regions. Both "interest-based" groups have a historic mistrust of state and federal governments that infringe on their lifeways.

Indigenous nations stake their cultural survival not on being absorbed into a colonial state, but on boundaries and sovereign institutions that protect their distinctive identities. Most scholars assume a stark choice between national self-determination and a common state citizenship or try to strike a compromise between the two. Fewer have explored ways to build common identities *outside* the state framework, by constructing or using common place-based identities.

Indigenous-settler environmental alliances are an example of an intercultural movement constructed not around a common state citizenship, but around a common "place membership." The symbolic frame of "place membership" is based on people living in a particular naturally or culturally significant place rather than within a particular political boundary.

This chapter looks at some of the best known examples of Indigenous-settler alliances, to document their histories, but also to illustrate larger themes of conflict and cooperation. Case studies can be treated as prototypes: of alliances that did not happen, of alliances that were initially successful but later floundered, of alliances that achieved success in some areas but not others, and of alliances that seemed to meet most of their goals and even began to expand beyond environmental issues.

Efforts to improve relations only at the governmental or large-scale geographic level are often unsuccessful in improving relations at the social- or local-scale level. "Top-down" approaches tend to increase resentment at the base of both communities against their respective leaders.[19] "Bottom-up" relations are certainly more difficult and complex, but people-to-people ties can result in deeper and longer lasting cooperation, particularly at a local scale, such as a single watershed or mountain range.

"People-to-people" relations are not simply an alternative to "government-to-government" relations, but each can form a parallel track that strengthens the other. If alliances overlook Native sovereignty, they can undermine tribal political strengths and threaten future cooperation. On the other hand, tribal leaders who rely only on legal precedents or political contacts may lose on issues if they ignore the need to shape local public opinion.

Mining and Sacred Lands in the Northern Plains

In eastern Montana and western South Dakota, some of the earliest alliances in the U.S. between Native nations and white ranchers confronted mining corporations, and later bombing ranges and toxic wastes. They defended lands viewed as "sacred" by Native peoples, and culturally or environmentally significant by white agriculturalists.[20]

Early Montana Alliances

In the 1970s, Northern Cheyenne tribal members joined with white ranchers to oppose plans for new coal mines and coal-fired power plants, using the U.S. Environmental Protection Agency's "Treatment-As-State" plan, allowing the tribe to declare Class I air status under the Clean Air Act. After the alliance won some victories, the alliance contracted and the historic economic animosities between the communities returned.[21] Yet the 1970s alliance made new alliances easier to form against new coal projects in the 1980s–2000s.[22] In general, the most traditionalist and pro-treaty Native activists were the first to build bridges to their white neighbors.

Alliances among tribes, rural environmentalists, and settler farmers/ranchers tended to gain more than alliances between tribes and urban environmental groups alone, as shown by the very different results in Montana alliances against gold mines in the sacred ranges of the Little Rocky Mountains (next to the Fort Belknap Reservation) and the Sweetgrass Hills, held sacred by several tribes.[23] Third-generation wheat farmer Richard Thieltges commented on the successful fight to protect the Sweetgrass Hills, "Farmers-ranchers, Native Americans and environmentalists are three sides of a natural alliance. We are the only people who truly have to bear the burden of what's happened to the land. So the mining industry tries to drive wedges between us."[24]

Black Hills Alliance

In South Dakota in the 1970s–1980s, similar alliances grew between Lakota (Sioux) tribal members and white ranchers against coal and uranium mining plans, despite intense racial tension between Native and white communities. It was in fact some of the most "militant" Native activists, who strongly valued the Black Hills as a sacred Lakota homeland, who initiated the Black Hills Alliance (BHA) in 1979.[25] Rancher Marvin Kammerer said, "I've read the Fort Laramie Treaty, and it seems pretty simple to me; their claim is justified. There's no way the Indians are going to get all of that land back, but the state land and the federal land should be returned to them. Out of respect for those people, and for their belief that the hills are sacred ground, I don't want to be a part of this destruction."[26] The Black Hills International Survival Gathering drew 11,000 people to Kammerer's ranch in July 1980, jumpstarting the Midwest movement against nuclear and fossil fuel energy development.[27]

The BHA successfully stopped uranium exploration in the Black Hills by 1981, and the subsequent Cowboy and Indian Alliance (with the poignant acronym of "CIA") halted a proposed bombing range near Hot Springs in 1987.[28] In the same year, a congressional bill failed to return Black Hills federal lands to the tribes.[29] A second Cowboy Indian Alliance stopped a proposed coal train from the Powder River coal basin to the Mississippi River by 2009. The Black Hills Clean Water Alliance has fought renewed uranium exploration in the 2010s–2020s.[30]

BHA organizer Madonna Thunder Hawk observed that Black Hills settler residents came to understand that the treaty could help to prevent uranium mining: "They realized how helpless they were in the face of eminent domain. But Indian people had treaty rights—they could *stop* things!"[31] Black Hills rancher Cindy Reed said of the bombing range project defeat, "This not Indian versus white. It's a land-based ethic versus a profit-oriented motive. This is a beautiful place. There's no reason to begin to ruin it."[32]

Keystone XL Pipeline

In South Dakota and Nebraska in 2013, a third Cowboy Indian Alliance brought together Lakota and other tribes with white ranchers and farmers to stop the proposed Keystone XL oil pipeline from the Alberta tar sands region.[33] White farmers and ranchers opposed the pipeline company's claim of eminent domain by stressing their right to their private property tracts, which originally were homesteads taken from the tribes.[34]

As they worked together against the pipeline, the tribes convinced some white neighbors to protect sacred burial sites on their lands. Ihankton-wan Nakota elder Faith Spotted Eagle observed, "We come from two cultures that clashed over land, and so this is a healing for the generations."[35] Rancher Paul Seamans, chair of Dakota Rural Action, said that the Lakota "feel the government should step up and do what's right by them on the 1868 Treaty ... They're not after the deeded land. They would like the government to recognize that they've been screwed, and ... to have the federal and state lands back ... After being around them and listening to their point of view, I get to thinking, 'hey, if I was Indian, I would be doing the same exact damn thing that they're doing'."[36]

In Nebraska, the pipeline company TransCanada tried to buy off some ranchers and farmers by moving the Keystone XL route away from their lands—but these landowners did not give up the fight and continue to work with others who are still directly affected, including Native communities. In a meeting to form the alliance, Bold Nebraska director Jane Kleeb remembered "that moment one of the ranchers stood up and said: 'I finally understand how you feel having your land taken away,' and one of the tribal leaders stood up and said, 'welcome to the tribe.'... There was this amazing connection ... We're all in this together in the fight."[37]

In 2014, the "CIA" drew water protectors to Washington DC for a large, visible march that attracted national media attention. When President Obama initially stopped Keystone XL the following year, Spotted Eagle commented, "We stood united in this struggle, Democrat, Republican, Native, Cowboy, Rancher, landowners, urban warriors, grandmas and grandpas, children, and through this fight against KXL we have come to see each other in a new better, stronger way." Although President Trump tried to restart the permit process as one of his first acts in January 2017, President Biden reversed the decision as one of his first acts in January 2021. The pipeline was stopped once again, with the support of a growing climate justice movement.

Standing Rock

The series of alliances led up to Standing Rock's resistance to the Dakota Access Pipeline (DAPL), which received unprecedented national and international attention in 2016–17. The Native water protectors who started the Standing Rock camps in April 2016 were always very open to unity with ranchers and farmers along the 1,172-mile route of the "black snake" carrying Bakken-fracked shale oil through North and South Dakota, Iowa, and Illinois, asking "everyone who farms or ranches in the local area, and everyone who cares about clean air and clean drinking water stand with us."[38] They were backed by former Keystone XL opponents from South Dakota and Nebraska, and by Iowa farmers who strongly opposed DAPL, resulting in many protests and dozens of arrests.

But in contrast, North Dakota farmers and ranchers were rarely visible in the fight to stop DAPL. Most local landowners gave in to the "eminent domain" confiscation of their property for the $3.8 billion pipeline, even if they mistrusted the company's promises of safety. Indigenous Environmental Network organizer and Cheyenne River tribal member Joye Braun stated, "When this proposed pipeline breaks, as the vast majority of pipelines do, over half of the drinking water in South Dakota will be affected. How can rubber-stamping this project be good for the people, agriculture, and livestock? It must be stopped ... with our allies, both native and non-native."[39]

One local landowner who visited the water protector camps said, "The first thing I thought about when I heard about the Bakken pipeline was that beautiful black soil that my grandmother taught me to love.... She'd always point it out to me when she'd see that beautiful topsoil.... [I]t hurts see it trenched and piled up and eroded the way it has been."

The pipeline companies pit Native and white communities against each other. The route originally was proposed to cross under the Missouri River near Bismarck, but the company rejected the route as jeopardizing drinking water of the residents in the state capital. In a classic case of a racialized "shell game" at work, the route was diverted southward to cross the Missouri just north of the boundary of the Standing Rock Reservation.[40]

Three reasons may explain the relative lack of visible rural white participation in North Dakota to stop DAPL. First, the oil fracking industry has become so powerful in the state that fatalistic private landowners assume that would lose any legal battle. Second, the DAPL permitting and construction were "fast-tracked" (in contrast to the drawn-out process around permitting Keystone XL), perhaps because pipeline companies realized that delays could allow a fourth Cowboy Indian Alliance to solidify. Third, state government and media accounts of the controversy tended to demonize and criminalize the Native opposition, causing local ranchers and residents to "view the protests with a mix of frustration and fear."[41] Concrete, war zone-style highway checkpoints isolated the camps from Native supporters but also had the (perhaps calculated) effect of preventing white North Dakotans from joining or even seeing the camps.

Trump also gave the green light to DAPL in January 2017, and the camps were violently cleared the following month. Nevertheless, the stand at Standing Rock inspired or strengthened other "water protector" alliances around the continent.[42]

Figure 3.2 Water protectors blockading and locking down bulldozers that had cut a swath through North Dakota ranchlands and tribal sacred sites for the Dakota Access Pipeline, north of the Standing Rock Sioux Reservation, on September 6, 2016. Permission by author.

Series of Alliances

The Northern Plains case studies represent a series of oppositional alliances that have made steady but uneven progress, making later alliances easier to form but rarely extending relations beyond environmental issues. Each environmental alliance represented two steps forward in relations, which was followed by one step backward as tensions resurfaced, followed by another alliance that took relations again two steps forward. Native activists used a "carrot-and-stick" strategy of strongly asserting their tribal rights while building bridges to white ranchers/farmers on common land-based values that they hold in common.

Around the world, sacred places (such as Jerusalem) tend to be viewed as sources of religious contention and exclusion. Yet "sacred" sites such as the Black Hills and Sweetgrass Hills have been the sources of the strongest alliances, even if their sacredness is understood in very different ways. Non-Native residents culturally value the land more than corporate or governmental "outsiders," and so can make some connections with much more deeply rooted Native values. When Natives and non-Native neighbors begin to construct such local-scale common territorial identities, they will inevitably confront entrenched majority racial identities, as the next set of case studies demonstrates.

Military Projects and Environmental Racism in Nevada and Wisconsin

When tribes in Nevada and Wisconsin allied with white agriculturalists to oppose bombing ranges and low-level flight ranges, the strength of their alliances was of concern to the U.S. military. Yet the Western Shoshone in Nevada, and Ho-Chunk (formerly Winnebago) in southern Wisconsin, could not convince many of their non-Native allies to stand by them when the military singled them out. The alliances achieved some limited successes but did not extend beyond the immediate environmental threat, and sharp differences remained on local resource control.

In Nevada, the Western Shoshone had worked with some white ranchers and townspeople in the late 1970s to successfully stop the proposed MX missile system.[43] When the military dropped the gargantuan plan, the alliance ended. Western Shoshone opposition to nuclear weapons testing on the tribe's treaty-ceded territory was not joined by white ranchers. In the 1980s, the alliance revived to oppose new military plans for low-level flight ranges and expanded bombing ranges, with mixed results.[44]

The Western Shoshone did not receive support from their white allies for their treaty rights, even though those rights would have provided a tool to stop the military projects. As the tribe's strong treaty claims were turned down in federal court (partly due to national security considerations), the tribe had fewer legal tools to offer their allies against the low-level jet flight plans. The alliances opposed "alien" military projects but did not substantially alter relations between Nevada tribes and their white neighbors.

In Wisconsin, the Ho-Chunk Nation joined with white farmers in the mid-1990s to oppose Air National Guard low-level flights and a bombing range expansion.[45] The flights were planned all over the farming valleys in the southwestern region of the state, while the more localized bombing range expansion primarily affected nearby Ho-Chunk communities.[46] Ho-Chunk Nation legislator Ona Garvin told white landowners about "the reason we had empathy for those people was the government policies that the Indian nations always had to follow, which meant the loss of land. Now it's the Department of Defense that's taking your land. So we understand where you are.... That's what really hit them ... That did a lot more for cultural understanding than if we had sat there and talked to them until we were blue in the face."[47]

In the end, however, the Air National Guard met the demands of white farmers not to expand the flight ranges but continued to pursue expansion of the bombing range. Some of the white farmers claimed victory and dropped their demand to oppose bombing range expansion, effectively selling out their Native allies.[48]

Dividing Alliances

The case studies of low-level military jet flights in Nevada and Wisconsin represent oppositional alliances that did not extend beyond the "outside" threat and failed to build a lasting improvement in relations between Native and non-Native communities. But by the 2010s, the Ho-Chunk also reclaimed former treaty lands around a former dam project and a closed munitions base and co-manages them with a state agency. This joint ownership of protected natural areas was grudgingly accepted by environmental groups and local non-Natives.[49]

Lessons from the military projects case studies reflect larger critiques around interracial "unity" environmental justice strategies. U.S. alliances between whites and people of color are often formed around the assumption that so-called minority communities will set aside their particular concerns for the common universal good. Yet mere "unity" is not adequate if it is applied to unequal players, and "lowest-common-denominator" politics will generally not succeed in building closer community ties.

Standing in the way of equality in the U.S. is the institution of white "privilege," or what can more accurately be termed white "advantage." Racism has not been used only to repress so-called minority groups but has served as a mechanism of social control of the white majority, deflecting from its members' other interests.[50] U.S. whites often exhibit a dual consciousness of their own racial self-interest, and their loyalty to more universal values, preventing or splitting interracial alliances.[51]

Companies or government agencies can use an ostensibly geographical "shell game" to shift environmental burdens away from white communities. By accepting short-term self-preservation, white communities are preventing long-term solutions to environmental problems. Alliances can be divided if white participants do not defend the interests of both communities, instead of accepting "out of sight, out of mind" outcomes, and failing to prioritize their "place membership" over their racial advantages.

Fishing and Mining in Northern Wisconsin

The concept of "insiders" and "outsiders" forms the core of two major conflicts in the recent history of northern Wisconsin, over tribal off-reservation treaty rights, and over corporate proposals to mine metals in the lands ceded by the treaties. In both cases, the main point of contention was fish.[52]

Spearfishing Conflict

The spearfishing conflict began in 1983 after federal courts recognized Ojibwe (Chippewa) treaty rights to harvest resources outside Wisconsin reservations. Anti-treaty groups told white sport fishers and business owners that the Ojibwe would harm the fishery and cause an economic disaster in the tourism-dependent region. Thousands of white residents protested at northern boat landings in the late 1980s and early 1990s against spearfishing (with signs saying "Save a Walleye—Spear an Indian"), harassed spearers and their families, and in some cases physically attacked them with thrown objects, bat, and vehicle assaults, and pipe bombs and sniper fire.[53]

The Ojibwe responded that the spearers harvested only 3 percent of the highly prized walleye. They welcomed support from non-Native "Witnesses for Nonviolence" to monitor and document the harassment and violence.[54] The Ojibwe bands asserted that the state government was "scapegoating" the Ojibwe for declining fish stocks and appealed to northern whites' historic mistrust of state resource agencies.[55]

By 1992, the anti-treaty movement dramatically declined in strength and influence. Northern whites had gained some understanding of fish biology and Native cultural traditions or had been intimidated by a federal court injunction against racial harassment. But in addition, some sport fishers began to see new mining plans as endangering the fishery, or at least posing a greater threat than Ojibwe spearfishing.[56]

Tribes presented their treaty rights, and their on-reservation sovereign rights, as legal obstacles to the mining plans. Instead of arguing over the fish, white anglers began to cooperate with tribes to protect the fish, recognizing some of the same tribal rights that they had earlier fought as useful for their own interests. Red Cliff Ojibwe spearfisher Walter Bresette had predicted during the treaty rights crisis that northern Wisconsin whites would realize that environmental and economic problems are "more of a threat to their lifestyle than Indians who go out and spear fish ... we have more in common with the anti-Indian people than we do with the state of Wisconsin."[57]

Yet the emerging alliance developed in different ways in different parts of the treaty-ceded territory. In the Lac Courte Oreilles Reservation area where the Ojibwe did not assertively practice off-reservation spearfishing, the alliance failed to stop the Flambeau copper mine from opening near Ladysmith in 1993.[58] Bresette concluded of the loss, "Where you don't have Indian rights, non-Indians lose."[59] Yet the areas where the Ojibwe had strongly pushed their spearfishing (between Lac du Flambeau and Mole Lake reservations) were the same areas where the strongest alliance was built with sport fishers. Lac du Flambeau spearing leader Tom Maulson says that the treaty rights conflict had offered "an education on everybody's part as to what Indians were about. It needed a conflict to wake them up."[60]

Crandon Mine

In 1976, Exxon Minerals proposed a huge zinc-copper mine upstream from the wild rice beds of the Mole Lake Reservation, and the trout-rich Wolf River.[61] Sportfishing group leader Bob Schmitz says that a "mutual love" of the Wolf River brought together angling groups and tribal members to fight the mine.[62]

The Midwest Treaty Network formed the Wolf Watershed Educational Project (WWEP) in 1995 to organize Wolf River communities downstream from the mine site. In the meantime, Mole Lake, Menominee, Forest County Potawatomi, and Stockbridge-Munsee came together as the Niiwin tribes to stop the mine. The WWEP began a series of speaking tours to form small local anti-mine groups. At the 22 towns visited in the first speaking tour, representatives of tribes, environmental groups, and sportfishing groups spoke, drawing about 1,100 people. Instead of sending the speakers only to speak separately to their own constituencies, the WWEP showed all three parts of the alliance at each of the communities, to model their cooperation, and some sport fishers heard a Native person speak for the first time in their lives.[63]

Mining companies had usually been able to portray mainstream, urban environmental activists as yuppies or hippies who do not care about rural jobs. The Crandon mine companies tried to pit Native Americans against white residents, environmentalists against union members, and rural northerners against urban residents. But they failed each time to divide Wisconsinites by race, by class, or by region. What mining companies faced along the Wolf River was something new—an environmental movement that was rural-based, multiracial, middle-class and working-class—and made up of many youth and elderly people.

The WWEP did not just address mining companies' environmental threats, but also their threats to rural cultures and local government democratic institutions, and their "boom-and-bust" economic disruptions. The movement drew from four strands in Wisconsin history: progressive populism, rural environmental ethics, resentment of northern residents against state government agencies, and the perseverance of Native nations to protect their treaty rights and tribal sovereignty. Alliance meetings were held every month for nine years, at different reservations and border towns, making decisions in a democratic way, rather than letting a few leaders, experts, or lawyers decide.

The grassroots movement made a strong an impression on the global mining industry. Toronto's *North American Mining* journal claimed that "The increasingly sophisticated political maneuvering by environmental special interest groups has made permitting a mine in Wisconsin an impossibility," and *Mining Voice* called mine opponents' websites "barbarians at the gates of cyberspace." *Mining Environmental Management* described the WWEP as "just one example of what is becoming a very real threat

Figure 3.3 Native and non-Native members of the Wolf Watershed Educational
Project celebrate their 2003 victory over the Crandon mine project, at
the Mole Lake Sokaogon Chippewa tribal headquarters in northern
Wisconsin. Permission by author.

to the global mining industry." And for many years, the Vancouver-based
Fraser Institute rated Wisconsin at or near the bottom of its annual "mining investment attractiveness score" because of the state's "well-publicized
aversion to mining."[64]

In October 2003, the 28-year fight to stop the proposed Crandon mine
came to a sudden and dramatic end, as the Mole Lake and Forest County
Potawatomi tribes gained ownership and divided the 6,000-acre mine
site.[65] The tribes bought the land at a "rummage sale" price (without the
mineral deposit value), partly because the grassroots movement had driven
away potential corporate partners for the mining company. The victory
led to greater economic cooperation between tribal and local governments,
and stronger landowner protection of sacred places and wild rice beds.

Bad River and Menominee

Just as in the Northern Plains case studies, the formation of an alliance
can make future alliances easier to form. In 2010, the Bad River Ojibwe
Tribe opposed the planned Penokees iron ore mine, which was backed
by Gov. Scott Walker despite its location upstream from wild rice beds
and pristine Lake Superior. By 2015, a local alliance (originally modeled
on the WWEP) brought together Ojibwe and their neighbors to stop the
proposed mine.[66] Bad River Chairman Mike Wiggins Jr. said that if local
people "acknowledge our differences as human beings but acknowledge
the commonality of our home ... we can all turn our attention outward
from ourselves, and start to love the big lake."[67]

In the same year, the Menominee Nation and its allies came out in opposition to the planned Back Forty zinc-gold mine just across the Menominee
River border in Michigan. The Menominee saw the mine as a threat to

sacred burial sites, and non-Natives saw the mine as a threat to the river ecosystem.[68] By 2021, one mining company relinquished its permits, and another company tried to restart the permit process.[69] Even though the Menominee Nation had not won its treaty rights battle in the courts, it maintained strong moral and cultural leadership to stop the proposed mine.

Conflict Engendered Cooperation

The counterintuitive outcome of the northern Wisconsin case studies is that the areas where the treaty conflicts had been the most intense is where the later environmental cooperation developed to the deepest and most successful extent and even extended environmental gains into cultural and economic cooperation. The assertion of tribal legal powers helped to equalize the reservation with white "border towns." The assertion of cultural traditions helped educate non-Natives about previously invisible tribal cultures. Tribal economic powers (through gaming) began to be viewed as benefiting settler communities. Anti-Indian prejudice continued to exist, but organized anti-Indian groups were soundly defeated.

Lessons from the Wisconsin fishing and mining case studies reflect larger issues of how people are defined as "outsiders" or "insiders" in a place. "Geographies of exclusion" are based on social/racial definition of place, which identifies a landscape with the group that lives there (or "should" live there). In this view, tribal fishers were portrayed seen as "outsiders" transgressing on white land, with anti-treaty protesters chanting "White Man's Land!," and the ironic demand "Indians Go Home!".

Geographies of inclusion, on the other hand, are based on a territorial definition of place, which identifies all people who live there with the land. In this view, Native and non-Native neighbors harvesting can define mining companies as new and more threatening "outsiders." Natives and rural whites at odds over resources saw each other as "outsiders" transgressing social boundaries, yet in the face of a threat by mining company "outsiders," they could start to see each other as "insiders" in a territorially defined community. Mole Lake tribal member Frances Van Zile describes this shift in consciousness when she says that many local white residents now "accept Mole Lake as part of home. It's not just my community. It's everybody's home.... when it's your home you try to take as good care of it as how can, including all the people in it."[70]

Fish, Water, and Dams in the Pacific Northwest

In the Pacific Northwest, salmon have been a keystone species since time immemorial, and their paths of migration have long defined the region as "home" for both Indigenous and settler populations. As the Nisqually treaty rights leader Billy Frank, Jr. stated, "We know our watersheds, we know our neighbors and for centuries, we've known the needs of salmon. This is our homeland. This is where we live. We aren't going anywhere."[71]

Colonial Geographies

The Pacific Northwest treaties of 1854–55 ceded 64 million acres of Native nations' homelands, in return for access to tiny reservations and their hunting, fishing, and gathering grounds.[72] Settler capitalism harmed the region's biodiverse native forests, cold streams, and river-mouth estuaries, where salmon had thrived. Logging increased stream temperatures and exposed the soil to erosion, causing spring floods that silted up the gravel beds where salmon lay their eggs. Farmers straightened meandering streams, installed fish-blocking culverts in irrigation channels, eliminated beaver dams, and erected dikes to turn wetlands into farms and cattle pastures. Industrialization paved over precious habitat and erected hydroelectric dams that blocked salmon runs, raising water temperatures in stagnant reservoir pools, and concentrating toxins and invasive species.

Northwest Native Resurgence

After a series of 1960–70s "fish wars" between Northwest tribes and state government-backed white fishers, a federal court in the 1974 Boldt Decision reaffirmed tribal treaty rights to harvest fish outside the reservations. In the 1980s, the federal courts took on "Phase II" of the Boldt treaty litigation, in which tribes demanded a voice in regulating logging, dams, and agricultural practices that blocked or harmed salmon.[73] The tribes and the state instituted a program of co-management of off-reservation fisheries and other natural wealth, enshrined in the 1989 Centennial Accord.[74]

Co-management began to normalize a role for Northwest treaty tribes in managing, protecting, and restoring their original treaty-ceded territories, which often correspond to watersheds.[75] Washington State organized Water Resource Inventory Areas based on watersheds rather than political boundaries.[76] Even some commercial and sportfishing groups came to support treaty rights to protect habitat, such the 2007 Martinez Decision, a key federal District Court ruling directing Washington State to eliminate culverts that block fish passage.[77]

Removing Barriers

Through leading environmental justice and restoration, and strengthening their political, economic, and cultural self-determination, tribal nations are providing models of resilience and regeneration for non-Native communities.[78] Two dams on the Elwha River of the northern Olympia Peninsula were successfully removed in the 2010s, allowing salmon to return to areas where they had been blocked for a century.[79] The dam removals revealed a submerged tribal sacred place, and restoring a beach and shellfish-gathering grounds for the Lower Elwha Klallam Tribe.[80]

The Elwha success inspired the removal of barriers to salmon runs on several other rivers in Washington, and four dams each on the beleaguered Snake River in eastern Washington and the Klamath River in southern Oregon. Tribally led alliances have also tried to offset the negative effects of existing dams, by building new fish-passage facilities, repairing other parts of the watersheds, and preventing new flood-control dams.[81]

Tribally led initiatives are seeking to respond to prevent floods through more holistic ecosystem management.[82] The Tulalip Tribes are relocating beavers to bring back beaver dams to rehydrate the upper reaches of the Snohomish River, storing spring snowpack runoff for release during summer low flows.[83]

Watershed Restoration

In the Nisqually River watershed, the Nisqually Tribe is recognized as the "lead entity" in creating Nisqually River Council watershed stewardship plans for local, state, and federal agencies, which have put more than three-quarters of the Nisqually River mainstem in protected ownership.[84] In the western half of the Nisqually Estuary, the U.S. Fish & Wildlife Service pulled dikes out of the Nisqually National Wildlife Refuge (renamed after Billy Frank Jr. after his 2014 passing), in collaboration with the Tribe.[85] Nisqually acquired the eastern half of the estuary from white farmer Kenneth Braget, when he gifted the land upon his death in 2006, and the Tribe has pulled out dikes to allow tides to again wash over the former cattle pasture and turned his farmhouse into a cultural center and tribal garden.[86]

Standing in the Nisqually Delta on a misty afternoon, one can observe an estuary landscape that is healing. After decades of being diked to create pasture, tidal flows are again allowed to bring salt water and aquatic species into old restored channels, like blood circulating back into blocked veins and capillaries in a human body.[87] Riparian vegetation is being brought back to prevent erosion, and sacred springs are being protected and returned to tribal ownership.[88]

After decades of being straightened to drain the wetlands, upstream tributaries of the Nisqually River are being remeandered, and log jams are being installed to create pools for salmon to rest on their long journey back home from the ocean. After decades of declining runs, at least some salmon are returning to the Nisqually watershed, because their habitat is finally being restored, and two hatcheries are restocking the river.[89] The landscape is slowly being healed and made more resilient, through tribal partnerships with non-Native residents and agencies.[90]

In the interior Northwest Plateau region, the Umatilla Tribes have also seen success in its Umatilla Basin Project collaboration with Oregon farmers and ranchers, in restoring riparian areas and bringing water and fish back to their basin.[91] Decades of work remains to be done, given the precarious state of Pacific Northwest watersheds which continue to face multiple ongoing ecological threats[92].

Indigenous Climate Resilience

In the 21st century, these threats are being intensified by the climate crisis, which causes sudden spring melts of snowpack and glaciers, generating floods and scouring out salmon egg nests, followed by summer low flows that are unhealthy for salmon. Using the template of habitat restoration, Northwest tribes are working on climate change adaptation with local non-Native governments that have usually opposed tribal sovereignty.[93]

The Swinomish Tribe was one of the first Native nations to develop a climate change adaptation plan in the Skagit River Delta.[94] The Tulalip Tribes defused a longstanding source of conflict between dairy farmers and tribal fishers over cattle waste in the Snohomish watershed's salmon streams by converting the waste into biogas energy.[95] Instead of waiting for Congress or the U.N. to act on climate change, tribes and their allies are starting to stitch together a patchwork of local, watershed-based solutions for climate resilience.

Several coastal tribes (such as Quinault and Quileute) face increasing flooding, storm surges, and potential sea-level rise, worsening the threat of tsunamis, so are moving infrastructure to higher ground.[96] Indigenous nations are becoming leaders in disaster resilience, drawing on a strong sense of community to meet catastrophes such as climate change and the pandemic.[97]

Fossil Fuel Wars

Many of the tribes have also taken the lead in unlikely alliances with rural white communities in opposing new port terminals that would ship coal and oil overseas, targeting the companies' "Achilles heel" of fossil fuel shipping.[98] In partnership with local and state allies, the Lummi Nation used its treaty fishing rights to stop plans for the Cherry Point coal terminal in 2015.[99] Lummi Nation Chairman Tim Ballew II described the victory as "a celebration of the power of treaty rights to protect all of us, to preserve our lands and waters for everyone who calls this place home."[100]

The Quinault Nation and non-Native neighbors in Grays Harbor County halted plans for three Bakken oil terminals in 2017.[101] Tribes and

Figure 3.4 The Quinault Indian Nation hosts the "Shared Waters, Shared Values" rally against a proposed Grays Harbor oil terminal, at Hoquiam City Hall on July 8, 2016. Quinault President Fawn Sharp and Vice President Tyson Johnston are joined by representatives of the Quileute, Makah, and Lummi tribes, Washington fishing association representatives, and local environmentalists (credit: Zoltán Grossman).

their allies have stopped a dozen other oil, coal, and natural gas projects along the Pacific Northwest coast "Thin Green Line," blocking the industry's shipping chokepoint.[102] Even conservative mayors opposed the long trains, which endanger their towns with either coal dust or explosive Bakken oil. In Idaho and eastern Oregon, tribal councils have joined direct actions to stop the "heavy haul" of equipment megaloads from coastal ports to the Alberta tar sands, forming even stronger cross-cultural alliances than in more "progressive" regions.[103]

The hotspots for the fossil fuel wars have been small working-class cities (such as Aberdeen-Hoquiam, Coos Bay, and Longview), where white residents had been at odds with tribes since the "fish wars" of the 1960s–70s, and with anti-logging environmentalists since the "spotted owl wars" of the 1990s. Urban-based environmental groups have not been as successful in these former timber towns (and Obama/Trump-voting counties), as the local cross-cultural anticorporate populist alliances.[104] As Quinault President Fawn Sharp said, "the relationships we have with our neighbors arose out of a relationship of much division, strife, and conflict, but through that ... they've come to know who we are."[105]

In British Columbia, similar alliances have stopped or challenged numerous oil pipelines from the Alberta tar sands, such as the Northern Gateway and Trans Mountain pipelines.[106] As in the U.S., it has been more difficult for alliances to stop fracked-gas pipelines promoted by the false discourse of "clean energy."[107] Canadian industry is even more reliant

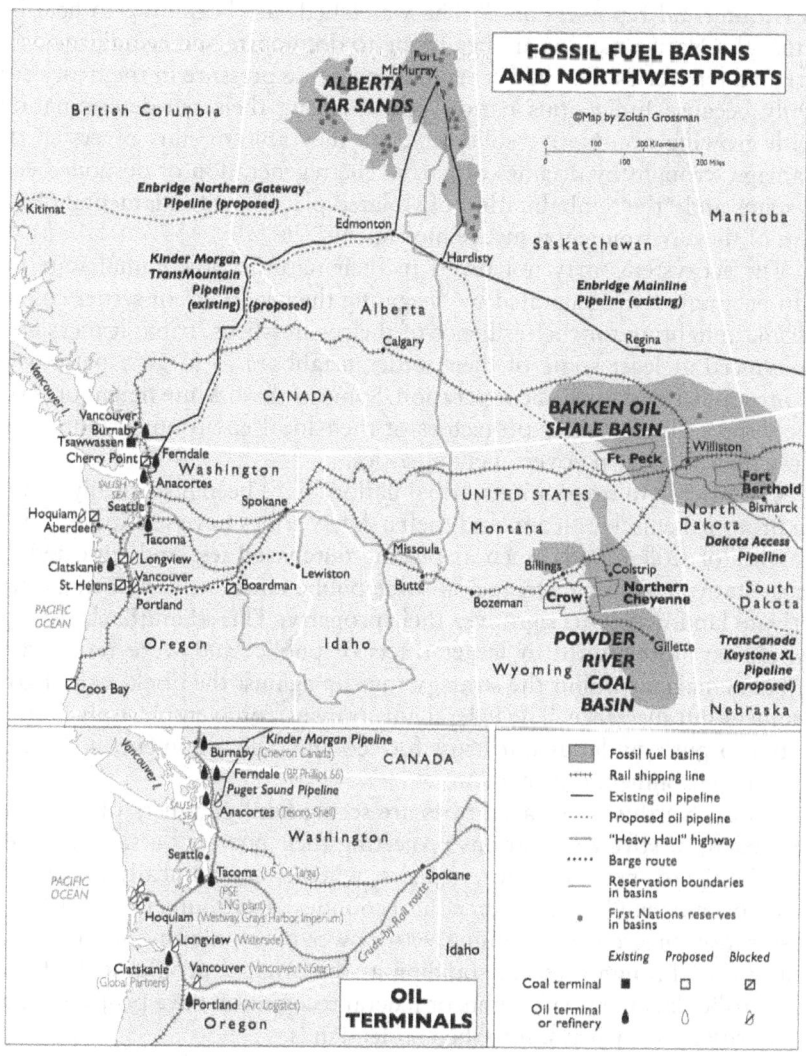

Figure 3.5 Fossil fuel basins and Northwest Ports. Permission by author.

than U.S. industry on an "extractivist mindset" to feed the global energy market.[108] With tiny reserve land bases, First Nations have needed to creatively assert their *de facto* authority in stolen Crown (federal) lands.[109]

Projecting Power

It is the leadership role of the tribal nations, using a creative mixture of Indigenous and Western knowledge systems, that has led this process of environmental repossession.[110] The watersheds are beginning to heal because the Native nations are beginning to decolonize and reindigenize the original territories they had ceded under intense pressure in the treaty era. Only because Indigenous nations are asserting their self-determination, with growing non-Native solidarity, are they able to start reversing the damage wrought by decades of harm. The regeneration of despoiled ecosystems and "the reinhabitation of violated places is a fundamental objective of the environmental justice movement."[111]

The ecosystems may not revert to their natural, precolonial state but can be gradually regenerated by "negating the negation" of settler colonialism. In fighting for the resilience of their watersheds, tribal leaders have convinced at least some of their settler neighbors to forge a path away from conflict and toward cooperation. Some of these white neighbors even view the tribes as better protectors of their local environment and economy than their own government agencies.

Constricted to reservations, tribal nations have been historically unable to protect themselves from land thefts. A few tribes have reacquired small patches of stolen lands, such as putting purchased fee lands into federal trust status, regaining acreage on some public lands, or convincing a few private landowners to sign over their property. This #LandBack strategy would be more fruitful if larger tracts of public land were returned to tribal management, but the strategy runs up against the brick wall of government intransigence. The federal government, for example, only permits tribes to put new lands into trust for economic development, not sacred site or environmental protection.

Pacific Northwest Treaty Tribes are seeing faint glimmers of a parallel approach, of sovereign nations projecting their powers outside their reservations into their original territories, whether economically (as leading investors and employers in some rural counties), environmentally (through restoration projects and gaining veto power over polluting industry), or culturally (through a moral standing as the original people of the land). Even tribes lacking treaty rights or federal recognition have projected their authority in other ways, into their stolen lands.

The priority of Indigenous nations and movements is and always will be for #LandBack. But at the same time, Native nations can gain "Power

Back" even before regaining direct property ownership, by becoming "lead entities" in managing the realities on the ground in their original homelands, developing alliances with non-Native communities whose own futures are better served by respecting tribal authority, and building community resilience to face the crises that colonial capitalism has inflicted on the land and people.

Conclusions

Participants in these "unlikely" environmental justice alliances do not simply tell a story about relations between Native peoples and rural whites against a temporary threat to their livelihoods or values. They also shed some light on how the differences between people do not have to undermine the similarities between them and can under certain conditions even reinforce the similarities. The Native Americans' assertion of treaty and sovereign rights was in the short term a barrier to intercultural communication but actually helped in the long term to facilitate cooperation with their rural white communities that have faced similar environmental (and sometimes economic) injustices. The concept of environmental justice can be enlarged to encompass settler communities that seek to correct power disparities with the Indigenous nations on whose land they have settled.

Cooperation would have certainly been possible without prior conflict, and conflicts do not inevitably lead to collaborative projects. But certain conflicts—in a particular form and met with a particular response—serve as an embryo from which cooperation can emerge. Prospects for cooperation can be embedded within conflicts and, under certain circumstances, even harsh conflict may ironically serve as an opportunity for improving relations.

In this context, environmental justice can be seen not simply as the absence of environmental injustices, but the presence of positive relationships that allow for the regeneration of common ecosystems into the future. Restoring the land and water is necessary for the healing of divisions between peoples, and healing the divisions between peoples is necessary for restoring the land and water.

Environmental justice alliances have not caused rural white and Native communities to fundamentally overcome their historic divisions. No matter how much their rights have been violated, white rural residents have been afforded an advantageous position relative to Native peoples and have not experienced the same levels of land dispossession, cultural domination, and outright genocide as Native nations.

Environmental justice alliances do, however, pose a significant challenge to common assumptions about the inevitability of conflict and the impossibility of lasting cooperation. Instead of accepting that their white neighbors will always be obstacles to environmental justice, some Native leaders have made a serious effort to enlist their solidarity, by getting them to recognize that Indigenous sovereignty strengthens a common defense of shared local places. Even in the Trump-Biden era of political polarization, these tribal leaders have become more effective in reaching populist-minded whites than urban progressives or mainstream "Big Green" organizations.

The Indigenous-settler "unlikely alliances" needed certain circumstances or preconditions to succeed. First, they must build a sense of a "common place," or a common bond to the landscape. The geographic setting obviously establishes the communities' priorities, based on their proximity to a perceived environmental threat. But their perception of the landscape's sacredness or cultural significance also makes a construction of an alliance between them more likely. The sense of a "common place" also forms the basis of mutual cultural education, as both communities learn about their neighbors' land ethics.

Second, the alliances are founded out of a sense of a "common purpose," in legal, political, or economic fields. A common political adversary, usually an outside corporation or government agency, provides an enemy to focus the anger or resentment of community members outward, instead of only at each other. In some case studies, the tribal gaming economy helps even the playing field between the communities, even if the core of the Indigenous resurgence is cultural.

Third, the alliances can also be built of a sense of "common understanding"—a more difficult concept to grasp or define. Some members of the Native and rural white communities in resource conflict were forced to search for a way out, and an alliance seemed like a convenient vehicle. Conflict can lead to cooperation in this manner only if the players consciously seek common goals.

It is remarkable that widely disparate regions of North America experienced the development of Indigenous-settler alliances over a rather short period. As economic globalization makes peoples (and places) more and more similar, and ethnic nationalism seeks to emphasize the differences between peoples (and places), the intercultural alliances recognize difference and similarity as mutually reinforcing conditions.

The Particular and the Universal

The intercultural alliances offer important insights into the relationship between "particularism" and "universalism." "Particularism" asserts the particular *differences* between ethnic/racial groups, or other groups based on other social identities. (In the U.S., particularism is often termed "identity politics," an inadequate term that omits Native political sovereignty over territory).[112] "Universalism" asserts common ground, or the *similarities* between distinct groups that claim inherent differences. For example, "universalism" can be based on common class consciousness, a common state citizenship, or a human tie to the common environment or climate.

The unlikely alliances challenge the idea that "particularism" (such as Native rights) is always in contradiction to "universalism" (such as environmental protection). The assertion of Indigenous political strength does *not* necessarily weaken the idea of joining with non-Natives to defend the land and can even strengthen it. The stories of these alliances may identify ways to weave together the assertion of differences between peoples with the goal of finding common-ground similarities between them. (I am perhaps drawn to this hope because of my own Hungarian background, with a Jewish father whose family was decimated by genocide, and a Catholic mother whose family valued its cultural identity, and my attempts to navigate between the fear and celebration of ethnic pride.)

Environmental justice alliances based solely on the "universalist" similarities of living on a common planet, and that assert that "we are all in the same boat," tend to fail without respecting particularist differences. The idea of "why can't we all just get along" (like "United We Stand") is often used to suppress or sideline marginalized voices. The "same-boat approach ... does not necessarily confront ongoing territorial dispossession and risks to health, economic vitality, lives, psychological well-being, and cultural integrity that Indigenous people experience."[113]

This overemphasis on unity makes alliances more vulnerable, since authorities may try to divide them by meeting the demands of the (relatively advantaged) white members. A few alliances (such as against low-level military flights) floundered because the white "allies" declared victory and went home and did not keep up the fight to also win the demands of their Native neighbors. "Unity" is not enough when it is a unity of unequal partners; Native leadership needs to always be directly involved in the decision-making process.

Environmental justice alliances can help level the playing field between the communities, in effect using "particularism" as a prerequisite to "universalism." In order to achieve "unity," the majority needs to recognize and respect difference and understands how doing so can benefit universal values. The challenges "posed to environmental racism from within aggrieved communities of color contain the potential to create struggles that unite the particular and the universal in new ways."[114]

Many Indigenous-led movements (such as the Zapatistas in Mexico, or protesters against corporate globalization in Bolivia) have mixed particularist appeals to end racist colonial oppression, with universalist class-based appeals to members of the dominant ethnic group. Native/non-Native environmental alliances are an example of a North American movement that (consciously or not) has creatively negotiated the tensions between "particularity" and "universality" and has attempted to interweave them by backing Indigenous sovereignty to protect the land and water for everyone.

Ironically, the areas of the most intense treaty conflicts (such as western South Dakota and northeastern Wisconsin) developed the earliest and strongest tribal alliances with white farmers, ranchers, and fishers. In these areas some white residents have begun to see the tribes as more effective guardians of local economies than their own local, state, and federal governments. Where tribal nations strongly asserted their sovereignty, they were able to build a bridge to their neighbors, and where they held back, the alliances were not as successful.

Using their sovereign powers and federal trust responsibility, tribal nations can draw federal agencies and courts into the fray in a way that local and state governments cannot. Indigenous peoples offer a strong cultural anchor to the movement that makes it less willing to compromise. Tribes cannot simply move away from risks, because they have harvesting rights only within their ancestral territory, and so are fixed in their place-based identity and political sovereignty. The alliances are promoting empathy and solidarity to replace the oppression and entitlement of settler colonialism, and moving toward decolonization on the ground.

Non-Native Responsibilities

The continued existence of Native nationhood today undermines the claims of settler colonial states to the land.[115] Unlikely alliances can help chip away at the legitimacy of colonial structures, even among the settlers themselves. By asserting their treaty rights and sovereignty, Indigenous nations are benefiting not only themselves, but also their treaty partners. Since Europeans in North America are more separated in time and place from their Indigenous origins, they need to respectfully ally with Native nations to help find their own path to what it means to be a human being living on the Earth—without appropriating Native cultures.

Settlers need to rethink assumptions about the certainty of our presence on Native lands and understand how "embracing difficult uncertainty can be an integral part of undoing settler privilege and a step toward decolonization."[116] Indigenous nations will be more willing to remain engaged in alliances if they experience "respect for Indigenous knowledges, control of knowledge mobilization, intergenerational involvement, self-determination, continuous cross-cultural education, and early involvement."[117]

It is not the role of non-Natives to dissect Native cultures, but to study Indigenous-settler relations, and white attitudes and policies. The responsibility of non-Natives is to help remove the barriers and obstacles to

Native sovereignty in their own governments and communities. This is most effectively done not merely as an individual ally working to reverse personal prejudice, but as part of a collective alliance creating new structures of relationships.

Settler neighbors can begin to look to Indigenous nations for models to make their own communities more socially just, more ecologically resilient, and more hopeful. As Red Cliff Ojibwe organizer Walt Bresette once told Wisconsin non-Natives fighting a proposed mine, "We can't go back where we came from ... and the fish can't go back where they came from.... Those of you who have come need to make this part of your place ... have to defend it with your heart and your soul."[118]

Notes

1 Jill M. Bystydzienski and Steven P. Schacht, *Forging Radical Alliances Across Difference: Coalition Politics for the New Millennium* (London/New York: Rowman & Littlefield, 2001).
2 James D. Fearon and David D. Laitin, "Explaining Intercultural Cooperation," *American Political Science Review* 90 (1996): 715–35.
3 Sissel A. Waage, "(Re)claiming Space and Place through Collaborative Planning in Rural Oregon," *Political Geography* 20 (2001): 839–57.
4 Nella Van Dyke and Holly J. McCammon (eds.), *Strategic Alliances: Coalition Building and Social Movements* (Minneapolis: University of Minnesota Press, 2020).
5 Thomas Beamish and Amy Luebbers, "Alliance Building Across Social Movements," *Social Problems*, November 2009.
6 Laura Pulido, *Environmentalism and Economic Justice: Two Chicano Struggles in the Southwest* (Tucson: University of Arizona Press, 1996).
7 Andrew Hurley, *Environmental Inequalities Class, Race, and Industrial Pollution in Gary, Indiana, 1945–1980* (Chapel Hill: University of North Carolina Press, 2009).
8 Leilani Nishime and Kim Hester (eds.), *Racial Ecologies* (Seattle: University of Washington Press, 2018).
9 Michelle Gawerc, "Coalition-Building and the Forging of Solidarity across Difference and Inequality," *Sociology Compass*, February 2021.
10 Fred Rose, *Coalitions across the Class Divide: Lessons from the Labor, Peace, and Environmental Movements* (Ithaca, NY: Cornell University Press, 1999).
11 Jonathan Clapperton and Liza Piper (eds.), *Environmental Activism on the Ground: Small Green and Indigenous Organizing* (Calgary, AB: University of Calgary Press, 2019).
12 Leah Horowitz and Michael Watts (eds.), *Grassroots Environmental Governance*: Community Engagements with Industry (London/New York: Routledge, 2019).
13 Shannon Elizabeth Bell, *Fighting King Coal: The Challenges to Micromobilization in Central Appalachia* (Cambridge; London: MIT Press, 2016).
14 Anne Bonds and Joshua Inwood, "Beyond White Privilege: Geographies of White Supremacy and Settler Colonialism," *Progress in Human Geography* 40, no. 6 (2015): 715–33.

15 Steven C. Beda, "Collaboration, Not Fighting, Is What the Rural West Is Really About," *The Conversation*, October 25, 2018.
16 Vicken Hillis, Kate A. Berry, Briana Swette, Lauren Porensky, Clare Aslan, and Sheila Barry (eds.), "Unlikely Alliances and their Implications for Resource Management in the American West," *Environmental Research Letters* 15, no. 4 (January 2020): 1–15.
17 Zoltán Grossman, *Unlikely Alliances: Native Nations and White Communities Join to Defend Rural Lands* (Seattle: University of Washington Press, 2017).
18 Lynne Davis, *Alliances: Re/Envisioning Indigenous-non-Indigenous Relationships* (Toronto: University of Toronto Press, 2010).
19 Donald L. Horowitz, *Ethnic Groups in Conflict* (Berkeley: University of California Press, 1985), 677.
20 Zoltán Grossman, "Cowboy and Indian Alliances in the Northern Plains," *Agricultural History* 77, no. 2 (2003): 355–89.
21 Gail Small, "The Search for Environmental Justice in Indian Country," *News From Indian Country*, March 1994.
22 Northern Plains Resource Council (NPRC), "Derail the Tongue River Railroad," December 2000.
23 Heather Abel, "Montana on the Edge: A Fight Over Gold Forces the Treasure State to Confront Its Future." *High Country News*, December 22, 1997.
24 Richard Thieltges, Sweet Grass Hills Protective Association, Helena, Mont., interview, August 6, 1997.
25 Black Hills Alliance, *Keystone to Survival: The Multinational Corporations and the Struggle for Control of Land* (Minneapolis, MN: Haymarket Press, 1981).
26 Peter Matthiessen, "High Noon in the Black Hills," *New York Times Magazine*, July 13, 1980.
27 Black Hills Clean Water Alliance, "Black Hills Alliance Archive," https://bhcleanwateralliance.org/about/black-hills-alliance/bha-archive
28 Kurt Chandler, "Clash Over a Canyon: The Meaning of Sacred," *Minneapolis Star-Tribune*, August 9, 1987.
29 Wayne King, "Bradley Offers Bill to Return Land to Sioux," *New York Times*, March 11, 1987.
30 Black Hills Clean Water Alliance, "Know the Issues," https://bhcleanwateralliance.org
31 Madonna Thunder Hawk, Black Hills Alliance, Pine Ridge, S.D., interview, June 26, 1999.
32 Bob Secter, "Indians, Ranchers Oppose Black Hills Weapons Tests," *Los Angeles Times*, August 30, 1987.
33 Brandon B. Derman, *Struggles for Climate Justice: Uneven Geographies and the Politics of Connection* (London: Palgrave Macmillan, 2020), 164–76.
34 Kai Bosworth, *Pipeline Populism: Grassroots Environmentalism in the Twenty-First Century* (Minneapolis: University of Minnesota Press, 2022).
35 Faith Spotted Eagle, Ihanktonwan Nakota/Dakota, Yankton, S.D., interview, March 20, 2014.
36 Paul Seamans, Dakota Rural Action, Draper, S.D., interview, April 3, 2014.
37 Jane Kleeb, Bold Nebraska executive director, Hastings, Neb., interview, March 20, 2014.
38 Camp of the Sacred Stones, "Tribal Citizens Rise Up against Bakken Oil Pipeline," March 29, 2016.
39 Quoted in Camp of the Sacred Stones, "Protection of Sacred Sites Leads to Clash with Dakota Access Private Security," September 4, 2016.

40 Carl Sack, "A #NoDAPL Map," *Huffington Post,* November 2, 2016.
41 Jack Healy, "Neighbors Say North Dakota Pipeline Protests Disrupt Lives and Livelihoods," *New York Times,* Sept. 13, 2016
42 Nick Estes and Jaskiran Dhillon (eds.), *Standing with Standing Rock: Voices from the #NoDAPL Movement* (Minneapolis: University of Minnesota Press, 2019). Winona LaDuke, *To Be a Water Protector: The Rise of the Wiindigoo Slayers* (Halifax, NS: Fernwood Publishing, 2020).
43 Dagmar Thorpe, "The MX Missile and the Western Shoshone: The Destruction of a People," *Western Shoshone Sacred Land Association,* Spring 1981.
44 Jane Braxton Little, "Military Watchdog," *Audubon,* December 2000. Jennifer Allen, "U.S. Navy Crashed and Burns in Austin, NV," *Western Shoshone Defense Project,* Spring 1997.
45 Susan Lampert Smith, "Opposition to Military Flyovers Taking Off," *Wisconsin State Journal,* October 8, 1995.
46 Citizens Opposed to Range Expansion, "Don't Bomb Our Forests!," October 1996.
47 Ona Garvin, Ho-Chunk Nation, Pittsville, Wis., interview, August 22, 1997.
48 Tomah Monitor Herald, "National Guard drops flight plan," *Tomah Monitor Herald,* April 15, 1996.
49 Zoltan Grossman, "Ho-Chunk Environmental Alliances and the Effects of White Racial Advantage," *Wisconsin Geography* (2002), https://wisconsingeography.files.wordpress.com/2013/06/ho.pdf
50 Theodore W. Allen, *The Invention of the White Race. Volume Two: The Origin of Racial Oppression in Anglo-America* (London/New York: Verso, 1997).
51 Noel Ignatiev, *How the Irish Became White* (London/New York: Routledge, 1995).
52 George Lipsitz, "Walleye Warriors and White Identities: Native Americans' Treaty Rights, Composite Identities and Social Movements," *Ethnic and Racial Studies* 31, no. 1 (2008), 101–22.
53 Ronald N. Satz, "Chippewa Treaty Rights: The Reserved Rights of Wisconsin's Chippewa Indians in Historical Perspective," *Transactions of the Wisconsin Academy of Sciences, Arts, and Letters* 79 (1991).
54 Rick Whaley & Walter Bresette, *Walleye Warriors: An Effective Alliance Against Racism and for the Earth* (Philadelphia, PA: New Society Publishers, 1994).
55 Rennard Strickland, *et al.,* "Keeping Our Word: Indian Treaty Rights and Public Responsibilities. A Report on a Recommended Federal Role Following Wisconsin's Request for Federal Assistance," report to Senate Committee on Indian Affairs, 1990.
56 Nathan Seppa, "Old Foes Now Allies: Indians, Sports Fishermen Join to Oppose Mine," *Wisconsin State Journal,* February 11, 1994.
57 Marv Balousek, "Indians, Environmentalists Forge Alliance," *Wisconsin State Journal,* August 28, 1988.
58 Al Gedicks, *The New Resource Wars: Native and Environmental Struggle Against Multinational Corporations* (Boston, MA: South End Press, 1993), 83–162.
59 Walter Bresette, Midwest Treaty Network, Red Cliff Ojibwe, interview, September 17, 1997.
60 Tom Maulson, Wa-Swa-Gon Treaty Association, Chairman of Lac du Flambeau Ojibwe Band of Lake Superior Chippewa, interview, July 12, 1997.
61 Al Gedicks and Zoltán Grossman, "Native Resistance to Multinational Mining Corporations in Wisconsin," *Cultural Survival Quarterly,* March 2001,

https://www.culturalsurvival.org/publications/cultural-survival-quarterly/native-resistance-multinational-mining-corporations

62 Bob Schmitz, Wolf River Watershed Alliance, White Lake, Wis., interview, June 13, 1997.

63 Grossman, *Unlikely Alliances*, 239-257.

64 Al Gedicks, *Resource Rebels: Native Challenges to Mining and Oil Corporations* (Boston, MA: South End Press, 2001), 127–58.

65 Ron Seely, "State Tribes' Influence Had Broadened over Years," *Wisconsin State Journal,* November 2, 2003.

66 Grossman, *Unlikely Alliances*, 257–67.

67 Mike Wiggins Jr., chairman of Bad River Band of Lake Superior Chippewa, Odanah, Wisconsin, interview, October 14, 2014.

68 Al Gedicks, "Fight Over Proposed Mine by Menominee River has Brought Together Unlikely Allies," *Earth Island Journal,* April 28, 2021.

69 Menominee Indian Tribe of Wisconsin, "Back Forty Mine," https://www.menominee-nsn.gov/GovernmentPages/Initiatives/Back40Mine/Back40Mine.aspx; No Back 40 Mine, "News," http://www.noback40.org

70 Frances Van Zile, Mole Lake Mining Impact Committee, Sokaogon Chippewa Community, Wisconsin, interview, July 18, 1999.

71 Billy Frank Jr., *Tell the Truth: The Collected Columns of Billy Frank Jr.* (Olympia, WA: NWIFC, 2015), 149, https://nwtreatytribes.org/tellthetruth

72 Alexandra Harmon, *The Power of Promises: Rethinking Indian Treaties in the Pacific Northwest* (Seattle: University of Washington Press, 2008).

73 Lewis Kamb, "Boldt Decision 'Very Much Alive' 30 Years Later," *Seattle Post-Intelligencer,* February 11, 2004.

74 Governor's Office on Indian Affairs, "Centennial Accord," State of Washington (1989), https://goia.wa.gov/relations/centennial-accord

75 Northwest Indian Fisheries Commission. "Fisheries Management," https://nwifc.org/about-us/fisheries-management

76 Washington State Department of Ecology, "Watershed Look-up," https://ecology.wa.gov/Water-Shorelines/Water-supply/Water-availability/Watershed-look-up

77 Dan Von Seggern, "Culvert Case Update: A Victory for Tribal Treaty Rights," *Center for Environmental Law & Policy*, July 12, 2018, https://celp.org/2018/07/12/culvert-case-update

78 Dina Gilio-Whitaker, *As Long as Grass Grows: The Indigenous Fight for Environmental Justice, from Colonization to Standing Rock* (Boston, MA: Beacon Press, 2019).

79 Lynda Mapes, *Elwha: A River Reborn* (Seattle: Mountaineers Books, 2013). John Gussman and Jennifer Plumb, *Return of the River*, documentary film, 2014, http://www.elwhafilm.com

80 Northwest Treaty Tribes, "New Sand Habitat Attracting More Life near Elwha River," *Northwest Treaty Tribes*, April 8, 2014, https://nwtreatytribes.org/new-sand-habitat-attracting-life-near-elwha-river

81 Zoltán Grossman and Alexander McCarty, eds., *Removing Barriers: Restoring Salmon Watersheds through Tribal Alliances,* Conceptualizing Place: Pacific Northwest Native Art & Geographies (Olympia, WA: The Evergreen State College, 2021), https://sites.evergreen.edu/removingbarriers

82 Chehalis Basin Strategy. "The Chehalis Basin Strategy," https://chehalisbasinstrategy.com

83 Jason Schilling, "Beavers Relocated to Improve Salmon Habitat," *Northwest Treaty Tribes*, September 10, 2014.

84 Nisqually River Council, *Nisqually Watershed Stewardship Plan* (2020), https://nisquallyriver.org/resources/nwsp

85 Nisqually Delta Restoration, "Nisqually Delta Restoration," http://deltaresto ration.nisquallyriver.org

86 Timothy W Ransom, *For the Good of the Order: The Braget Farm and Land Use in the Nisqually Valley* (Centralia, WA: Gorham Printing, 2000).

87 Bellamy Pailthorp, "Biologist David Troutt 'This Is an Amazing, Productive Ecosystem'," KNKX News, September 11, 2017, http://apps.knkx.org/SalishSea/nisqually

88 Andy Hobbs, "Olympia's McAllister Springs Site Will Be Returned to Nisqually Tribe," *The Olympian*, January 13, 2017.

89 Nisqually Tribe, "Salmon Recovery Program," http://www.nisqually-nsn.gov/index.php/administration/tribal-services/natural-resources/salmon-recovery-program

90 Beth Rose Middleton, *Trust in the Land: New Directions in Tribal Conservation* (Tucson: University of Arizona Press, 2011), 185–94.

91 Jennifer L. Phillips, Jill Ory, and André Talbot, *Anadromous Salmonid Recovery in the Umatilla River Basin, Oregon* (Portland, OR: CRITFC, 2000).

92 Northwest Indian Fisheries Commission, *State of Our Watersheds: A Report by the Treaty Tribes of Western Washington* (Olympia, WA: NWIFC, 2020), https://nwifc.org/publications/state-of-our-watersheds

93 Zoltán Grossman and Alan Parker, *Asserting Native Resilience Pacific Rim Indigenous Nations Face the Climate Crisis* (Corvallis: Oregon State University Press, 2012).

94 Swinomish Climate Change Initiative, "About," https://www.swinomish-climate.com

95 Julia-Grace Sanders, "From Poop to Power: Manure from 2,300 Cows May Run 600 Homes," *Bellingham Herald*, December 14, 2020.

96 Valerie Volcovici, "A U.S. Tribe's Uphill Battle against Climate Change," *Reuters*, April 13, 2020. Quileute Tribe, "Quileute Tribe Move to Higher Ground," Move to Higher Ground, https://mthg.org

97 Zoltán Grossman, "The Resilience Doctrine: Indigenous Nations Understand Disaster Resilience," *CounterPunch*, February 3, 2021, https://www.counterpunch.org/2021/02/03/the-resilience-doctrine-indigenous-nations-understand-disaster-resilience

98 Zoltán Grossman, "The Achilles Heel of the Fossil Fuels Monster," *Works in Progress*, December 10, 2012, https://olywip.org/the-achilles-heel-of-the-fossil-fuels-monster

99 Maggie Allen, Stoney Bird, Sara Breslow, and Nives Dolšak, "Stronger Together: Strategies to Protect Local Sovereignty, Ecosystems, and Place-Based Communities from the Global Fossil Fuel Trade," *Marine Policy* 80 (2017): 168–76.

100 Tim Ballew II, "Cherry Point Victory Shows Treaty Rights Protect Us All," *Bellingham Herald*, May 14, 2016.

101 Zoltán Grossman, "Quinault Nation Builds Bridges to Stop Grays Harbor Oil Terminal," *Works in Progress*, December 9, 2016, https://olywip.org/quinault-nation-builds-bridges-stop-grays-harbor-oil-terminal

102 Eric de Place, "The Thin Green Line: The Northwest Faces off Against Titanic Coal and Oil Export Schemes," *Sightline Institute*, March 20, 2014, https://www.sightline.org/2014/03/20/the-thin-green-line; Robert McClure, "How Cascadia's Climate Activists Fought Off Fossil Fuels and Succeeded," *Crosscut*, January 18, 2021, https://crosscut.com/environment/2021/01/how-cascadias-climate-activists-fought-fossil-fuels-and-succeeded

103 Corrie Grosse, *Working Across Lines: Resisting Extreme Energy Extraction* (Berkeley: University of California Press, 2022).

104 Zoltán Grossman, "In 2017, Fusing Identity and Class Politics in 'Trumpland'," *Common Dreams,* January 3, 2017, https://www.commondreams.org/views/2017/01/03/2017-fusing-identity-and-class-politics-trumpland

105 Fawn Sharp, president of Quinault Indian Nation and Affiliated Tribes of Northwest Indians, Taholah, Washington, interview, October 29, 2015.

106 Paul Bowles and Henry Veltmeyer, *The Answer is Still No: Voices of Pipeline Resistance* (Halifax, NS: Fernwood Press, 2014). Toban Black, Stephen D'Arcy, Tony Weis, and Joshua Kahn Russell, eds. *Line in the Tar Sands: Struggles for Environmental Justice* (Oakland, CA: PM Press, 2014): 488–510.

107 Fiona MacPhail and Paul Bowles, "Fractured Alliance: State-Corporate Actions and Fossil Fuel Resistance in Northwest British Columbia, Canada," *Journal of Political Ecology* 28 (2021).

108 Anna J. Willow, *Understanding ExtrACTIVISM: Culture and Power in Natural Resource Disputes* (London/New York: Routledge, 2018). Paul Bowles and Gary Norman Wilson (eds.), *Resource Communities in a Globalizing Region: Development, Agency, and Contestation in Northern British Columbia* (Vancouver: University of British Columbia Press, 2016).

109 Shiri Pasternak, *Grounded Authority: The Algonquins of Barriere Lake against the State* (Minneapolis: University of Minnesota Press, 2017).

110 Katie Big-Canoe and Chantelle A. M. Richmond, "Anishinabe Youth Perceptions about Community Health: Toward Environmental Repossession," *Health & Place* 26 (March 2014): 127–35.

111 Devon G. Pena, "Endangered Landscapes and Disappearing Peoples?: Identity, Place, and Community in Ecological Politics," in *The Environmental Justice Reader: Politics, Poetics, and Pedagogy,* ed. Joni Adamson et Al., (Tucson: The University of Arizona Press, 2002), 65.

112 Charles Taylor, *Multiculturalism: The Politics of Recognition* (Princeton, NJ: Princeton University Press, 1994).

113 Kyle Powys Whyte, "White Allies, Let's Be Honest About Decolonization.," *Yes!* Magazine, April 3, 2018.

114 George Lipsitz, "Unexpected Affiliations: Environmental Justice and the New Social Movements," *Works and Days* 24, nos. 1–2 (2006): 27.

115 Audra Simpson, *Mohawk Interruptus: Political Life across the Borders of Settler States* (Durham, NC: Duke University Press, 2014).

116 Eva Mackey, *Unsettled Expectations: Uncertainty, Land and Settler Decolonization* (Halifax, NS: Fernwood Publishing, 2006).

117 Nicholas J. Reo, Kyle P. Whyte, Deborah McGregor, M.A. (Peggy) Smith, and James F Jenkins, "Factors that Support Indigenous Involvement in Multi-Actor Environmental Stewardship," *AlterNative* 1, no. 11 (2017): 58–68.

118 Protect the Earth, "In His Own Voice: Speaking for the Generations," *Superior Broadcast Network,* 1999, https://www.youtube.com/watch?v=q4Vtzj5LwUU

Bibliography

Abel, Heather. "Montana on the Edge: A Fight over Gold Forces the Treasure State to Confront its Future." *High Country News,* December 22, 1997.

Allen, Maggie, Stoney Bird, Sara Breslow, and Nives Dolšak, "Stronger Together: Strategies to Protect Local Sovereignty, Ecosystems, and Place-Based Communities from the Global Fossil Fuel Trade." *Marine Policy* 80 (2017): 168–76.

Allen, Theodore W. *The Invention of the White Race. Volume Two: The Origin of Racial Oppression in Anglo-America.* London/New York: Verso, 1997.

Ballew, Tim II. "Cherry Point Victory Shows Treaty Rights Protect Us All." *Bellingham Herald*, May 14, 2016.

Beamish, Thomas, and Amy Luebbers, "Alliance Building Across Social Movements." *Social Problems*, November 2009.

Beda, Steven C. "Collaboration, Not Fighting, Is What the Rural West is Really About." *The Conversation,* October 25, 2018.

Bell, Shannon Elizabeth. *Fighting King Coal: The Challenges to Micromoblization in Central Appalachia.* Cambridge, MA; London: MIT Press, 2016.

Big-Canoe, Katie, and Chantelle A. M. Richmond. "Anishinabe Youth Perceptions About Community Health: Toward Environmental Repossession." *Health & Place* 26 (March 2014): 127–35.

Black Hills Alliance. *Keystone to Survival: The Multinational Corporations and the Struggle for Control of Land.* Minneapolis, MN: Haymarket Press, 1981.

Black Hills Clean Water Alliance. "Black Hills Alliance Archive." https://bhclean wateralliance.org/about/black-hills-alliance/bha-archive/

Black, Toban, Stephen D'Arcy, Tony Weis, and Joshua Kahn Russell, eds. *Line in the Tar Sands: Struggles for Environmental Justice.* Oakland, CA: PM Press, 2014.

Bonds, Anne, and Joshua Inwood, "Beyond White Privilege: Geographies of White Supremacy and Settler Colonialism." *Progress in Human Geography* 40, no. 6 (2015): 715–33.

Bosworth, Kai. *Pipeline Populism: Grassroots Environmentalism in the Twenty-First Century.* Minneapolis: University of Minnesota Press, 2022.

Bowles, Paul, and Henry Veltmeyer. *The Answer Is Still No: Voices of Pipeline Resistance.* Halifax, NS: Fernwood Press, 2014.

———, and Gary Norman Wilson, eds. *Resource Communities in a Globalizing Region: Development, Agency and Contestation in Northern British Columbia.* Vancouver: University of British Columbia Press, 2016.

Bystydzienski, Jill M., and Steven P. Schacht. *Forging Radical Alliances Across Difference: Coalition Politics for the New Millennium.* London/New York: Rowman & Littlefield, 2001.

Clapperton, Jonathan, and Liza Piper, eds. *Environmental Activism on the Ground: Small Green and Indigenous Organizing.* Calgary, AB: University of Calgary Press, 2019.

Davis, Lynne. *Alliances: Re/Envisioning Indigenous-non-Indigenous Relationships.* Toronto: University of Toronto Press, 2010.

De Place, Eric. "The Thin Green Line: The Northwest Faces Off against Titanic Coal and Oil Export Schemes." Sightline Institute, March 20, 2014. https://www.sightline.org/2014/03/20/the-thin-green-line/.

Derman, Brandon B. *Struggles for Climate Justice: Uneven Geographies and the Politics of Connection.* London: Palgrave Macmillan, 2020.

Estes, Nick, and Jaskiran Dhillon, eds. *Standing with Standing Rock: Voices from the #NoDAPL Movement.* Minneapolis: University of Minnesota Press, 2019.

Fearon, James D., and David D. Laitin. "Explaining Intercultural Cooperation." *American Political Science Review* 90 (1996): 715–35.

Gawerc, Michelle. "Coalition-Building and the Forging of Solidarity Across Difference and Inequality." *Sociology Compass, February 2021.*

Gedicks, Al. *The New Resource Wars: Native and Environmental Struggle Against Multinational Corporations.* Boston, MA: South End Press, 1993.

_____. *Resource Rebels: Native Challenges to Mining and Oil Corporations.* Boston, MA: South End Press, 2001.

_____. "Fight Over Proposed Mine by Menominee River Has Brought Together Unlikely Allies." *Earth Island Journal,* April 28, 2021. https://www.earthisland. org/journal/index.php/articles/entry/fight-proposed-mine-menominee-river-unlikely-allies/.

Gilio-Whitaker, Dina. *As Long as Grass Grows: The Indigenous Fight for Environmental Justice, from Colonization to Standing Rock.* Boston, MA: Beacon Press, 2019.

Governor's Office on Indian Affairs. "Centennial Accord." State of Washington (1989). https://goia.wa.gov/relations/centennial-accord.

Grossman, Zoltan. "Ho-Chunk Environmental Alliances and the Effects of White Racial Advantage." *Wisconsin Geography* (2002). https://wisconsingeography. files.wordpress.com/2013/06/ho.pdf

_____. "Cowboy and Indian Alliances in the Northern Plains." *Agricultural History* 77, no. 2 (2003): 355–89.

_____. "The Achilles Heel of the Fossil Fuels Monster." *Works in Progress,* December 10, 2012. https://olywip.org/the-achilles-heel-of-the-fossil-fuels-monster/.

_____. "Quinault Nation Builds Bridges to Stop Grays Harbor Oil Terminal." *Works in Progress,* December 9, 2016. https://olywip.org/quinault-nation-builds-bridges-stop-grays-harbor-oil-terminal/

_____. *Unlikely Alliances: Native Nations and White Communities Join to Defend Rural Lands.* Seattle: University of Washington Press, 2017.

Grossman, Zoltán. "The Resilience Doctrine: Indigenous Nations Understand Disaster Resilience." *CounterPunch,* February 3, 2021. https://www.counterpunch. org/2021/02/03/the-resilience-doctrine-indigenous-nations-understand-disaster-resilience

_____, and Alexander McCarty, eds. *Removing Barriers: Restoring Salmon Watersheds Through Tribal Alliances.* Conceptualizing Place: Pacific Northwest Native Art and Geographies. Olympia, WA: The Evergreen State College, 2021. https://sites.evergreen.edu/removingbarriers/

_____, and Alan Parker, eds. *Asserting Native Resilience Pacific Rim Indigenous Nations Face the Climate Crisis.* Corvallis: Oregon State University Press, 2012.

Grosse, Corrie. Working Across Lines: Resisting Extreme Energy Extraction. Berkeley: University of California Press, 2022.

Harmon, Alexandra. *The Power of Promises: Rethinking Indian Treaties in the Pacific Northwest.* Seattle: University of Washington Press, 2008.

Healy, Jack. "Neighbors Say North Dakota Pipeline Protests Disrupt Lives and Livelihoods." *New York Times,* September 13, 2016.

Hillis, Vicken, Kate A. Berry, Briana Swette, Lauren Porensky, Clare Aslan, and Sheila Barry, eds., "Unlikely Alliances and their Implications for Resource Management in the American West." *Environmental Research Letters* 15, no. 4 (January 2020): 1–13.

Horowitz, Donald L. *Ethnic Groups in Conflict*. Berkeley: University of California Press, 1985.

Horowitz, Leah, and Michael Watts, eds. *Grassroots Environmental Governance: Community Engagements with Industry*. London/New York: Routledge, 2019.

Hurley, Andrew. *Environmental Inequalities Class, Race, and Industrial Pollution in Gary, Indiana, 1945–1980*. Chapel Hill: University of North Carolina Press, 2009.

Kamb, Lewis. "Boldt Decision 'Very Much Alive' 30 Years Later." *Seattle Post-Intelligencer*, February 11, 2004.

LaDuke, Winona. *To Be A Water Protector: The Rise of the Wiindigoo Slayers*. Halifax, NS: Fernwood Publishing, 2020.

Lipsitz, George. "Unexpected Affiliations: Environmental Justice and the New Social Movements." *Works and Days* 24, nos. 1–2 (2006): 25–44.

_____. "Walleye Warriors and White Identities: Native Americans' Treaty Rights, Composite Identities and Social Movements." *Ethnic and Racial Studies* 31, no. 1 (2008): 101–22.

Mackey, Eva. *Unsettled Expectations: Uncertainty, Land and Settler Decolonization*. Halifax, NS: Fernwood Publishing, 2006.

Mapes, Lynda. *Elwha: A River Reborn*. Seattle: Mountaineers Books, 2013.

McClure, Robert. "How Cascadia's Climate Activists Fought Off Fossil Fuels and Succeeded." *Crosscut*, January 18, 2021. https://crosscut.com/environment/2021/01/how-cascadias-climate-activists-fought-fossil-fuels-and-succeeded

Menominee Indian Tribe of Wisconsin, "Back Forty Mine." https://www.menominee-nsn.gov/GovernmentPages/Initiatives/Back40Mine/Back40Mine.aspx

Middleton, Beth Rose. *Trust in the Land: New Directions in Tribal Conservation*. Tucson: University of Arizona Press, 2011.

Nishime, Leilani, and Kim Hester, eds., *Racial Ecologies*. Seattle: University of Washington Press, 2018.

Nisqually Delta Restoration. "Nisqually Delta Restoration." https://www.nisqually deltarestoration.org/

Nisqually River Council. *Nisqually Watershed Stewardship Plan* (2020). https://nisquallyriver.org/resources/nwsp

Nisqually Tribe, "Salmon Recovery Program." http://www.nisqually-nsn.gov/index.php/administration/tribal-services/natural-resources/salmon-recovery-program

No Back 40 Mine, "News." http://www.noback40.org

Northwest Indian Fisheries Commission. *State of Our Watersheds: A Report by the Treaty Tribes of Western Washington*. Olympia, WA: NWIFC, 2020. https://nwifc.org/publications/state-of-our-watersheds/

Pasternak, Shiri. *Grounded Authority: The Algonquins of Barriere Lake against the State*. Minneapolis: University of Minnesota Press, 2017.

Phillips, Jennifer L., Jill Ory, and André Talbot. *Anadromous Salmonid Recovery in the Umatilla River Basin, Oregon*. Portland, OR: CRITFC, 2000.

Pulido, Laura. *Environmentalism and Economic Justice: Two Chicano Struggles in the Southwest*. Tucson: University of Arizona Press, 1996.

Ransom, Timothy W. *For the Good of the Order: The Braget Farm and Land Use in the Nisqually Valley*. Centralia, WA: Gorham Printing, 2000.

Reo, Nicholas J., Kyle P. Whyte, Deborah McGregor, M.A. (Peggy) Smith, and James F Jenkins. "Factors That Support Indigenous Involvement in Multi-Actor Environmental Stewardship." *AlterNative* 1, no. 11 (2017): 58–68.

Rose, Fred. *Coalitions across the Class Divide: Lessons from the Labor, Peace, and Environmental Movements.* Ithaca, NY: Cornell University Press, 1999.

Sack, Carl. "A #NoDAPL Map." *Huffington Post,* November 2, 2016.

Satz, Ronald N. "Chippewa Treaty Rights: The Reserved Rights of Wisconsin's Chippewa Indians in Historical Perspective." *Transactions of the Wisconsin Academy of Sciences, Arts, and Letters* 79 (1991).

Secter, Bob. "Indians, Ranchers Oppose Black Hills Weapons Tests." *Los Angeles Times,* August 30, 1987.

Simpson, Audra. *Mohawk Interruptus: Political Life across the Borders of Settler States.* Durham, NC: Duke University Press, 2014.

Small, Gail. "The Search for Environmental Justice in Indian Country." *News from Indian Country,* March 1994.

Swinomish Climate Change Initiative. "About." https://www.swinomish-climate.com

Van Dyke, Nella, and Holly J. McCammon, eds. *Strategic Alliances: Coalition Building and Social Movements.* Minneapolis: University of Minnesota Press, 2020.

Volcovici, Valerie. "A U.S. Tribe's Uphill Battle against Climate Change." *Reuters,* April 13, 2020.

Von Seggern, Dan. "Culvert Case Update: A Victory for Tribal Treaty Rights." *Center for Environmental Law & Policy,* July 12, 2018. https://celp.org/2018/07/12/culvert-case-update

Waage, Sissel A. "(Re)Claiming Space and Place Through Collaborative Planning in Rural Oregon." *Political Geography* 20 (2001): 839–57.

Whaley, Rick, and Walter Bresette. *Walleye Warriors: An Effective Alliance Against Racism and for the Earth.* Philadelphia, PA: New Society Publishers, 1994.

Whyte, Kyle Powys. "White Allies, Let's Be Honest About Decolonization." *Yes! Magazine,* April 3, 2018.

Willow, Anna J. *Understanding ExtrACTIVISM: Culture and Power in Natural Resource Disputes.* London/New York: Routledge, 2018.

Part II

Indigenous Movements and Environmental Justice in the United States, Canada, Mexico, and the Caribbean

4 Environmental Justice in Hawai'i and Oceania

Kyle Kajihiro

I am this land, and this land is me.[1]

(Pualani Kanaka'ole Kanahele)

We sweat and cry salt water, so we know that the ocean is really in our blood.[2]

(Teresia Teaiwa)

Introduction

This chapter provides a brief survey of key environmental justice (EJ) issues in Hawai'i and the U.S.-affiliated Pacific Islands of Guåhan/Guam, the Commonwealth of the Northern Mariana Islands, and the Republic of the Marshall Islands (RMI). Although EJ issues are widespread in this region, they are not usually framed as problems of EJ. However, the EJ implications of regional environmental issues become clearer if we understand the historical drivers of environmental change: imperialism, militarization, war, settler colonialism, extractivist capitalism, tourism, and global climate change.

This chapter examines the history of EJ issues in Hawai'i, which are rooted in U.S. imperialism, settler colonialism, and the rise of sugar capitalism in the islands. While there are numerous examples of EJ issues in Hawai'i, this chapter focuses on the impacts of industrial agriculture on land and water use, the developmentalist economics of the post-statehood period, and the widespread EJ impacts of the U.S. military.[3] It concludes with a brief discussion of EJ issues in the Mariana Islands and the Marshall Islands, two U.S.-affiliated Pacific Island groups, whose environmental issues are closely tied to political-economic, geopolitical, and historical developments emanating from Hawai'i.

DOI: 10.4324/9781003214380-7

Historical Roots of Environmental Injustice in Hawai'i

To understand the historical context of EJ issues in Hawai'i and the Pacific, we must briefly review the history of how the United States came to occupy Hawai'i. The Hawaiian Kingdom was founded in 1795 when Hawai'i Island ruler Kamehameha I conquered the islands of Maui, Lāna'i, Moloka'i, Kaho'olawe, and O'ahu. In 1810, Kaua'i and Ni'ihau voluntarily joined the Hawaiian Kingdom, consolidating the Hawaiian archipelago under a single government. By 1843, Great Britain, France, and the United States recognized the Hawaiian Kingdom as an independent and sovereign state.

When Captain James Cook arrived in Hawai'i in 1778, he marveled at the health and vitality of the Kanaka Maoli people and the scope of their social and cultural achievements. Scholars estimate that the Kanaka Maoli population at the time of western contact was around 683,200.[4] However, due to introduced diseases and the disruption of Indigenous lifeways, by 1900 the Kanaka Maoli population collapsed to 30,546,[5] a 96% decline. This drastic decrease led to the abandonment of many agricultural sites and a dramatic reduction in food production.

While Kanaka Maoli-introduced plant and animal species and landscape modifications certainly caused environmental changes, the most radical environmental changes came with the penetration of capitalist economic relations, first with extractivist ventures related to whaling and sandalwood, but later with the advent of industrial sugar production. Deforestation due to the overharvesting of forest products and feral livestock caused erosion and the sedimentation of reefs.

The Māhele (1848) and subsequent Kuleana Act (1850) were a series of reforms in the land tenure system instituted by the Hawaiian Kingdom to partition land, assign land use rights, and record land titles. While the original intent was to codify and protect Indigenous title and use rights to land, the long-term outcome of these changes was to dispossess the majority of Kānaka Maoli (Native Hawaiians) and consolidate foreign land ownership. These EJ issues were exacerbated by imperialist interventions and the consequent loss of Indigenous control over their country.

The primary aim of the United States in Hawai'i was to acquire for a naval coaling station the sprawling inlet and estuary on the island of O'ahu known to Kānaka Maoli as Ke Awalau o Pu'uloa (today commonly known as "Pearl Harbor"). In 1873, General John Schofield, posing as a tourist, transmitted secret reconnaissance reports about suitable sites for a U.S. naval base. He reported that "Pearl River ... is the key to the Central Pacific Ocean, it is the gem of these islands, valueless to them because they cannot use it, but more valuable to the United States than all else the islands have to give."[6] What Schofield dismissed as "valueless" was the complex and sophisticated mariculture and agriculture food system that Kānaka Maoli had built at Ke Awalau o Puʻuloa.[7]

In 1887, white businessmen forced King Kalākaua, under threat of violence, to ratify a new constitution, commonly known as the "Bayonet Constitution." This partial *coup d'état* dramatically concentrated the power of the white minority elites while disenfranchising the majority of poor Kānaka Maoli and Asian immigrant workers. The new settler-dominated government approved a commercial treaty with the United States to waive tariffs on Hawaiian sugar exports. In exchange, the United States obtained exclusive rights to build a naval coaling station at Ke Awalau o Puʻuloa. In the ensuing Hawaiian political unrest sparked by this affront to Hawaiʻi's sovereignty, Kalākaua's successor Queen Liliʻuokalani attempted to rescind the Bayonet Constitution. The white sugar planters and the U.S. minister countered by landing U.S. troops in support of a settler *coup d'état*. Despite successful political resistance by Hawaiian nationals to block annexation by the United States, the Spanish American War in 1898 gave annexationists an opportunity to take the Hawaiian Islands under the rubric of military necessity. As the U.S. military seized and transformed the land and sea of Ke Awalau o Puʻuloa into its naval station, Kanaka Maoli ʻāina momona (abundant lands) was radically altered. The U.S. occupation of Hawaiʻi enabled the United States expansion into an overseas empire.

The United States transformed Hawaiʻi into a key military outpost in the Pacific, which made Hawaiʻi a target at the start of the Pacific War. After the United States defeated Japan in World War II, the United Nations placed Hawaiʻi on the list of non-self-governing territories—colonies eligible for decolonization. In 1959, following a questionable statehood referendum in which U.S. military personnel and settlers were allowed to vote on Hawaiʻi's political status, the U.N. removed Hawaiʻi from the list. The United States passed legislation admitting Hawaiʻi as the 50th state of the United States, which brought a wave of foreign investment and economic growth, as well as an influx of settlers. Statehood may have created new political and economic opportunities for many of the Asian settler groups in Hawaiʻi who were previously subjected to racist discrimination, but it added new layers of multiethnic settler oppression to Kānaka Maoli, whose land and culture came under increased development pressure.

Industrial Agriculture and Environmental Justice

In the aftermath of the U.S. annexation of Hawaiʻi in 1898, the dramatic growth of the sugar industry radically transformed the environment and caused significant EJ impacts in Hawaiʻi, which continue to have lingering effects long after sugar's demise.[8] The systematic transformation of

Hawaiian landscapes by industrial agriculture contributed to the loss of native ecosystems and the creation of landscapes dominated by alien species.

With the growth of the sugar industry, the struggle to control water use became one of the major EJ issues in Hawai'i. Kānaka Maoli revere wai (fresh water) as a sacred source of life. One of the primary elemental deities in the Hawaiian religious system is Kāne, who can take the form of sunlight and heat and all forms of fresh water, which poetically describe what western scientists call the hydrological cycle. The Hawaiian word for wealth or abundance is waiwai, which has water as its root.

After seizing power in Hawai'i, the sugar planters acquired vast parcels of land to grow sugar and pineapple. But these thirsty crops required substantial irrigation. In 1879, the first artesian well in Hawai'i was drilled to irrigate 'Ewa Plantation on O'ahu. The discovery of underground water sources fueled a growth in sugar cultivation. To satisfy the needs of the burgeoning sugar industry, a number of sugar plantations also embarked on massive water diversion projects, involving an extensive series of tunnels through the hearts of mountains and elaborate networks of irrigation ditches. These ditch systems siphoned water from the wet windward valleys of the island to agricultural lands on sunny leeward plains. As a result, many windward streams dried up, disrupting fragile stream ecologies and their community of native species, and driving Kanaka Maoli kalo (taro) farmers to ruin. Hawai'i's sugar capitalists, at one point the most efficient and profitable in the world, dominated water rights throughout the territorial period.

In 1916, the O'ahu Sugar Company constructed a 25-mile Waiāhole ditch system, including a 3-mile tunnel through the Ko'olau mountains to divert 30 million gallons per day of windward O'ahu water to irrigate sugar on the 'Ewa plains. Waiāhole and Waikāne streams practically disappeared.

After Hawai'i's admission as a state in 1959, locally selected judges replaced federal appointees and brought a deeper appreciation of Hawai'i's historical, cultural, and political context to their interpretation of the laws, which reshaped jurisprudence in Hawai'i.[9] The Hawai'i Supreme Court of Chief Justice William Richardson, a Kanaka Maoli, ruled in *McBryde Sugar Company v. Robinson* (1973) that private water users may have certain use rights to water but did not have property interests in water. Importantly, the court concluded that water is a public trust under the stewardship of the government. This legal definition of water as a public trust, which is now foundational in Hawai'i's water code, derives from the Kanaka Maoli relationship to wai as a sacred natural element. Despite 129 years of settler colonial domination, many Kanaka Maoli values and concepts have been retained within the laws of the State of Hawai'i.

The 1978 Hawai'i constitutional convention introduced a number of progressive reforms, including language affirming the State's duty to protect natural resources and the environment, with specific provisions for the protection of water resources for the benefit of the people. Subsequent legislation established the Water Code and the Commission on Water Resource Management (CWRM) to govern the equitable and sustainable allocation of water.

In the 1980s and 1990s, as competition from cheaper foreign sources of sugar brought an end to the era of sugar in Hawai'i, the former sugar corporations, many of which had merged with transnational corporations, turned to real estate and tourism development. These housing and industrial developments also required access to water.

With the economic demise of sugar, and the waning of sugar's political power, Kanaka Maoli kalo farmers and environmentalists in windward O'ahu saw an opportunity to seek the return of waters to windward streams. Through a lengthy campaign of grassroots organizing and litigation, Kanaka Maoli groups, environmentalists, and other residents of Waiāhole-Waikāne squared off against major landowners, development interests, and state officials. The Hawai'i Supreme Court ruled in 2000 that the State had a public trust obligation to govern water use for several public purposes. These public trust purposes included protection of the environment, Kanaka Maoli rights, and other domestic uses, all of which have priority over commercial uses. The decision resulted in the restoration of a significant amount of water to windward streams.

There are currently two major fights over water allocation on the island of Maui, Nā Wai 'Ehā and East Maui. In the Nā Wai 'Ehā (four great waters) case, Hui o Nā Wai 'Ehā and Maui Tomorrow Foundation petitioned the CWRM in 2004 to restore streamflows in the four streams after the Wailuku Sugar Company ended sugar production and began selling water to the Hawaiian Commercial & Sugar Company (HCSC) and other developers. After a long legal battle, in 2021, the CWRM allocated 14% for kalo cultivation, 28% for reasonable and beneficial uses, such as diversified agriculture, and 7% for municipal water use but retained only 51% of the streamflow.[10]

In the East Maui Water case, Sierra Club of Hawai'i sued the State of Hawai'i Board of Land and Natural Resources, Alexander and Baldwin, and the East Maui Irrigation Company to seek a reduction in water diversion from east Maui streams. In 2021, the State Environmental Court ordered Alexander and Baldwin, and the East Maui Irrigation Company to reduce its water diversion from 45 to 25 million gallons per day (mgd).

The liberal use of toxic pesticides and herbicides in Hawai'i's agricultural sector has become another EJ issue in Hawai'i. While the sugar industry experimented with a number of biological pest control methods, some of which turned out to be environmentally disastrous, the pineapple industry preferred pesticides such as heptachlor and dibromochloropropane, which are carcinogenic and neurotoxic. Despite the EPA's cancelation of heptachlor use in the United States in 1976, Hawai'i pine-apple growers continued using it for several more years until their stocks were depleted. In the 1980s, a public health scare swept Hawai'i when high levels of heptachlor were detected in local milk, the result of dairy cows being fed pesticide-laden pineapple tops. According to the environmental poet W.S. Merwin, in 1985, "Ten times as much poison is used in Hawai'i per square mile, and three times as much per capita, as in any state on the mainland."[11] In 1977, a spill of ethylene dibromide at the Del Monte pineapple plantation at Kunia, O'ahu contaminated a nearby well which supplied drinking water to plantation workers. The Environmental Protection Agency (EPA) listed the 3,000-acre site on the National Priorities List (NPL, "Superfund") in 1994. After some remediation, the EPA delisted portions of the site and allowed limited economic redevelopment.

Post-World War II Transformations

World War II introduced dramatic changes to the political economy of Hawai'i. During the period of martial law in Hawai'i from 1941 to 1944, the U.S. military supplanted the power of the white settler oligarchy and took large tracts of land for training and military base development. The wartime influx of troops and military workers changed the population demographics and prompted the growth of new industries to meet their needs. Meanwhile, large numbers of second-generation Japanese settlers joined the U.S. military during the war and obtained college educations on the G.I. Bill. At the same time, workers on the plantations and docks organized powerful unions which won major sugar plantation and longshore strikes and built the multiethnic political base of the Democratic Party. This Democratic wave swept the elections of 1954 and replaced the decades-old white-dominated Republican Party rule in Hawai'i. The new multiethnic Democratic Party machine in Hawai'i implemented some important liberal reforms but failed to deliver on many of its promises of social and economic equality. Instead, the Democratic Party aligned itself with the interests of the growing tourism, construction, and military sectors to sustain economic growth and maintain its hold on power.

Cold War military policies brought steady military spending to Hawai'i. With the advent of cheaper air travel, mass tourism grew to become Hawai'i's main economic driver, overtaking industrial agriculture. The post-statehood development boom drove up the cost of living and displaced many agricultural areas and rural communities to make way for sprawling suburbs, golf courses, and resorts. But it also spurred many affected communities to fight back.

From 1970 to 1971, a historic anti-eviction struggle took place at Kalama Valley in east O'ahu. It was the first in a wave of protest movements in reaction to the frenzied development of the 1960s and 1970s. The land owner at Kalama Valley, the Bishop Estate, undertook a luxury housing development project which threatened to displace a number of hog farmers. Students and activist groups rallied to defend the farmers who chose to resist eviction. Inspired by the Black Panthers and other Third World liberation struggles, these activists formed Kokua Kalama Valley (which they later renamed Kokua Hawai'i). In the ensuing months' long occupation of the valley, three dozen Kokua Hawai'i members were arrested. While they were unable to stop the evictions, the Kalama Valley movement galvanized activists across the islands and sparked numerous anti-eviction and anti-development struggles, including protests against a resort development at Nukoli'i on Kaua'i; Chinese immigrant anti-gentrification organizing in Honolulu's Chinatown; Filipino plantation workers fighting to keep their homes at Ota Camp, a plantation camp in Waipahu; and Okinawan, CHamoru, and Kanaka Maoli farmers fighting eviction for a suburban housing development project in Waiāhole-Waikāne on the windward side of O'ahu. Dozens of land struggles, many of them led by Kānaka Maoli, took place in Hawai'i between 1970 and 1999.[12] These struggles were multiethnic and primarily working class in composition. However, in many cases, the political strategy and organizing skills came from students at the University of Hawai'i (UH), especially the newly formed Ethnic Studies Program.

While the Kalama Valley occupation was multiethnic and primarily framed as a class struggle, political differences emerged. The people of color leaders called on haole (white settler) members to step back from visible leadership positions. Kanaka Maoli leaders began to differentiate their uniquely Indigenous claims to land and sovereignty from the social and economic justice demands of the other working-class people of color. In these nascent assertions of indigeneity, Haunani-Kay Trask, a renowned Kanaka Maoli activist and scholar, saw Kalama Valley as the birthplace of the modern Hawaiian movement.[13]

Another controversial industry to grow in the wake of the sugar industry was the agrochemical industry, which exploited Hawai'i's mild year-round semi-tropical climate to produce genetically modified seed corn.[14] The concerns about the health effects of herbicides and pesticides used on seed crops have sparked grassroots efforts to restrict or heavily regulate pesticide and herbicide use on genetically modified crops.[15]

Environmental Justice and the U.S. Military

Militarization is a major source of environmental injustice in Hawai'i and Oceania, the vast region encompassing Melanesia, Micronesia, Polynesia, and Australia. Insular East Asia and Southeast Asia, including Japan,

Okinawa, and the Philippines, are also heavily affected by militarization. As described above, imperial geopolitics and military strategy drove the United States to overthrow and occupy Hawaiʻi. When the United States purported to annex Hawaiʻi in 1898 as a wartime measure during the Spanish American War, it began a dramatic militarization of Hawaiʻi's landscape, building a string of new military installations to encircle Oʻahu "with a ring of steel,"[16] including Pearl Harbor, Fort Shafter, Schofield Barracks, Fort DeRussy, and Fort Ruger. Military land takings peaked during World War II, when Hawaiʻi was under martial law. Today, the military occupies 142 sites and controls approximately 228,639 acres of land in Hawaiʻi.[17] The military is most concentrated on Oʻahu where it occupies roughly 94,000 acres, or 24.6% of the land. The majority of the military's land holdings in Hawaiʻi—approximately 68%—are former Government and Crown Lands of the Hawaiian Kingdom, which have been designated as Hawaiian Trust Lands or "Ceded" lands.

With the transformation of Ke Awalau o Puʻuloa into Pearl Harbor, the once productive Kanaka Maoli food system was lost and the area has become a toxic Superfund site.[18] Today, the Department of Defense has identified approximately 1,000 contamination sites at 115 installations in Hawaiʻi.[19] For many years, the U.S. military operated with few environmental regulations on its activities. But the rise of the environmental movement and the passage of new environmental and cultural protection laws in the 1960s and 1970s created political opportunities for groups to hold the military accountable.

Kahoʻolawe: The (Re)Birth of Aloha ʻĀina

One of the formative EJ social movements to emerge in Hawaiʻi in the 1970s was the movement to protect the island of Kahoʻolawe from military live-fire training. Located approximately 7 miles south of Maui, and in the rain shadow of the massive Haleakalā volcano, Kahoʻolawe is the smallest and one of the driest of the main Hawaiian Islands. Many Kānaka Maoli consider Kahoʻolawe to be a sacred manifestation of Kanaloa, a Polynesian deity of the deep sea and long-distance voyaging. In the 19th and 20th centuries, overgrazing by feral goats, which were originally a gift from British captain Vancouver in 1792, resulted in severe deforestation and erosion. However, the most devastating environmental harm came with the military occupation of the island. Within days of the bombing of Pearl Harbor in 1941, the U.S. military seized the island, removed the resident cattle ranchers, and turned the entire island into a live-fire training area. This severely damaged cultural sites and left the island littered with unexploded ordnance. In 1965, the U.S. Navy conducted Operation Sailor Hat on Kahoʻolawe. This three-shot series of simulated nuclear blasts, each using 500 tons of TNT, left a deep water-filled crater in the lava rock as a monument to the violence.

On January 4, 1976, a group of Kanaka Maoli activists staged a bold occupation of Kaho'olawe to press their demands for federal legislation to compensate Native Hawaiians for historical wrongs committed by the United States. Although most of the flotilla was turned back by the Coast Guard, a group of nine activists made it to the island. Of the nine, two young Kanaka Maoli men, Emmett Aluli and Walter Ritte, evaded capture for three days and explored the island's endangered cultural sites. Profoundly moved by the experience, they came away with a new mission to stop the bombing of Kaho'olawe and restore the island's environmental and cultural treasures. The Protect Kaho'olawe 'Ohana (PKO) was born.[20]

The PKO engaged in a series of nonviolent direct actions to disrupt naval exercises, lengthy bouts of litigation and political lobbying, and the revitalization of Kanaka Maoli cultural practices on the island. In 1976, the PKO sued the U.S. Navy under the National Environmental Policy Act (NEPA) and the National Historic Preservation Act (NHPA). In a 1980 Consent Decree and Order, the Navy recognized the PKO as traditional stewards of Kaho'olawe and agreed to complete a more thorough environmental impact study to include extensive cultural and archaeological surveys, remove the invasive goats, and importantly to permit the PKO regular access to Kaho'olawe to engage in cultural activities and restoration practices.[21]

In 1977, tragedy struck when two PKO leaders—George Helm and Kimo Mitchell—disappeared at sea during an attempted rescue of two other PKO members stranded on Kaho'olawe during a Navy training event. Suspicions still circulate that the two men were assassinated. While their loss was a tragic blow to the young organization, the event helped to galvanize public support for the PKO.[22]

After years of persistent grassroots organizing by the PKO, in 1990, President George H.W. Bush ordered an end to military training on Kaho'olawe. Through special legislation, Congress appropriated $400 million for the cleanup and restoration of the island. The State of Hawai'i established the Kaho'olawe Island Reserve Commission (KIRC) to receive the land from the Navy in trust for a future Kanaka Maoli national government and to manage ongoing environmental and cultural site restoration of the newly designated cultural reserve. However, the ten-year cleanup effort was only partially successful. Only 10% of the island was cleared to a level considered safe for unrestricted land use. The remaining 90% of the island, as well as the submerged lands surrounding the island, are still considered hazardous due to the presence of unexploded munitions.[23]

Nevertheless, the significance of the movement to protect Kaho'olawe goes far beyond the achievements on the island itself. Kaho'olawe became a piko (navel, umbilicus) for the rising Hawaiian sovereignty movement. The PKO revived a 19th-century Hawaiian independence slogan—"aloha 'āina"—meaning a deep love and political commitment to Hawai'i's land and people. It is a core ethical and political principle that has come to define and animate the Kanaka Maoli struggle.

The concept of aloha 'āina is key to understanding EJ issues in Hawai'i. Although the word "aloha" has been abused and exploited commercially by the settler state and the tourism industry, the philosophy and practice of aloha involves a profound sense of love, care, and reciprocal responsibilities. The word "'āina" means that which feeds and refers to a living land as kin, in a mutual relationship of care with humans. Unlike capitalistic conceptions of land as an inanimate thing which can be partitioned, owned, and developed as a form of property, 'āina is understood to have a living ancestral relationship to humans. Aloha 'āina has become an ideological framework for Kanaka Maoli cultural resurgence and political independence activism.

Mākua Valley: Protecting the Parents

The idea and spirit of aloha 'āina rekindled by Kaho'olawe spread to other sites throughout Hawai'i. One site where aloha 'āina took root is Mākua, a valley on the Wai'anae coast of O'ahu, near the western tip of the island. It is shaped like a deeply curved bowl bounded by steep cliffs on all sides except for the gradually sloping valley floor which descends to a broad white sand beach. The valley was once the site of a fishing village and dryland agricultural complex, with at least three heiau (temple sites) and other sacred structures, agricultural terracing, and storied landscape features. The name "Mākua" means "parents." Some Kānaka Maoli consider Mākua to be a sacred place, associated with creation stories, supernatural beings, healing powers, and the passage of the spirits of the deceased into the ancestral realm. In the legendary epic story of Hi'iaka, the sister of the volcano goddess Pele, Hi'iaka battles a destructive kupua, a supernatural shapeshifting being, who had killed a girl from Mākua because she had rejected his affections. In the mo'olelo, Hi'iaka brings the girl to Mākua beach to resuscitate her by chanting the following prayer:

E ka pua o ka 'ilima e,
Hōmai ana ho'i he ola
E Mākua i ka nu'a o ke kai-e
Ha'awi mai ana ho'i ua ola-e
E ola ku'u kama i ka hu'a o ke kai-e
A ola ho'i iā Kāne i ka wai ola-e

Oh blossom of the 'ilima
Let life descend
Oh, Mākua of the ocean swells
Grant life
That my child of the frothy sea may live
That life may be gained by the living waters of Kāne[24]

The chant refers to 'ilima (*Sida fallax*), an indigenous plant in arid climates with vivid orange flowers associated with the island of O'ahu. It calls on the power of the sea, the realm of Kanaloa, and the life-giving waters of Kāne, a deity of heat and fresh water. Hi'iaka revives the girl, then defeats the deadly intruder, and is honored with a feast before continuing with her journey. This story suggests that Mākua is regarded as a place of life and healing.

In the late 19th and early 20th centuries, Mākua sustained a small ranching and farming community, with its own railroad station, church, and a scattering of homes. During World War II, when Hawai'i was under martial law, the U.S. military seized around 6,000 acres of government land in Mākua and the surrounding area and condemned 28 private parcels.

Kanaka Maoli families who had lived in Mākua for generations were forcibly evicted by the military to make way for training. In 2001, one kupuna (elder) named Walter Kamana testified at a public meeting about the childhood trauma of being evicted by the military and watching his church being bombed:

I was small, used to run when the plane come in. The plane had no respect for people living in the valley. Only had a small little church. You ever seen your church get bombed one Sunday? I seen that, small boy. I seen my church get taken away by a bomb.[25]

Eviction survivors and their descendants later became a force campaigning for the return of the land.[26]

The Territory of Hawai'i issued a revocable permit stipulating that the Army could use the public lands until six months after the end of World War II and that it needed to restore the land prior to returning the land. However, when the war ended, the Army continued training as it lobbied to make Mākua into a permanent training area. After extending the original permit in 1946 and again in 1953, the State of Hawai'i and the U.S. government in 1964 signed a $1-for-65-years lease of 782 acres of Hawaiian trust land (i.e. the former Government and Crown lands of the Hawaiian Kingdom) covering the lower portion of Mākua. At the same time, 3,236 acres of Hawaiian trust land in the upper portions of the valley were "set aside" for Army use by Presidential executive order.[27]

Beginning in the 1960s, Kanaka Maoli families had begun to build temporary shelters outside the training area on Mākua beach for seasonal habitation, fishing, and recreation. Later, houseless families joined the community to establish a permanent settlement there. The State conducted several evictions of these encampments between 1960 and 1996, the most notable involving anti-eviction protests and arrests in 1983 and 1996.

In 1976, within months of the first activist landing on Kahoʻolawe, Waiʻanae community members organized a solidarity rally at Mākua. This was the beginning of the movement to recover Mākua from the U.S. military. PKO leaders met on the beach at Mākua and vowed that they would support an end to the bombing of Mākua after the campaign to protect Kahoʻolawe was successful. In the 1980s, Ivanhoe Naiwi, a survivor of the original evictions, led efforts to call for the cleanup and return of Mākua. Then, in 1992, Kānaka Maoli and settler activists in Waiʻanae formed the group Mālama Mākua to fight the Army's proposed open burn/open detonation site in Mākua. Mālama Mākua is notable as one of the first groups in Hawaiʻi to specifically confront military toxins within an EJ framework. The Army withdrew its permit application after meeting strong community opposition and scrutiny by regulators. Mālama Mākua turned its attention to address other military impacts on Mākua. In 1998, Mālama Mākua sued the U.S. government under NEPA to compel the Army to conduct a thorough environmental impact statement for its activities in the valley. However, following the 9/11 attacks on the United States by operatives of Al Qaeda, Mālama Mākua and the Army reached a settlement agreement which would allow for some military training with restrictions and additional conditions. The 2001 settlement agreement limited the number of days, types of training, and areas of training. It required the Army to (1) complete a comprehensive environmental impact study, including supplemental archaeological and cultural studies and environmental toxicological studies within three years; (2) clear surface ordnance from the lower portion of the valley and at prioritized cultural sites; and (3) permit cultural access for Mālama Mākua two days per month and overnight access twice a year for annual Makahiki ceremonies to open and close the season of Lono, a deity of agriculture. In 2004, when the Army failed to meet the deadlines for its environmental studies, the court imposed an injunction on live-fire training. There has not been live-fire training in Mākua since then. Meanwhile, thousands of community members have been able to access and participate in cultural activities in Mākua.[28]

Around 2020, the Army began environmental impact studies for its proposed retention of approximately 30,000 acres of leased Hawaiian trust lands in Mākua, Kahuku, and Poamoho on Oʻahu, and Pōhakuloa on Hawaiʻi island, after the leases expire in 2029. Numerous groups are organizing to terminate these leases and hold the Army to cleaning up and restoring these lands.

The Red Hill Fuel Leak and Water Crisis

The Red Hill Bulk Fuel Storage Facility consists of 20 enormous underground fuel tanks constructed within a ridge of the Ko'olau mountain range near Ke Awalau o Pu'uloa. In 1938, during the tense years preceding the U.S. entry into World War II, planning began for an underground fuel storage facility at Pearl Harbor which would be hidden and protected from attack.[29] Construction of the tanks began in 1940 and was completed in 1943. The facility was classified as secret until 1995.

Seven miles of underground tunnels, pipelines, and rail tracks connect the tank farm to the pumphouse and distribution points at Joint Base Pearl Harbor-Hickam, 2.5 miles away. The tanks are constructed out of welded quarter-inch thick steel plates clad in concrete. Each tank is 250 feet tall by 100 feet in diameter and has a capacity of 12.5 million gallons.[30] Altogether the facility can hold up to 255 million gallons of fuel. It is the military's largest fuel storage facility and considered a national strategic fuel reserve. However, these aging tanks located at mere 100 feet above the Pearl Harbor aquifer pose an imminent threat to the source of 77% of the drinking water on O'ahu.

In January 2014, 27,000 gallons of jet fuel leaked from a newly repaired tank. While community groups called for Red Hill to be permanently shut down, the Navy, business leaders, and powerful politicians blocked legislation to close the tanks or implement more stringent safety measures. Instead, the Navy and regulators signed an Administrative Order on Consent (AOC), a regulatory settlement agreement requiring the Navy to improve safety measures while continuing to operate.[31] The AOC gave the Navy 22 years to implement a secondary containment solution, which the Honolulu Board of Water Supply and community groups strongly condemned.

In the ensuing legal and administrative wrangling that followed, the public learned shocking facts about failures and leaks going back to the very beginning of the facility. One technical report cites a former Red Hill worker who, between 1943 and 1945, witnessed the spill of approximately 1.3 million gallons of fuel in the tunnels which escaped into Hālawa Stream.[32] In 1948, an earthquake off of O'ahu caused a release of tens of thousands of gallons of fuel. In the lifetime of the Red Hill facility, the Navy has documented approximately 73 leaks, resulting in estimated releases totaling between 200,000 gallons and 1.5 million gallons of fuel into the environment.[33] The Navy's own risk analysis study predicted that the probability of a leak of between 1,000 and 30,000 gallons is 80.1% over the next five years, and 96% over the next ten years.[34]

In May 2021, the Navy reported a "small" leak and claimed to have recovered all of the fuel.[35] But, as Oʻahu residents learned several months later, the Navy grossly underestimated the true size of the spill. On Saturday evening, November 20, 2021, residents of Foster Village and the Āliamanu Military Reservation reported strong diesel or gasoline odors.[36] Unbeknownst to the residents, deep beneath their neighborhood, a Navy worker had crashed their cart into a pipe which released approximately 19,000 gallons of jet fuel on the floor of the tunnel. This fuel originated with the May 2021 leak and was taken up by the drainage pipe system unknown to Navy operators until the crash occurred. The fuel escaped into a floor drain directly into the inundation gallery for the Navy's water system.[37] Pumps then distributed the fuel-tainted water to 93,000 people on the Navy's water system. Initially, the Navy reported that this second leak was contained and that the water was safe to drink. But, by November 28, residents of military housing began complaining on social media that their homes smelled "like a gas station" and that they were experiencing severe rashes, headaches, diarrhea, and vomiting from drinking or bathing in the tap water.[38]

More than 2,000 people were hospitalized with severe reactions. Pets vomited and grew lethargic. One dog died. Viral videos on social media showed tap water with an oily sheen. In one video, a flame put to the water's surface sparked and crackled. The contamination threw schools, businesses, and the lives of thousands of families into turmoil as their water turned into poison. For five months, more than 4,000 military families were displaced to temporary housing. Businesses were shuttered. Schools were forced to use bottled water.

When the Navy stopped pumping water from its Red Hill water shaft, the Board of Water Supply shut down three of its wells in the vicinity as a precautionary measure to prevent drawing contaminated water into its water system. The Board of Water Supply has said that it is not sure when, or even if it could ever safely restart its three inactive wells. This has precipitated a water shortage on Oʻahu.[39]

In the Fall of 2021, Hawaiʻi Peace and Justice and the Sierra Club of Hawaiʻi began collaborating on a more focused and systematic campaign to shut down Red Hill. They convened an initial organizing committee of key leaders and activists from different parts of the island, representing different organizations and sectors, including Kanaka ʻŌiwi, environmentalists, labor, interfaith, and peace and social justice groups. This organizing committee later grew into the Oʻahu Water Protectors (OWP), which formed around three points of unity: "1. Water is life. Access to clean drinking water is a basic human right. 2. The Red Hill fuel tanks are a threat to O'ahu's drinking water and must be retired as soon as possible. 3. We demand that local, state, and federal officials take urgent action to shut down the Red Hill fuel tanks in order to protect drinking water on O'ahu."[40] Centering their politics on the life-giving power of water enabled the group to move fluidly between numerous decentralized actions and flexible tactics without losing focus. It also allowed the group to embrace a diversity of political tendencies.

The November 2021 leak sparked a wave of intense activism, including the formation of a new Kanaka Maoli coalition, Ka'ohewai, the bamboo water carrier, dedicated to the protection of the sacred waters of Kāne. Their first action was a nonviolent direct action aiming to culturally reframe the terms of the debate from military priorities to the vital necessity of water for life on O'ahu. In the pre-dawn darkness on December 12, 2021, Ka'ohewai led a group of approximately 80 kia'i (protectors), most of them Kānaka 'Ōiwi to conduct a nonviolent direct action at the entrance of the Pacific Fleet Command headquarters. Activists brought stones from their home territory of the island, convened and built a ko'a, a type of stone altar typically used to attract resources, such as fish, or in this case, clean water and water protectors. Ka'ohewai leaders announced that the structure would remain as a site for ceremony and other gatherings for as long as it was needed until the Red Hill tanks were no longer a threat to the wai. The site, which was dedicated to Kāne, a deity of fresh water and heat, has become a gathering space for water protectors to conduct ceremonies and political gatherings.[41]

After weeks of the Navy refusing to shut down the tanks, on March 7, 2022, Secretary of Defense Lloyd Austin III unexpectedly issued a memorandum directing the Secretary of the Navy, in coordination with the Commander of the United States Indo-Pacific Command, "to take all steps necessary to defuel and permanently close the Red Hill Bulk Fuel Storage Facility."[42] The Navy has submitted a defueling plan, but it claims that it may take up to two years to safely defuel and permanently close the facility. As of this writing, approximately 100 million gallons of fuel remain in the tanks and continue to threaten the aquifer. Community groups and local government officials continue to apply pressure to ensure that the military follows through with the defueling and closing of the facility.

Wai'anae: An Environmental Justice Hotspot

As land development continues to expand into rural and agricultural parts of O'ahu, such as the 'Ewa district, polluting activities have been pushed further west into the Wai'anae district, one of Hawai'i's poorest areas with a large Kanaka Maoli population. The Campbell Industrial Park and former Barbers Point Naval Air Station (Kalaeloa) near Wai'anae host some of the worst polluting facilities, including oil refineries, power generation plants, incinerators, recycling facilities, and industrial toxic waste disposal sites. A few miles away, the Waimānalo Gulch Landfill, the only municipal landfill on the island, is already over capacity and prone to contaminated runoff during torrential rains. A few miles down the coast, the Kahe Point Power Plant is known to produce hazardous air emissions. In Lualualei Valley, the PVT Landfill, Oʻahu's only construction and demolition landfill, which receives hazardous materials such as asbestos and other contaminants, is already over capacity. The company wants to expand into an adjacent parcel, which is opposed by many in the community.[43]

In 2010, a development company, Tropic Land LLC sought to rezone a 96-acre parcel of land in Lualualei from agricultural to urban in order to develop a light industrial park on the flank of a ridge known to be sacred to the demigod Māui. This would have broken up the agricultural district, making it more susceptible to urbanization in the future. A coalition led by the Waiʻanae Environmental Justice Working Group and the Concerned Elders of Waiʻanae campaigned against the rezoning of this land. In an EJ victory, the Land Use Commission eventually rejected the petition to rezone the parcel. Then in 2019, MAʻO Organic Farms and its non-profit the Waiʻanae Community Re-Development Corporation purchased the parcel to return it to agricultural production and youth leadership development programs.[44]

Approximately a third of Waiʻanae land is occupied by the military, including Mākua and Lualualei valleys. The Naval Magazine Lualualei and Naval Radio Transmitting Facility occupy nearly 10,000 acres of land, much of it fertile agricultural land. There are contamination sites on the base. Residents have long been concerned about potential health hazards from the powerful radio transmitters. Chemical weapons and conventional munitions have been dumped off the Waiʻanae coast, some of which wash up on the beach during large surf events.

Protecting Mauna a Wākea (Mauna Kea)

Another ongoing EJ issue on Hawaiʻi island involves the struggle over astronomy development on Mauna a Wākea. Rising above Hawaiʻi island at an elevation of 13,803 feet, Mauna a Wākea (mountain of Wākea) is considered to be a sacred mountain by many Kānaka Maoli, born from the mating of Papahānaumoku (Papa who births islands) and Wākea (deity of the expanse of the sky). It is the realm of multiple deities related to water forms.

In 1960, the UH built the first 2.2-meter telescope on Mauna a Wākea within a designated conservation district. In 1968, the Department of Land and Natural Resources leased the summit of Mauna a Wākea to the UH for 65 years to create a Maunakea Science Reserve. Subsequently, the astronomy industry built a total of 13 telescopes. These telescopes have caused significant negative environmental impacts, including the destruction of habitat for endangered Hawaiian species such as the wekiu bug (*Nysius wekiuicola*), the desecration of Kanaka Maoli cultural sites, and the release of toxic contamination.

While Kānaka Maoli have opposed telescope construction from the start, opposition to new telescopes gained momentum around 2,000 when the University approved a revised master plan for Mauna a Wākea which allowed expansion beyond the cap of 13 telescopes established in the 1983 master plan. In 2000, the UH and NASA proposed to expand the Keck Telescope with an array of smaller outrigger telescopes. Mauna Kea Anaina Hou, the Royal Order of Kamehameha, Sierra Club Moku Nui chapter, KAHEA—The Hawaiian Environmental Alliance, and other groups opposed the expansion and engaged in legal challenges and protests. In 2006, NASA decided to withdraw its financial support for the project, which effectively killed the proposal.

However, within a few years, the UH and an international science consortium proposed to build a Thirty-Meter Telescope (TMT) on Mauna a Wākea, which, at the time, would have been the largest telescope in the world. A lengthy political and legal contest ensued after community groups intervened to oppose the project at multiple points in the approval process. A key legal setback for the public trust doctrine came when the Hawai'i Supreme Court cited the degradation principle to conclude that the TMT would not cause substantial and adverse impacts. According to William Crowell, a J.D. candidate at the William S. Richardson School of Law at the UH:

> The degradation principle, as depicted in this case, is the concept that cultural and natural resources lose legal protection when the 'āina (land) has previously suffered a "substantial adverse impact" (i.e., degradation) and, therefore, any future impact cannot be considered substantial and adverse.[45]

The Supreme Court decision seemed to clear the way for TMT construction. Then, in October 2014, a group of Kānaka Maoli interrupted a groundbreaking ceremony at the construction site on the summit of Mauna a Wākea. When developers began to move construction equipment up the mountain in June 2015, hundreds of kia'i (protectors) blocked the winding access road in multiple rows up the mountain. Despite dozens of arrests by police, construction was halted indefinitely. The kia'i set up a checkpoint at the Mauna Kea Visitors Center.

After several more years of legal and political wrangling, in July 2019, TMT set out to resume construction. This time kia'i established a pu'uhonua (place of refuge) and a roadblock at the entrance to Mauna Kea Access

Road. They named the encampment Puʻuhonua o Puʻuhuluhulu, a place of refuge at the base of a small forested hill named Kīpuka Puʻuhuluhulu.

As police arrived to clear the road for construction equipment to pass, some kiaʻi chained themselves to a cattle grate, and kūpuna (elders) formed a line across the road, backed by a line of women, and behind them, other kiaʻi. Scenes of the police arresting kūpuna went viral and stoked anger across Hawaiʻi and around the world. Thousands rushed to join the action, and the puʻuhonua grew into a miniature city, complete with a kūpuna tent, a medic tent, a kitchen, recycling and trash disposal facilities, portable toilets, a university, a media center, a command center, public safety teams, a donations operation and free store, an information booth and orientation protocol, and a ceremonial space where prayers and ceremonies were conducted three times a day.[46]

In 2019, the TMT corporation halted construction indefinitely. With no immediate threat of construction and the onset of the COVID-19 pandemic, in March 2020, the leaders of the Puʻuhonua decided to disband the encampment. In 2022, despite opposition from the UH, the State legislature created a new entity to govern Mauna a Wākea, with representation from Kanaka Maoli cultural practitioners. Some see this as a move toward greater accountability to Kanaka Maoli values while others are skeptical about the intentions of this new entity. At the same time, the National Science Foundation announced that it is considering a proposal to fund the TMT project and initiated a new environmental review process, which, as of this writing, is currently underway.

The Kū Kiaʻi Mauna movement generated solidarity actions around the world and inspired similar protectors' movements in Hawaiʻi. Residents of Waimānalo, a community with a large Kanaka Maoli population, successfully blocked the construction of a city-sponsored sports facility at Hūnānāniho in Waimānalo on Oʻahu. And in Kahuku, on Oʻahu's North Shore, the community rallied to block the construction of an industrial wind farm just a few hundred yards from schools and residences. Unfortunately, the Kū Kiaʻi Kahuku group was unable to stop the construction of the wind turbines. However, these activists have continued to organize on other issues, stopping a controversial missile defense radar facility proposed for Kahuku, and now mobilizing to demand an end to a $1-for-65-year Army lease of Hawaiian trust land in the same vicinity.

EJ Issues in U.S.-Affiliated Pacific Islands

This last section will briefly discuss EJ issues at two U.S.-affiliated Pacific Islands: the Mariana Islands, which includes the U.S. unincorporated territory of Guam (also referred to by its Indigenous CHamoru name, Guåhan) and the Commonwealth of the Northern Mariana Islands, and the RMI.

The Mariana Islands (Guåhan/Guam and the Northern Mariana Islands)

The Mariana Islands are an archipelago in the Western Pacific Ocean, including Guåhan/Guam and the Commonwealth of the Northern Mariana Islands, and forming an arc approximately 1,565 miles (2,519 km) from South to North. The Indigenous CHamoru people have inhabited the archipelago for thousands of years. In 1521, the Portuguese explorer Ferdinand Magellan was the first European to encounter the Mariana Islands. Spain claimed the Mariana Islands in 1667 in honor of the Spanish Queen Mariana of Austria. From 1565 to 1815, the islands were an important link in the Manila Galleon Trade Route between the Philippines and South America. In 1898, when Spain lost the Spanish American War, it ceded Guåhan, the Philippines, and Puerto Rico to the United States, and in 1899, it sold the Northern Mariana Islands and the Caroline Islands (Palau, Yap, Chuuk, Pohnpei, Kosrae) to Germany. Following Germany's defeat in World War I, the League of Nations granted Japan a mandate over the Northern Mariana, Caroline, and Marshall Islands.

A series of U.S. Supreme Court cases known as the "Insular Cases" established differential political statuses for America's new colonies, including Guåhan and Hawai'i. With its long history of white settler colonialism, the Court deemed Hawai'i to be an "incorporated territory" as defined in the Northwest Ordinances, which held out the possibility of its eventual incorporation into the United States. However, due to the "alien" racial composition of Guåhan, Puerto Rico, the Philippines, and American Sāmoa, which was acquired by the United States in 1900, the Court defined these islands as "unincorporated territories," possessions of the United States but "foreign ... in a domestic sense."[47]

After Japan attacked U.S. military forces on Guåhan on December 8, 1941, it occupied the island until August 1944, when U.S. Marines recaptured the island. That year, the United States also captured Saipan and Tinian in the Northern Mariana Islands. Following the U.S. victory over Japan in World War II, Hawai'i and Guåhan were placed on the United Nations list of non-self-governing territories, colonies eligible for self-determination. Guåhan remains on the United Nations List as an unincorporated territory of the United States. In 1947, the United Nations placed the Northern Mariana Islands in the Trust Territory of the Pacific Islands (TTPI) under the sole trusteeship of the United States. In 1978, the Northern Mariana Islands formalized its political status as a commonwealth of the United States. While still subordinate to the United States, commonwealth status conferred more autonomy and fewer rights under the U.S. system than Guåhan or Hawai'i.

Guam/Guåhan is frequently referred to as the "tip of the spear" of U.S. military power in the Pacific. Joint Region Marianas includes the facilities of Naval Base Guam and Andersen Air Force Base and training areas in the Northern Mariana Islands. There are 39 U.S. military sites in Guåhan occupying 37,141 acres. In the Northern Mariana Islands, the U.S. military controls four military sites, including 15,000 acres of leased land on Tinian, facilities on Saipan, and a bombing range on Farallon De Medinilla island. On Guam, reports that the carcinogenic herbicide Agent Orange was used during the Vietnam War have prompted an investigation into possible contamination and health impacts.[48]

Both Guåhan and the Northern Marianas are experiencing a massive military expansion as part of the United States' "pivot" to the Pacific. After the 1995 rape of a 12-year-old Okinawan schoolgirl by U.S. Marines, the United States and Japan agreed to reduce the U.S. military footprint in Okinawa by moving 8,000 III Marine Expeditionary Force personnel and their approximately 9,000 dependents to Guåhan. In April 2012, after fiscal, legal, and political setbacks in the Guam Buildup and fierce resistance to the Futenma Replacement Facility in Henoko, Okinawa, the United States and Japan agreed to loosen the timeline and reduce the proposed number of Marines relocating to Guåhan down to 5,000 (mostly rotational/without family members). Additionally, 2,700 Marines would move to Hawai'i, 800 to the continental United States, and 1,300 on a rotational basis to Australia.

From 2007 to 2010, the U.S. Navy prepared an Environmental Impact Statement for the Guam and CNMI Military Relocation (the Guam Buildup) under the NEPA. Plans included building infrastructure for the relocating Marine Corps unit and opening a live-fire training range at Pågat Point. The group We Are Guåhan led the campaign to oppose the military buildup. A successful public awareness campaign and legal challenge under the NHPA caused the military to drop plans for a training range at Pågat.

In 2010, the Navy completed an environmental review process for expanding the Mariana Island Range Complex (MIRC EIS/OEIS 2010), which encompasses 497,469 square nautical miles of air- and sea-space surrounding the entire Mariana Island chain. In 2015, the Navy proposed to expand beyond the MIRC to create the Mariana Island Training and Testing (MITT EIS/OEIS), which would be a total of 984,601 square nautical miles. A major concern is the impact of Navy sonar activity on marine mammals. Additionally, in 2013, the Navy began environmental review for the Commonwealth of the Northern Mariana Islands Joint Military Training project, which seeks to expand facilities and training activities on Tinian Island and to acquire the entire island of Pagan for live-fire training. A number of groups have mobilized to oppose this expansion, including Guardians of Gani', Pagan Watch, Tinian Women's Association, and Alternative Zero Coalition, Save Pagan, and Our Islands Are Sacred. In 2022, the Marine Corps announced that it had dropped its proposal to train on Pagan.[49]

However, new plans have surfaced for a live-fire range in the ecologically and culturally sensitive limestone forests of Litekyan (Ritidian). The group Prutehi Litekyan—Save Ritidian is leading the campaign to stop this training range. Recently, construction activity has destroyed large swaths of native forest, disturbed CHamoru archaeological and cultural sites, and unearthed ancestral remains. The group is currently fighting a proposed military open burn/open detonation pit in the northern part of the island.

Nuclear Testing and Rising Seas in the Marshall Islands

The Marshall Islands are an archipelago of low atolls in Micronesia whose peoples have an ancient history of seafaring. These islands have endured German, Japanese, and U.S. colonialism. In the aftermath of World War II, the United Nations placed the Marshall Islands, Chuuk, Pohnpei, Yap, Kosrae, Palau, and the Northern Mariana Islands under the administration of the United States as the TTPI. Unlike other colonies administered by the United Nations as non-self-governing territories during their decolonization processes, the TTPI was designated a strategic trust under the sole administration of the United States. As trustee, the United States curtailed full independence for these islands and sought to perpetuate military control and economic dependency. While the Northern Marianas opted for commonwealth status under U.S. sovereignty, which gave its residents U.S. citizenship, the RMI, the Republic of Palau, the Federated States of Micronesia (FSM) consisting of Chuuk, Pohnpei, Yap, and Kosrae, became nominally independent states in "free association" with the United States. Under their Compacts of Free Association (COFA), the United States provides economic aid and controls defense, while COFA citizens enjoy unrestricted access to the United States. In return, the United States has unrestricted military access to the lands and seas of COFA states.[50]

Two major EJ issues in the Marshall Islands are the legacy of nuclear testing and the effects of climate change on sea-level rise. After World War II, the United States implemented its nuclear weapons testing program in the Marshall Islands atolls of Bikini and Enewetak and later relocated the residents of Kwajalein in order to build a missile test range there. Between 1946 and 1958, the United States detonated 67 atomic and nuclear bombs in the Marshall Islands, 23 in Bikini, and 44 at Enewetak.[51] On March 1, 1954, the United States conducted its largest nuclear detonation, "Castle Bravo," at Bikini Atoll in the Marshall Islands. The explosion yielded approximately 15 megatons of TNT, 1000 times more powerful than the atomic bomb dropped on Hiroshima in 1945. The radioactive fallout covered an area approximately 7,000 square miles, contaminating the land and inhabitants of Rongelap, Utirik, and Ailinginae, U.S. military observers on Rongerik Atoll, and the Japanese fishing vessel, "The

Lucky Dragon." Soon after the strange "snow" fell on Rongelap, the people became very sick with radiation burns and poisoning. Radiation was later detected as far away as Europe and India. According to the Atomic Heritage Foundation, "This was the worst radiological disaster in U.S. history and caused worldwide backlash against atmospheric nuclear testing."[52] The U.S. commanders were aware that wind conditions could pose a contamination risk for the inhabitants of nearby atolls but proceeded with the test anyway. A week after the blast, the United States began a medical study, code-named Project 4.1, on the effects of radiation on the Marshallese survivors. However, this research was done without informed consent of the Marshallese survivors.

Under the 1986 U.S.-RMI COFA, the United States agreed to continue providing health care for the 176 Marshallese directly affected by the tests and established a $150 million trust fund to pay compensation to the affected islanders. When the Marshall Islands Nuclear Claims Tribunal trust fund was depleted in 2011, the RMI government petitioned for additional funding pursuant to the "changed circumstances" clause in the COFA. The petitioners argued that new revelations that the extent of radioactive fallout was much wider than previously disclosed and that updated radiation safety standards constituted changed circumstances to warrant additional compensation. However, the United States declined to allocate additional monies to the trust fund. Researchers continue to find elevated levels of radiological contamination in Bikini and Enewetak and urge the U.S. government to conduct a more thorough cleanup of the nuclear fallout.[53] On Runit Island in the Enewetak Atoll, the United States buried an estimated 93,000 cu yd of radioactive debris and encased it in a large concrete dome. However, the dome is beginning to break down and with rising sea levels, could release its radioactive contents.

In recent years, Marshall Islands have experienced severe flooding on "king tides," the effects of rising sea levels. The RMI government and grassroots Marshallese activists have been leading voices for curbing global carbon emissions in the climate talks. However, in general, EJ issues in Oceania have been ignored and neglected. This imperial attitude is epitomized by a quote from the former U.S. National Security Advisor Henry Kissinger, who in 1969, while debating the morality of taking Micronesian land for military purposes, said, "There are only 90,000 people out there. Who gives a damn?"[54]

Against Kissinger's sociopathic disregard for the peoples of Oceania, this chapter ends with excerpts from a poem by Marshallese poet and climate justice activist Kathy Jetñil-Kijiner. In 2014, at the United Nations Climate Summit in New York, she moved the participants to tears and received a standing ovation with this challenge and message of hope to world leaders. The poem is written as a letter to her infant daughter,

Matafele Peinam, "a seven month old sunrise of gummy smiles ... thighs that are thunder and shrieks that are lightning ... so excited for bananas, hugs and our morning walks past the lagoon." She paints a picture of the looming threat to her beautiful home: "that lucid, sleepy lagoon lounging against the sunrise/men say that one day/that lagoon will devour you," and recounts predictions that generations of Marshallese "will wander rootless/with only a passport to call home."

Apologizing to the Carteret islanders of Papua New Guinea and the Taro islanders of the Solomon Islands, who have already lost their homes to rising seas, she consoles her daughter with a promise to fight for the planet. "We are drawing the line here," she says. "Because baby we are going to fight." Jetñil-Kijiner condemns the imperial gaze that ignores the plight of places like the Marshall Islands, Kiribati, Tuvalu, the Maldives—and one might add Hawai'i and Guåhan—as if they did not exist. And she praises global EJ movements "who see us." She reassures her daughter that "there are thousands out on the street/marching with signs/hand in hand/chanting for change NOW/they're marching for you, baby/they're marching for us/because we deserve to do more than just/survive/we deserve/to thrive."

In a final act of faith in our common humanity, Jetñil-Kijiner places the fate of her sleeping daughter in our hands:

dear matafele peinam,

you are eyes heavy
with drowsy weight
so just close those eyes, baby
and sleep in peace

because we won't let you down
you'll see[55]

Jetñil-Kijiner gently slyly inverts Kissinger's cynicism into a challenge: "Who gives a damn?" Indeed. The question is, will we rise to meet that challenge?

Notes

1 Pualani Kanaka'ole Kanahele, "I Am This Land, and This Land Is Me," *Hulili* 2, no. 1 (2005): 23.
2 Quoted in Epeli Hau'ofa, "The Ocean in Us," in *Voyaging Through the Contemporary Pacific*, ed. David J. Hanlon and Geoffrey M. White (Lanham, MD: Rowman & Littlefield Publishers, 2000), 32.
3 Melody Kapilialoha MacKenzie, Susan K Serrano, and Koalani Laura Kaulukukui, "Environmental Justice for Indigenous Hawaiians: Reclaiming Land and Resources," *Natural Resources and Environment* 21, no. 3 (Winter 2007): 32–42.

4   David A Swanson, "A New Estimate of the Hawaiian Population for 1778, the Year of First European Contact," *Hūlili: Multidisciplinary Research on Hawaiian Well-Being* 11, no. 2 (2019): 203–22.

5   Swanson, "A New Estimate of the Hawaiian Population for 1778, the Year of First European Contact," 222.

6   M. John Schofield to William Sherman, "Schofield to Sherman," February 15, 1873, DSO-11472-49951-Hawaiian Islands (1872–93), Library of Congress.

7   John N. Cobb, "Commercial Fisheries of the Hawaiian Islands," Report U.S.F.C. (Washington, DC: United States Fishing Commission, 1901); William Kenji Kikuchi, "Hawaiian Aquacultural System" (Ph.D., United States – Arizona, The University of Arizona, 1973).

8   For a political ecology of sugar in Hawai'i see Carol A. MacLennan, *Sovereign Sugar: Industry and Environment in Hawai'i* (Honolulu: University of Hawai'i Press, 2014).

9   D. Kapua'ala Sproat, "A Question of Wai: Seeking Justice through Law for Hawai'i's Streams and Communities," in *A Nation Rising: Hawaiian Movements for Life, Land, and Sovereignty*, ed. Noelani Goodyear-Ka'ōpua, Ikaika Hussey, and Erin Kahunawaika'ala Wright, Narrating Native Histories (Durham, NC: Duke University Press, 2014): 199–219.

10 Kehaulani Cerzo, "State Decision Reached in Na Wai 'Eha Water Case," *Maui News*, July 1, 2021, https://www.mauinews.com/news/local-news/2021/07/state-decision-reached-in-na-wai-eha-water-case/

11 William S. Merwin, "Hawai'i Wakes Up To Pesticides," *The Nation*, March 2, 1985, https://www.thenation.com/article/archive/hawaii-wakes-pesticides/

12 Luciano Minerbi, "Native Hawaiian Struggles and Events, A Partial List 1973–1993," *Social Process in Hawai'i* 35 (1994): 1–14; Davianna McGregor-Alegado, "Hawaiians: Organizing in the 1970s," *Amerasia Journal* 7, no. 2 (1980): 29–55.

13 Haunani-Kay Trask, "Birth of the Modern Hawaiian Movement: Kalama Valley, O'ahu," *Hawaiian Journal of History* 21, no. 13 (1987): 126–53.

14 Benjamin Schrager and Krisnawati Suryanata, "Seeds of Contestation: The Emergence of Hawai'i's Seed Corn Industry," in *Food and Power in Hawai'i: Visions of Food Democracy*, ed. Aya Hirata Kimura and Krisnawati Suryanata (Honolulu: University of Hawai'i Press, 2016): 138–55; Andrea Brower, "From the Sugar Oligarchy to the Agrochemical Oligopoly: Situating Monsanto and Gang's Occupation of Hawai'i," *Food, Culture & Society* 19, no. 3 (July 2, 2016): 587–614, https://doi.org/10.1080/15528014.2016.1208342

15 Christopher Pala, "Pesticides in Paradise: Hawai'i's Spike in Birth Defects Puts Focus on GM Crops," *The Guardian*, August 23, 2015, sec. US news, https://www.theguardian.com/us-news/2015/aug/23/hawaii-birth-defects-pesticides-gmo

16 William Addleman, 1946 History of the United States Army in Hawai'i 1849–1939 Hawai'i War Records Depository Collection, University of Hawai'i; Ian Lind, "Ring of Steel: Notes on the Militarization of Hawai'i," *Social Process in Hawai'i*, 31 (1984): 25–48.

17 U.S. Department of Defense, "Base Structure Report Fiscal Year 2018 Baseline," 2019, https://www.acq.osd.mil/eie/Downloads/BSI/Base%20Structure%20Report%20FY18.pdf; U.S. Indo-Pacific Command, "Hawai'i Military Land Use Master Plan (HMLUMP)," April 30, 2021.

18 US EPA, "Pearl Harbor Naval Complex Site Profile," accessed October 3, 2022, https://cumulis.epa.gov/supercpad/cursites/csitinfo.cfm?id=0904481

19 Lena Groeger et al., "Bombs in Your Backyard," *ProPublica*, November 30, 2017, https://projects.propublica.org/bombs/

20 "Protect Kaho'olawe 'Ohana," Protect Kaho'olawe 'Ohana, accessed October 3, 2022, http://www.protectkahoolaweohana.org/

21 Aluli et al v. Rumsfeld, Civ. No. 76-0380, No. 76-0380 (D. Hawai'i October 13, 1976); Consent Decree and Order, Aluli v. Brown, No. Civil No. 76-0380 (United States District Court for the District of Hawai'i December 1, 1980).

22 Rodney Morales (ed.), *Ho'iho'i Hou: A Tribute to George Helm and Kimo Mitchell* (Honolulu, Hawai'i: Bamboo Ridge Press, 1984).

23 Parsons-UXB Joint Venture, "Final Summary: After Action Report Kaho'olawe Island Reserve" (Naval Facilities Engineering Command Pacific Division, December 2004), http://kahoolawe.hawaii.gov/downloads/PUXBFinal%20 SummaryAfterActionReport.pdf

24 Kepā Maly and Institute for Sustainable Development, "Oral History Study: Ahupua'a of Mākua and Kahanahāiki, District of Wai'anae, Island of O'ahu," N62742-94-D 0006 D.O. 22 (U.S. Navy, PACDIV, June 1998).

25 "Makua Valley Public Meeting Held on January 27, 2001 (Condensed Transcript and Concordance)," Ralph Rosenberg Court Reporters, Inc., January 27, 2001, 56.

26 Marion Kelly and Sidney Michael Quintal, "Cultural History Report of Makua Military Reservation and Vicinity, Makua Valley, Oahu, Hawai'i," (Honolulu, Hawai'i: U.S. Department of the Army, U.S. Army Engineer Division, Pacific Ocean, April 1977).

27 State of Hawai'i Board of Land and Natural Resources, "State General Lease No. S-3848; U.S. Lease, Contract No. DA-94-626-ENG-79," August 17, 1964.; Lyndon B. Johnson, "Executive Order 11166 – Setting Aside for the Use of the United States Certain Public Lands and Other Public Property Located at the Makua Military Reservation, Hawai'i," *The American Presidency Project*, August 15, 1964, https://www.presidency.ucsb.edu/documents/executive-order-11166-setting-aside-for-the-use-the-united-states-certain-public-lands-and

28 "Citizens and Army Settle Lawsuit over Live-Fire Training at Makua, Hawai'i," *Earthjustice*, October 4, 2001, http://earthjustice.org/news/press/2001/citizens-and-army-settle-lawsuit-over-live-fire-training-at-makua-hawai-i

29 Historic American Engineering Record, National Park Service, "Historic American Engineering Record U.S. Naval Base, Pearl Harbor, Red Hill Underground Fuel Storage System," 2015, http://lcweb2.loc.gov/master/pnp/habshaer/hi/hi1000/hi1016/data/hi1016data.pdf

30 U.S. Fleet Forces Command, *Red Hill: The Building of an Enduring Monument (N1901-00-0001)*, 2001, https://www.youtube.com/watch?v=lIz8IstwnWU

31 "Administrative Order on Consent – Red Hill Bulk Fuel Storage Facility, Oahu, Hawai'i, EPA DKT NO. RCRA 7003-R9-2015-01; DOH DKT NO. 15-UST-EA-01; and Statement of Work (SOW)." (September 28, 2015), http://health.hawaii.gov/shwb/files/2015/09/Red-Hill-AOC_Final_29SEP151.pdf

32 ERC Environmental and Energy Services Co. (ERCE), "Red Hill Oily Waste Disposal Pit Remedial Investigation/Feasibility Study (RI/FS) – Work Plan" (Department of the Navy Pacific Division Naval Facilities Engineering Command, March 1992), 2–33. https://eha-cloud.doh.hawaii.gov/iheer/api/documents/147487/download

33 "Shut Down Red Hill," Sierra Club of Hawai'i, accessed October 9, 2022, https://sierraclubhawaii.org/redhill

34 ABS Consulting, "Quantitative Risk and Vulnerability Assessment Phase 1 (Internal Events without Fire and Flooding) – Red Hill Bulk Fuel Storage Facility NAVSUP FLC Pearl Harbor, HI (PRL) (INTERNAL REPORT NOT FOR PUBLIC RELEASE)," November 12, 2018.

35 William Cole, "Navy Confirms 1,000-Gallon Fuel Release at Red Hill," *Honolulu Star-Advertiser*, May 7, 2021, https://www.staradvertiser.com/2021/05/07/breaking-news/navy-confirms-1000-gallon-fuel-release-at-red-hill/

36 Sophie Cocke, "Odor from Red Hill Fuel Release Sparks 911 Calls," *Honolulu Star-Advertiser*, November 23, 2021, https://www.staradvertiser.com/2021/11/23/hawaii-news/odor-from-red-hill-fuel-release-sparks-911-calls/

37 Vice Chief of Naval Operations, "Command Investigation into the 6 May 2021 and 20 November 2021 Incidents at Red Hill Bulk Fuel Storage Facility," June 13, 2022.

38 Christina Jedra, "Navy Investigating 'Chemical Smell' in Military Housing Drinking Water," *Honolulu Civil Beat*, November 29, 2021, https://www.civilbeat.org/2021/11/navy-investigating-chemical-smell-in-military-housing-drinking-water/; Mahealani Richardson, "Alarming New CDC Survey Shows 'Worse Health' among Those Impacted by Red Hill Fuel Spills," *Hawai'i News Now*, November 9, 2022, https://www.hawaiinewsnow.com/2022/11/10/cdcdoh-survey-shows-worse-health-after-red-hill-fuel-spills/

39 Sophie Cocke, "Board of Water Supply Says 3 Oahu Wells May Never Reopen after Latest Navy Contamination," *Honolulu Star-Advertiser*, January 4, 2022, https://www.staradvertiser.com/2022/01/03/breaking-news/board-of-water-supply-says-3-oahu-wells-may-never-reopen-after-latest-navy-contamination/

40 O'ahu Water Protectors, "Ola I Ka Wai – Water is Life."

41 "Ka'ohewai Demands Shutdown of Kapūkaki (Red Hill) Fuel Tanks," *Kanaeokana* (blog), December 17, 2021, https://kanaeokana.net/shutdownredhill/; Ka'ohewai. Kāhea to the Lāhui! 'Anahulu at the Ko'a Starts Now, 2022, https://fb.watch/d-SvOaxOp1/

42 Lloyd J. Austin III to Senior Pentagon Leadership; Commanders of the Combatant Commands; Defense Agency and DOD Field Activity Directors, "Immediate Actions to Permanently Close the Red Hill Bulk Fuel Storage Facility at Joint Base Pearl Harbor-Hickam and to Redistribute Fuel in Accordance with INDO-PACOM Plans for Strategic Fuel Storage in the Pacific Region," March 7, 2022.

43 Anthony Makana Paris and Kamuela Werner, "Not in Anyone's Backyard," *Ka Wai Ola*, April 30, 2020, https://kawaiola.news/aina/not-in-anyones-backyard/; Kassandra Kometani, "Environmental Justice for Native Hawaiians: Preventing Landfill Expansion on the Wai'anae Coast – Environmental Amicus," *Environmental Law Education Center*, March 2, 2022, https://elecenter.com/1143/environmental-justice-for-native-hawaiians-preventing-landfill-expansion-on-the-waianae-coast/

44 Candace Fujikane, *Mapping Abundance for a Planetary Future: Kanaka Maoli and Critical Settler Cartographies in Hawai'i* (Durham, NC: Duke University Press, 2021), 84–85.

45 William N. K. Crowell, "Chipping Away at the Public Trust Doctrine: Mauna Kea and the Degradation Principle," *Asian-Pacific Law & Policy Journal* 21, no. 2 (Spring 2020): 3.

46 Personal observations by the author (2019).

47 "Downes v. Bidwell, 182 U.S. 244 (1901)," accessed March 26, 2018, https://supreme.justia.com/cases/federal/us/182/244/case.html

48 Anita Hofschneider, "Yale Report Says Guam Vets Likely Exposed To Agent Orange," *Honolulu Civil Beat*, May 12, 2020, https://www.civilbeat.org/beat/yale-law-report-says-guam-vets-likely-exposed-to-agent-orange/

49 Anita Hofschneider, "Military Won't Proceed with Marianas Bombing Range but the New Plan Is Unclear," *Honolulu Civil Beat*, April 8, 2022, https://

www.civilbeat.org/2022/04/military-wont-proceed-with-marianas-bombing-range-but-the-new-plan-is-unclear/
50 Congressional Research Service and Thomas Lum, "The Compacts of Free Association," in Focus, August 15, 2022, https://crsreports.congress.gov/product/pdf/IF/IF12194/1
51 "Marshall Islands | Atomic Heritage Foundation," Atomic Heritage Foundation, 2019, https://www.atomicheritage.org/location/marshall-islands
52 Ibid.
53 Ivana Nikolić Hughes and Hart Rapaport, "The U.S. Must Take Responsibility for Nuclear Fallout in the Marshall Islands," *Scientific American*, April 4, 2022, https://www.scientificamerican.com/article/the-u-s-must-take-responsibility-for-nuclear-fallout-in-the-marshall-islands/
54 Walter J. Hickel, *Who Owns America?* (Englewood Cliffs, NJ: Prentice-Hall, 1971), 208.
55 Kathy Jetñil-Kijiner, "United Nations Climate Summit Opening Ceremony – A Poem to My Daughter," *Kathy Jetñil-Kijiner*, September 24, 2014, https://www.kathyjetnilkijiner.com/united-nations-climate-summit-opening-ceremony-my-poem-to-my-daughter/

Bibliography

ABS Consulting. "Quantitative Risk and Vulnerability Assessment Phase 1 (Internal Events without Fire and Flooding) - Red Hill Bulk Fuel Storage Facility NAVSUP FLC Pearl Harbor, HI (PRL) (INTERNAL REPORT NOT FOR PUBLIC RELEASE)." 2018. Contract N62742-14-D-1884, Delivery Order 0028. Naval Facilities Engineering Command Pacific Division.
Addleman, William C. *History of the United States Army in Hawai'i 1849—1939*. Schofield Barracks: Hawaiian Division, 1939.
"Administrative Order on Consent - Red Hill Bulk Fuel Storage Facility, Oahu, Hawai'i, EPA DKT NO. RCRA 7003-R9-2015-01; DOH DKT NO. 15-UST-EA-01; and Statement of Work (SOW)." 2015. U.S. Environmental Protection Agency, Hawai'i State Department of Health. http://health.hawaii.gov/shwb/files/2015/09/Red-Hill-AOC_Final_29SEP151.pdf
Aluli et al v. Rumsfeld, Civ. No. 76-0380. 1976. D. Hawai'i.
Austin, Lloyd J., III. Letter to Senior Pentagon Leadership; Commanders of the Combatant Commands; Defense Agency and DOD Field Activity Directors. 2022. "Immediate Actions to Permanently Close the Red Hill Bulk Fuel Storage Facility at Joint Base Pearl Harbor-Hickam and to Redistribute Fuel in Accordance with INDOPACOM Plans for Strategic Fuel Storage in the Pacific Region," March 7, 2022.
Brower, Andrea. "From the Sugar Oligarchy to the Agrochemical Oligopoly: Situating Monsanto and Gang's Occupation of Hawai'i." *Food, Culture & Society* 19, no. 3 (2016): 587–614. https://doi.org/10.1080/15528014.2016.1208342
Cerzo, Kehaulani. "State Decision Reached in Na Wai 'Eha Water Case." *Maui News*, July 1, 2021. https://www.mauinews.com/news/local-news/2021/07/state-decision-reached-in-na-wai-eha-water-case/
"Citizens and Army Settle Lawsuit Over Live-Fire Training at Makua, Hawai'i." *Earthjustice*, October 4, 2001. https://earthjustice.org/news/press/2001/citizens-and-army-settle-lawsuit-over-live-fire-training-at-makua-hawai-i

Cobb, John N. "Commercial Fisheries of the Hawaiian Islands." Report U.S.F.C. United States Fishing Commission, 1901.

Cocke, Sophie. "Odor from Red Hill Fuel Release Sparks 911 Calls." *Honolulu Star-Advertiser*, November 23, 2021. https://www.staradvertiser.com/2021/11/23/hawaii-news/odor-from-red-hill-fuel-release-sparks-911-calls/

_____. "Board of Water Supply Says 3 Oahu Wells May Never Reopen after Latest Navy Contamination." *Honolulu Star-Advertiser*, January 4, 2022. https://www.staradvertiser.com/2022/01/03/breaking-news/board-of-water-supply-says-3-oahu-wells-may-never-reopen-after-latest-navy-contamination/

Cole, William. "Navy Confirms 1,000-Gallon Fuel Release at Red Hill." *Honolulu Star-Advertiser*, May 7, 2021. https://www.staradvertiser.com/2021/05/07/breaking-news/navy-confirms-1000-gallon-fuel-release-at-red-hill/

Congressional Research Service, and Thomas Lum. 2022. "The Compacts of Free Association." In Focus IF12194. https://crsreports.congress.gov/product/pdf/IF/IF12194/1

Crowell, William N. K. "Chipping Away at the Public Trust Doctrine: Mauna Kea and the Degradation Principle." *Asian-Pacific Law and Policy Journal* 21, no. 2 (2020): 1–47.

"Downes v. Bidwell, 182 U.S. 244 (1901)." n.d. Accessed March 26, 2018. https://supreme.justia.com/cases/federal/us/182/244/case.html

ERC Environmental and Energy Services Co. (ERCE). 1992. "Red Hill Oily Waste Disposal Pit Remedial Investigation/Feasibility Study (RI/FS) – Work Plan." N62742-88-D-0032 (DO 0012). Department of the Navy Pacific Division Naval Facilities Engineering Command. https://eha-cloud.doh.hawaii.gov/iheer/api/documents/147487/download

Fujikane, Candace. *Mapping Abundance for a Planetary Future: Kanaka Maoli and Critical Settler Cartographies in Hawai'i*. Durham, NC: Duke University Press, 2021.

Groeger, Lena, Ryann Grochowski Jones, and Abrahm Lustgarten. "Bombs in Your Backyard." *ProPublica*, November 30, 2017. https://projects.propublica.org/bombs/

Hau'ofa, Epeli. "The Ocean in Us." In *Voyaging Through the Contemporary Pacific*, edited by David J. Hanlon and Geoffrey M. White, 113–31. Lanham, MD: Rowman & Littlefield Publishers, 2000.

Hickel, Walter J. *Who Owns America?* Englewood Cliffs, NJ: Prentice-Hall, 1971.

Historic American Engineering Record, National Park Service. 2015. "Historic American Engineering Record U.S. Naval Base, Pearl Harbor, Red Hill Underground Fuel Storage System." HAER HI-123. http://lcweb2.loc.gov/master/pnp/habshaer/hi/hi1000/hi1016/data/hi1016data.pdf

Hofschneider, Anita. "Yale Report Says Guam Vets Likely Exposed to Agent Orange." *Honolulu Civil Beat*, May 12, 2020. https://www.civilbeat.org/beat/yale-law-report-says-guam-vets-likely-exposed-to-agent-orange/

Hughes, Ivana Nikolić, and Hart Rapaport. "The U.S. Must Take Responsibility for Nuclear Fallout in the Marshall Islands." *Scientific American*, April 4, 2022. https://www.scientificamerican.com/article/the-u-s-must-take-responsibility-for-nuclear-fallout-in-the-marshall-islands/

Jedra, Christina. "Navy Investigating 'Chemical Smell' in Military Housing Drinking Water." *Honolulu Civil Beat*. November 29, 2021. https://www.

civilbeat.org/2021/11/navy-investigating-chemical-smell-in-military-housing-drinking-water/

Jetñil-Kijiner, Kathy. "United Nations Climate Summit Opening Ceremony – A Poem to My Daughter." *Kathy Jetñil-Kijiner*, September 24, 2014. https://www.kathyjetnilkijiner.com/united-nations-climate-summit-opening-ceremony-my-poem-to-my-daughter/

Kanahele, Pualani Kanaka'ole.. "I Am This Land, and This Land Is Me." *Hulili* 2, no. 1 (2005): 21–30.

Ka'ohewai. Kāhea to the Lāhui! 'Anahulu at the Ko'a Starts Now. 2022. https://fb.watch/d-SvOaxOp1/.

kealaiwikuamoo. "Ka'ohewai Demands Shutdown of Kapūkaki (Red Hill) Fuel Tanks." *Kanaeokana* (blog), December 17, 2021. https://kanaeokana.net/shutdownredhill/

Kelly, Marion, and Sidney Michael Quintal. 1977. "Cultural History Report of Makua Military Reservation and Vicinity, Makua Valley, Oahu, Hawai'i." DACA84-76-C–0182. Honolulu: U.S. Department of the Army, U.S. Army Engineer Division, Pacific Ocean.

Kikuchi, William Kenji. *Hawaiian Aquacultural System.* Dissertation, Tucson: University of Arizona, 1973.

Kometani, Kassandra. "Environmental Justice for Native Hawaiians: Preventing Landfill Expansion on the Wai'anae Coast - Environmental Amicus." *Environmental Law Education Center.* March 2, 2022. https://elecenter.com/1143/environmental-justice-for-native-hawaiians-preventing-landfill-expansion-on-the-waianae-coast/

Lind, Ian. "Ring of Steel: Notes on the Militarization of Hawai'i." *Social Process in Hawai'i*, 31 (1984): 25–48.

MacKenzie, Melody Kapilialoha, Susan K Serrano, and Koalani Laura Kaulukukui. "Environmental Justice for Indigenous Hawaiians: Reclaiming Land and Resources." *Natural Resources and Environment* 21, no. 3 (2007): 32–42.

MacLennan, Carol A. *Sovereign Sugar: Industry and Environment in Hawai'i.* Honolulu: University of Hawai'i Press, 2014.

"Makua Valley Public Meeting Held on January 27, 2001 (Condensed Transcript and Concordance)." 2001. Ralph Rosenberg Court Reporters, Inc.

Maly, Kepā and Institute for Sustainable Development. 1998. "Oral History Study: Ahupua'a of Mākua and Kahanahāiki, District of Wai'anae, Island of O'ahu." BCH Project No. 442.0122. N62742-94-D-0006 D.O. 22. U.S. Navy, PACDIV.

"Marshall Islands | Atomic Heritage Foundation." Atomic Heritage Foundation. 2019. https://www.atomicheritage.org/location/marshall-islands

Merwin, William S. "Hawai'i Wakes Up To Pesticides." *The Nation*, March 2, 1985. https://www.thenation.com/article/archive/hawaii-wakes-pesticides/

Minerbi, Luciano. "Native Hawaiian Struggles and Events, a Partial List 1973–1993." *Social Process in Hawai'i* 35 (1994): 1–14.

Morales, Rodney, ed. *Ho'iho'i Hou: A Tribute to George Helm and Kimo Mitchell.* Honolulu, Hawai'i: Bamboo Ridge Press, 1984.

No author. "Protect Kaho'olawe 'Ohana." Protect Kaho'olawe 'Ohana. October 3, 2022. http://www.protectkahoolaweohana.org/

O'ahu Water Protectors. 2021. "Ola I Ka Wai – Water Is Life." https://oahuwaterprotectors.org/

Pala, Christopher. "Pesticides in Paradise: Hawai'i's Spike in Birth Defects Puts Focus on GM Crops." *The Guardian*, August 23, 2015. https://www.theguardian.com/us-news/2015/aug/23/hawaii-birth-defects-pesticides-gmo

Paris, Anthony Makana, and Kamuela Werner. "Not in Anyone's Backyard." *Ka Wai Ola*, April 30, 2020. https://kawaiola.news/aina/not-in-anyones-backyard/

Parsons-UXB Joint Venture. 2004. "Final Summary: After Action Report Kaho'olawe Island Reserve." Contract No. N62742-95-D-1369. Naval Facilities Engineering Command Pacific Division. http://kahoolawe.hawaii.gov/downloads/PUXBFinal%20SummaryAfterActionReport.pdf

Richardson, Mahealani. "Alarming New CDC Survey Shows 'Worse Health' among Those Impacted by Red Hill Fuel Spills." Hawai'i *News Now*, November 9, 2022. https://www.hawaiinewsnow.com/2022/11/10/cdcdoh-survey-shows-worse-health-after-red-hill-fuel-spills/

Schofield, M. John. Letter to William Sherman. "Schofield to Sherman," February 15, 1873. DSO-11472-49951-Hawaiian Islands (1872–93). Library of Congress.

Schrager, Benjamin, and Krisnawati Suryanata. "Seeds of Contestation: The Emergence of Hawai'i's Seed Corn Industry." In *Food and Power in Hawai'i: Visions of Food Democracy*, edited by Aya Hirata Kimura and Krisnawati Suryanata, 138–155. Food in Asia and the Pacific. Honolulu: University of Hawai'i Press, 2016.

"Shut Down Red Hill." Sierra Club of Hawai'i. October 9, 2022. https://sierraclubhawaii.org/redhill

Sproat, D. Kapua'ala. "A Question of Wai: Seeking Justice through Law for Hawai'i's Streams and Communities." In *A Nation Rising: Hawaiian Movements for Life, Land, and Sovereignty*, edited by Noelani Goodyear-Ka'ōpua, Ikaika Hussey, and Erin Kahunawaika'ala Wright, 199–219. Narrating Native Histories. Durham, NC: Duke University Press, 2014.

State of Hawai'i Board of Land and Natural Resources. "State General Lease No. S-3848; U.S. Lease, Contract No. DA-94-626-ENG-79." 1964.

Swanson, David A. "A New Estimate of the Hawaiian Population for 1778, the Year of First European Contact." *Hūlili: Multidisciplinary Research on Hawaiian Well-Being* 11, no. 2 (2019): 203–22.

Trask, Haunani-Kay. "Birth of the Modern Hawaiian Movement: Kalama Valley, O'ahu." *Hawaiian Journal of History* 21, no. 13 (1987): 126–53.

U.S. Department of Defense. "Base Structure Report Fiscal Year 2018 Baseline." 2019. https://www.acq.osd.mil/eie/Downloads/BSI/Base%20Structure%20Report%20FY18.pdf

US EPA. n.d. "Pearl Harbor Naval Complex Site Profile." Overviews and Factsheets. Accessed October 3, 2022. https://cumulis.epa.gov/supercpad/CurSites/srchsites.cfm

U.S. Fleet Forces Command, dir. 2001. *Red Hill: The Building of an Enduring Monument (N1901-00-0001).* https://www.youtube.com/watch?v=lIz8IstwnWU

Vice Chief of Naval Operations. 2022. "Command Investigation into the 6 May 2021 and 20 November 2021 Incidents at Red Hill Bulk Fuel Storage Facility."

5 Alaska Native Environmental Activism

Holly Miowak Guise

As Indigenous Alaskans have used the land sustainably for over 10,000 years, Alaska Native history is rich with stories of environmental preservation and environmental justice. Alaska stems from the Unangax̂ word "alaxsxaq" describing the land. The geography consists of rainforests in Southeast Alaska, 30 vast mountain ranges, including Mt. Denali representing the highest peak in North America, the expansive Yukon-Kuskokwim delta, and over 6,640 miles of coastline not including the coastline of 1,800 islands. The Native Peoples of Alaska include over 220 federally recognized tribes. Alaska has 20 distinct Indigenous languages from two main language groups of the Iñuit-Unangan and Ne-Dene.[1] While there are important differences between these nations and their ecologies, Indigenous Alaskans have tended to share a common history of living sustainably from the land, water, and natural ecology.

In recent centuries, the relationships between Native people and the Alaskan landscape have been marked by eras of colonialism, which may be defined as an empire and government imposed upon Indigenous lands and people without the consent of Indigenous people. From Russian contact, beginning in the 18th century and including the fur trade, up through US settler colonial occupation after 1867, sequential colonial projects have brought financial profits to colonial governments, working to the detriment of natural ecologies and Native peoples. Within harsh conditions of ongoing colonial violence against Alaskan people and their homelands, Alaska Natives have fought to protect and exercise their sovereignty—that is, their inherent right to self-governance, which does not disappear even after a colonial government imposes a new order.[2] In doing so, they have systematically linked together the protection of the environment with the protection of Indigenous governance of land and resources, thereby producing powerful conceptualizations and practices of environmental justice.

This chapter examines Alaska Native environmental activism in the 20th and 21st centuries. While Native efforts to preserve and protect the environment go back much longer, accelerating colonial incursions and

DOI: 10.4324/9781003214380-8

resource extraction projects catalyzed widespread Indigenous resistance and mobilization during this era. With the help of media such as newspapers, Native activists—including journalists, artists, and elders—organized grassroots movements emphasizing environmental preservation alongside the protection of Indigenous land rights and access to subsistence lifestyles that allowed Native peoples to continue harvesting land and sea mammals for sustainable food. The result was a multi-faceted Alaska Native environmental justice movement centered on the promotion of Indigenous sovereignty that played out within and across many different spaces, from press campaigns and artistic and literary works to community actions and public protests.

The Alaska Native environmental justice movement sometimes intersected with non-Native conservation activism, which focused increasing attention on Alaska in response to concerns about resource depletion and climate change. Important grassroots alliances emerged between Indigenous and Western activists; yet, there were also frequent clashes, as when Western environmentalists pressed for anti-Indigenous legislation that challenged Indigenous livelihood and limited subsistence activities.[3] Greenpeace's protest of Arctic whaling in Iñupiat communities provides one such example, and conservationists protesting Unangax̂ sealing provides another. Over time, the relationship between Native activists and non-Native conservationists evolved to amplify Indigenous approaches to environmental issues that position sovereignty as an essential element of environmental protection. In this way, Alaska Native activism played an important role in shaping a transnational anticolonial environmental justice movement that continues today. Ultimately, the success of environmental justice campaigns in Alaska evolved to include conservationists who worked alongside Indigenous communities to augment Indigenous voices for a sustainable environment. This chapter includes an analysis of Indigenous food sovereignty evidenced by the Barrow Duck-in, Unangax̂ sealing, and ivory carving, alongside organized Indigenous protests to colonial projects, including Project Chariot, the Rampart Dam, and the ecological crisis of the Exxon Valdez oil spill.

Organizing against Destruction: Protesting the Rampart Dam, Weapons Testing, and Nuclear Waste

In the decades following WWII, Alaska Natives faced an increasingly powerful US colonial state that attempted to reshape Alaska in accordance with Cold War norms of development and national security. Paths the US government pursued along these lines included the building of a megadam, atomic testing projects, and experiments using nuclear bombs to reshape the landscape, all of which threatened disastrous environmental and human impacts. When Alaska Natives found their homelands targeted by such campaigns, activists successfully mobilized broad opposition, with community political activists as well as newspapers and journalists performing central roles in voicing dissent and bringing people together. In the wake of these events, artists also played a central role in keeping their memory and meaning alive. This can be seen in the cases of the Rampart Dam, the US Atomic Energy Commission's planned underground testing of atomic bombs at Cape Thompson, and experiments to use nuclear bombs to create a harbor in Point Thomas.

Rampart Dam represents a pivotal moment of Alaska Native organizing to protect Indigenous lands, villages, and wildlife. Planned construction of this project reveals links between the US Army Corps of Engineers, corporations, politicians, and congressional representatives. Each of these organizations excluded the voices of Alaska's Indigenous peoples revealing the multi-colonial institutions' Native people had to challenge to protect their land.

In the 1950s, the US Army Corps of Engineers sought to dam the Yukon River and to construct the largest artificial reservoir in the world that would have spanned over 10,000 square miles.[4] This proposal stemmed from the desire to create artificial hydroelectric power. This dam was projected to cost an estimated $1.3 billion during a decade-long construction project with concrete at 530 feet high and the top reaching 4,700 feet.[5] With $100,000 in monetary support, Yukon Power for America, Inc. sought to lobby the Rampart project through Congress.[6]

Key leaders of the colonial territory of Alaska, including Senator Ernest Gruening, endorsed the plan for the dam proposed by the federal government. Gruening acknowledged his role in being an "outspoken advocate for the Rampart Canyon project from the moment I was elected to the Senate."[7] He offered contradictory sentiments in suggesting that the creation of swamplands from the construction of the Rampart reservoir would be "uninhabited except for seven small Indian villages on the river," and that "the flooding would, to all intents and purposes, destroy no property values."[8] Gruening remained unconcerned about Native villages, but perhaps his attitude would have been different if white towns resided in the region. As the US had a history of removing Native people from their original homelands, perhaps the pattern would have continued in relocating these seven Native villages. Gruening downplayed the importance of Native people and their subsistence relationship to the land in suggesting that Native people could acquire better living conditions away from their ancestral home villages and new employment working for the government and the construction of the dam. He bemoaned the role of conservationists in the national press while neglecting to mention the Gwich'in activists who protested the construction.

Native people adamantly protested the creation of an artificial reservoir in Alaska. Centering the voices of Gwich'in activist Sarah James, historian Finis Dunaway accounts for the oral history of Gwich'in activism dating back to the time when the military and Alaska Senator Gruening sought to flood the flats destroying seven Native villages.[9] Gwich'in activists mobilized forming Gwitchya Gwitchin Ginkhye (Yukon Flats People Speak) in 1964; Dunaway describes the Rampart Dam as galvanizing the Gwich'in to protect their homeland from environmental destruction. As Athabascan elder Alfred "Al" Wright described in an interview, the government eventually scrapped the large project. Wright listed the villages that would have been underwater otherwise, "It would've put Rampart, Stevens Village, Beaver, Fort Yukon, Circle City, Chalkyitsik all out under water. Oh, and Venetie would've been under water too."[10]

Alongside Indigenous voices who opposed the dam, conservationists, including University of Alaska faculty member Daniel W. Swift, who served as Vice President for the Alaska Conservation Society, and advocates from the US Fish and Wildlife Service (FWS) sought to terminate the construction plan. Leveraging their presence to a national audience using *The New York Times*, a reporter highlighted concerns by the FWS that 1,600 people would be displaced.[11] According to the reporter, FWS stated, "Nowhere in the history of water development in North America have the fish and wildlife losses anticipated to result from a single project been so overwhelming."[12] Dissent from this federal organization revealed how various government bureaucracies went against one another, and in this case it worked in favor of Gwich'in land protectors. While the FWS was a complicated entity that did not systematically center Indigenous interests, as will be explored below, Gwich'in land protectors' work to create temporary alliances with these stakeholders served to advance their aims here.

Academic researchers identified the Rampart Dam as destructive to wildlife preservation. Had the dam proceeded, it would have devastated the salmon population, the habitat for 1.5 million migratory birds, and relocated Native villages.[13] Scientists worried that the artificial reservoir would destroy the forests of the Yukon Flats and melt the permafrost.[14] With grant funding from the Natural Resources Council of America, researchers at the University of Michigan sought to identify problems with low-cost electricity and decisions involving Alaskan development. The study published as the Summary Report in 1966 failed to mention Indigenous Alaskans. Several categories related to ecology mentioned the detrimental effects on the fisheries, which would "negate thirty years of endeavor in waterfowl in North America", moose, mammals, and forests.[15]

Politicians who supported environmental concerns intervened to protect the flatlands. US Secretary of the Interior Stewart Udall opposed the dam's construction in 1967, and in 1980 President Jimmy Carter designated the space as the Yukon Flats National Wildlife Sanctuary, thereby protecting it from future exploitation. Here, protection for the Yukon Flats likely stemmed from a concern for the waterfowl, and not necessarily US politicians prioritizing Indigenous sovereignty over their homelands. While the establishment of the Yukon Flats National Wildlife Sanctuary did protect the ecology and subsistence lifestyle, the creation of a federally recognized national wildlife sanctuary depleted Indigenous sovereignty in title alone. Nonetheless, the Native villages remained in the Yukon Flats.

Another key environmental justice campaign, and one that was ultimately more successful in achieving its main aim, involved the building of opposition to US atomic testing programs in Alaska. In 1958, the US Atomic Energy Commission announced plans to test atomic bombs underground at Cape Thompson. This proposed manmade harbor by nuclear explosion epitomized Cold War era US governmental research that disregarded the impact on the local Indigenous population and the land. However, this colonial campaign met massive resistance catalyzed by a new Indigenous newspaper, the *Tundra Times*. Indigenous studies scholar Maria Sháa Tláa Williams (Tlingit) contends that the *Tundra Times* Native-run newspaper gave voice to Native peoples and garnered an Alaska Native solidarity movement since its founding in the 1960s.[16] In addition to uniting Alaska Native people during a tumultuous time of colonial litigation from the land claims era, the *Tundra Times* played a key role in shaping opposition to colonial projects that threatened the environment, beginning with the campaign against atomic testing.

In 1962, Howard Rock (Iñupiaq) founded the *Tundra Times* to unite Indigenous voices and to disseminate Indigenous perspectives in the news.[17] Born in Tikiġaq (English name: Point Hope), Rock attended the Bureau of Indian Affairs (BIA) federally run White Mountain School for high school. With an educational loan from the BIA, Rock attended the University of Washington to take art classes. When Rock returned to Tikiġaq in May 1961, he grew alarmed that the US Atomic Energy Commission planned to test atomic bombs. In addition to drafting a letter on behalf of the Village Council of Point Hope and mailing it to Secretary of Interior Stewart Udall in July of 1961, Rock sought to amplify Indigenous concerns using outside support.[18] Rock knew intertribal collaborations would be necessary to raise Native voices, so he invited LaVerne Madigan, the executive director of the Association on American Indian Affairs (AAIA) to access the growing concern in Point Hope.[19] Through networking and finding a donation from Henry Forbes of Massachusetts, Rock founded the *Tundra Times* and immediately set about using the paper to mobilize activism against atomic testing. This newspaper that began in October 1962 committed itself to representing an unbiased presentation of Native affairs. With an eight-page biweekly series that households could buy subscriptions to, the

Tundra Times ran investigative series on Native affairs across the state. In addition to community organizing in the Native village Tikiġaq, the *Tundra Times* helped to defeat the Atomic Energy Commission's Project Chariot in 1969.[20]

As the *Tundra Times*'s role in preventing atomic testing at Cape Thompson shows, journalism facilitated by Indigenous leadership gave voice to Native peoples and energized an Alaska Native solidarity movement, thereby playing a crucial role in fostering Alaska Native activism around issues related to sovereignty and the environment.[21] Alongside news sources like the *Tundra Times*, Native people had other unifying sources in journalism through *Village Voice* run by Rural Alaska Community Action Program, the *Mukluk Telegraph* administered by the Alaska Native Tribal Health Consortium, and news provided by the various 13 Alaska Native Corporations. Outside Alaska, other Indigenous-run presses have played a key role in unifying pan-Indigenous politics, issues, and concerns. *Indian Country Today*, founded in 1981 originally under *The Lakota Times*, has continued a legacy of serving as a free press for Indigenous peoples. Long before the era of social media, these news outlets, as well as Indigenous journalists working at other papers, galvanized popular opposition to colonial projects that threatened the environment across a broad range of contexts, including other attempted nuclear testing and experimentation campaigns.

Russia, the US, and the governments of the circumpolar north set up nuclear testing sites throughout the Arctic, and communications systems, including the White Alice Communications System and the Distant Early Warning (DEW) line that continue to contaminate the land. The protest against "Project Chariot" marks one of the most successful Indigenous movements to halt government testing and nuclear contamination in the Arctic. In Point Hope in Northern Alaska, the Iñupiat successfully coordinated a grassroots campaign to prevent the US from creating a manmade harbor in their town called "Project Chariot" that the federal government first proposed in 1958.[22] This manmade harbor at Cape Thompson, 32 miles from Point Hope and 40 miles from Kivalina intended to stimulate an economy in Northern Alaska using a massive atomic explosion. One *Anchorage Times* article addressed opposition by Iñupiat and opponents of nuclear power and in line with scientists and the Atomic Energy Commission by claiming that, "Nuclear power opens amazing new avenues for greater achievements. Mankind must explore them and the Alaska experiment may be the key."[23]

Even though the detonations that would have totaled 2.4 megatons never occurred, in an effort to study the rate of contamination the US still experimented on the Alaskan landscape. In 1962, the government brought materials from a nuclear explosion in Nevada to Point Hope for experimentation and then buried the radioactive material, thereby polluting the land and causing adverse health consequences for the local Iñupiat population. After these events, the government remained interested in the hypothetical construction of Project Chariot measuring guessed fallout. A report by geologist Arthur Piper prepared for the US Atomic Energy Commission in 1966 detailed these appraisals of radioactive materials absorbed by local water supplies.[24] Piper concluded that "...effects on the hydrologic environment could be substantial and could seriously handicap man's subsequent activities."[25]

As a key leader in the mobilization of Alaska Natives on the grassroots level and in the press, Charles Etok Edwardsen Jr. "Etok" vocally opposed continued proposals for radiation testing in Alaska. In 1993, Etok published a letter to President Bill Clinton as an opinion essay in the *Tundra Times*. His letter dated decades after proposed radiation testing shows how Alaska continued to be in potential jeopardy of radiation exposure into the late 20th century. Here, Etok claimed, "In Alaska, many of us are not happy with the prospect of ARCO altering the Earth's neutral atmospheric properties."[26] Etok's engagement with the press reveals how he used the resource to show how one could reach directly to the President to voice their concern and to share information widely for all others.

Despite the success preventing the construction of Project Chariot in Northern Alaska, made possible by unrelenting Iñupiat activism and the power of the press, between 1965 and 1971, the US conducted a series of nuclear blasts on Amchitka Island in the Aleutian Islands to test nuclear weapons. In 1971, just a couple years after the defeat of Project Chariot, a new "Project Cannikin" involved a 5-megaton blast with grave ecologic consequences, including loss of marine life, destroyed freshwater habitat long-term decline of the salmon runs in the area, and a landscape visibly disrupted by the blast.[27] Still, Indigenous resistance to these events continued, including through the medium of art, by which artists have meditated on and created historical reminders of such environmental atrocities.

For example, the atrocity of nuclear testing near the Iñupiat village of Point Hope is remembered by Iñupiat artist Ken "Ooyahtoona" Lisbourne. Lisbourne was born in 1950 and raised in Point Hope. Lisbourne became known for his art depicting joyful occasions like community gatherings and successful whale hunts. Alongside these celebrations of Iñupiat heritage and celebrating the maintenance of the subsistence lifestyle, he painted a few artworks that depicted the darker sides of colonial exploitation, including pollution from nuclear fallout.[28] In meeting the late Ken Lisbourne at the Alaska Federation of Natives in 2015, he mentioned that the image of Point Thompson depicted the people and animals that died from the fallout from government testing. Each body and animal carcass indicates a life lost to government exploitation of the land and local population.

His art calls attention to this past history that remains less known to the general US public and even within Alaska.

Sustaining Food Sovereignty: Asserting Rights to Traditional Native Resources

Other forms of environmental activism include Native people being able to utilize their own resources, including birds, animal pelts, and ivory, which they have a history of doing sustainably for many thousands of years. Alaska Native environmental activism has pursued this right in the face of colonial government interventions, as can be seen in the Barrow Duck-in and the fights to maintain sealing, ivory harvesting, and whaling rights. Importantly, Indigenous activists have often had to assert these rights not only in opposition to colonial laws, but in the face of Western conservation movements that attempted to usurp Native sovereignty.

The Barrow Duck-in represents Indigenous assertions of food sovereignty and resistance to government entities that sought to control Indigenous livelihoods and survival in their homelands. Due to the 1918 and 1937 Migratory Bird Treaty Conventions signed among the US, Canada, and Mexico, with these nations leaving out Indigenous representation, the federal government tried to prohibit the harvest of certain waterfowl. These hunting restrictions did not take into consideration the need for food and traditional hunting in the Indigenous community. Whaling is the main food source in Utqiaġvik, although one year the community only harvested two whales which equated to only a few meals per household. The Iñupiat turned to duck hunting to feed the community, as a traditional food. Following the arrest of two Iñupiat hunters in Utqiaġvik (also known by the English name Barrow) in 1961, the Iñupiat convened at the Point Barrow Conference on Native Rights.[29] This event was sponsored by the AAIA from November 15 to 18, 1961.[30] In asserting their aboriginal land rights, hunting rights, and by declaring these rights, the Iñupiat demonstrated Indigenous organizing power against colonial institutions. In their formal statement, they exclaimed, "We did not know before what our aboriginal rights are, but now we know. They are our Iñupiat Paitot. Our Iñupiat rights to own the land and minerals of our ancestors, to hunt and fish without restrictions over this land and the sea."[31] Here, in addition to hunting birds and placing them at the door of US federal officials, the Iñupiat created their own declaration of rights as a governing entity and as an act of defiance to go alongside their actions. The actions of hunting for food despite federal regulations demonstrated the Iñupiat exercise of Indigenous food sovereignty.

Sealing represents another arena in which Alaska Native peoples have fought to protect food sovereignty alongside labor rights. Seals have long been an important food source for Alaska Native people, particularly the Unangax̂ from the Pribilof Islands. With awareness through the *Tundra Times*, Alaska Native journalists sought to highlight the importance of Unangax̂ sealing that went unpaid and exploited.[32] Since 1867, the US FWS and private companies, including the Fouke Fur Company profited from Unangax̂ labor.

In response to federal control and exploitation of sealing rather than an Indigenous-focused effort, *Tundra Times* propelled Unangax̂ sealing rights forward by providing updates on the matter, with coverage by the paper centering Indigenous rights, including Unangax̂ labor rights, ultimately garnering pan-Alaska Native support. As scholar Barbara Boyle Torrey documents, the journalistic campaign that began in 1964 led to Alaska Senator Bob Bartlett visiting the Unangax̂ home island of St. Paul in 1965, and then holding the Bartlett Hearings to listen to Unangax̂. [33] This ultimately led to the Fur Seal Act of 1966 which allowed Indigenous peoples of Alaska to continue to harvest seal pelts, and it also established a townsite where Pribolobians could receive civil service retirement benefits and their previous labor would count toward benefits as employees of the federal government. While these monetary benefits were extended, so too did federal oversight in monitoring the sealing operations. The federal government could continue to conduct sealing operations in the Pribilof Islands while meeting a minimum requirement in acknowledging Unangax̂ as federal workers.

Yet over the next few decades, Indigenous sealing rights would come under attack again, this time by outside environmentalist groups. In the 1960s, US conservationists made sealing one of their top priorities to discontinue. Without taking into account the Indigenous tradition of sealing, which had occurred for several thousands of years on the Pribilof Islands even when the islands were not inhabited although sparsely visited according to oral tradition, the conservationists attacked Unangax̂ sealers. It would have been more helpful if the conservationists had defended Unangax̂ subsistence rights and labor rights as federal employees needing protection from worker exploitation as had occurred for decades prior dating back to 1867, rather than attacking the tradition of sealing. Instead, conservationists acted on behalf of the animals they never met, on the lands they never visited, and forever altered the lives of the Indigenous peoples who resided on the lands and stewarded the land for thousands of years. Precursing and paralleling the anti-whaling campaigns on Alaska's North Slope that included efforts by Greenpeace from the 1970s to 1980s, the mainstream Western conservationist movement attacked Unangax̂ seal harvesting and traditional way of life.

Pribolobian activists traveled to Washington, DC, to lobby Congress for their rights and to challenge Western conservationists seeking to create laws that would curb Indigenous food sovereignty. Mike Zacharof delivered a speech in Unangam Tunnu language which Larry Merculieff translated. Zacharof argued that halting the seal harvest would terminate employment for the entire island. He closed by saying, "Do these people know what would happen to us if the seal harvest is stopped or greatly cut down? Is the seal going to be extinct if we do not stop the harvest? Because if we cut the harvest, it will hurt an honest-to-goodness rare and endangered species- the Aleut people."[34] In 1974, the St. George Pribolobians lost their sealing rights and could only harvest seals for meat and they had to use a traditional boat to catch them. The problem was that Unangax̂ had never hunted seals from boats. So, Unangax̂ men demonstrated the absurdity of this activity and showed they could only catch one female seal hunting from the water. The government then decided to cap the number of seals harvested on St. George Island to 350 annually, which is not enough food for a village of 150 people who have relied on the food from the land for several generations. Like the Barrow Duck-In, Unangax̂ continue to advocate for their food sovereignty through the practice of sealing.

A similar chain of events has characterized whaling. In 1978, Greenpeace activists sought to halt Iñupiat and Yupiit communities from traditional whaling practices.[35] Such calls by external environmentalists to cease Indigenous subsistence practices for food and maintaining lifeways perpetuated colonial projects of cultural genocide in forcing assimilation and reliance on the Western economy. In advocating to protect the whales at all costs, environmentalists of the 1970s alienated Iñupiat and Yupiit in refusing to recognize the importance of the subsistence lifestyle to maintaining Indigenous livelihood.

Another flashpoint emerged around walrus hunting. While several governments and organizations have attempted to ameliorate the illegal and unsustainable global ivory trade, which involved the exploitation of elephant ivory from Africa and Asia, the global effort to tackle this issue acted in an anti-Indigenous manner. More specifically, Alaskan ivory comes from the walrus. Walrus are hunted traditionally by Iñuit in northern and western Alaska where either a male or female walrus produces ivory. Harvesting walrus means utilizing each piece of walrus for meat, blubber, and food preservation. In rural areas where walrus meat is harvested annually, ivory sales allow Native ivory carvers to secure a modest income while they maintain their traditional and sustainable lifestyle. Environmental historian Bathsheba Demuth has articulated that the empires of the US and the Soviets had targeted the walrus population of the Bering Sea, and that Indigenous peoples of Kamchatka and Alaska had been exercising food sovereignty sustainably before the arrival of these empires.[36]

In 2016, Alaska Natives circulated a petition to remove Alaskan ivory from the US laws banning the sale and trade of ivory. Susie Silook (St.

Lawrence Island Yupik) founded Sikuliiq, an Alaska Native artists' advocacy group aimed at protecting the Indigenous use and sale of walrus ivory.[37] Silook is a published author who has written on the role of ivory carving as applying healing and serving as an anti-depressant.[38] Her organization gathered approximately 1,000 signatures in a petition to President Barack Obama to protect the sale of Native arts. In addition to gathering signatures, Silook spoke at the Alaska Federation of Natives in 2016 not only indicating her advocacy on behalf of Alaska Natives, but also signifying widespread Native support to protect the sale of walrus ivory from anti-Indigenous state and federal bans.

Other Indigenous advocates for the protection of walrus ivory sales included Rosita Worl, Ph.D. (Tlingit), the President of the Sealaska Heritage Institute. In October 2016, Worl testified at the Field Hearing of the Fisheries, Water, and Wildlife Subcommittee describing the ban on walrus ivory as detrimental to Alaska Native artists. KINY Juneau radio reported that Worl and US Senator Dan Sullivan asked the digital commerce platform Etsy to reverse its ban on Alaska Native artists selling ivory work online and to reinstitute the accounts they suspended since Alaska Native artists are protected under the Marine Mammal Protection Act.[39] This shows how Worl partnered with political leaders and the press to push private corporations, like Etsy, to lift the ban on Alaska Native ivory.

Fighting Colonial Extraction: Defending Arctic Ecologies of Land, Water, and Animals

As a land rich in natural resources, Alaska has continuously been the target of colonial and corporate projects intent on extracting these resources in ways that destroy natural ecologies and threaten great harm to its human and non-human inhabitants. Thus, fighting resource extraction through the exertion of Indigenous sovereignty has been a central goal of Alaska Native environmental justice activism. This can be seen in campaigns to defend Alaska's land, water, and animals—as well as its Alaska Native inhabitants—against destruction wrought by drilling and mining.

The Arctic Wildlife Refuge in Northeastern Alaska, established in 1960, is on Gwich'in land, and Alaska Natives have worked hard to protect this endangered environment from colonial development and extraction. National awareness to protect the Arctic Refuge emerged from a traveling slide show with photographs by Lenny Kohm, in the late 1980s.[40] Kohm's slideshow featured photographs of the environment, delicate Arctic ecology, and his work was only made possible by his collaborations with Gwich'in activists. Such an alliance shows how Gwich'in activists strategically worked alongside other public-facing grassroots organizers like Kohm to spread awareness on protecting the Arctic.

Public awareness about protecting the Arctic was accelerated by the Exxon Valdez oil spill in March 1989, which resulted in 10.8 million gallons of oil from Prudhoe Bay polluting Alaskan waters. In addition to harming locals and fishermen in this region, this pollution severely impacted the hatcheries and fisheries, migratory birds, and sea mammals. Following the initial years after the spill, scientists and conservation biologists have continued to study the impact on the environment and ecology for decades.[41] During the initial cleanup, Alaska Native villages confronted waves of temporary workers and volunteers for the oil cleanup projects, and *Tundra Times* reported that this influx of quick changes and new work outside of subsistence activities adversely impacted Native communities.[42] These changes went alongside the inability to access traditional subsistence foods. Even five years after the oil spill, Native people could not eat shells polluted by the spill. A *Tundra Times* journalist interviewed Native people from the Chenega Corporation, including Chuck Totemoff who shared, "Subsistence in our area has dropped off dramatically since the oil spill."[43] John Christensen of Chenega also described the extra effort to find safe subsistence foods for his household, "We have had to travel farther, using better and safer equipment." Native people still remember the broken promises from Exxon, and they display these sentiments in public art. In 2006, Alaska Native carver and Cordova fisherman Mike Webber created a ridicule pole of Exxon for the Native Village of Eyak. Webber explained in his art and public history, "Ridicule poles are erected to publicly humiliate a person, typically a wealthy person, for not paying his debt in full. I completed the pole for the 18th anniversary of the oil spill. I carved a hole in the heart area, meaning that when Exxon said they would make us whole, they ended up putting a hole in us, a hole in our hearts."[44]

After several decades following the Exxon Valdez oil spill in the Prince William Sound, dAXunhyuu (Eyak) activists have been advocating to restore Indigenous lands. For example, geographer Jen Rose Smith (dAXunhyuu) has written about Indigenous kelp knowledge and conservation farming.[45] Such practices demonstrate the efforts of Indigenous mobilization to promote Indigenous-led farming and fishing as well as to protect the sea and land from future catastrophes.

By sharing stories across the US and globe, documentary films have contributed to awareness on the need to protect the Arctic from oil drilling, especially the Arctic National Wildlife Refuge. *The Last Great Wilderness* (1990), *Oil on Ice* (2004), and *The Arctic Our Last Great Wilderness* (2021) are three such films. Together these films sought to amplify Athabascan voices in protecting the Arctic. In the early millennium, the Anchorage Museum hosted film screenings, including *Oil on Ice*. *The Arctic Our Last Great Wilderness* is narrated by Princess Daazhraii Johnson (Neets'aii Gwich'in from Arctic Village). She founded the Fairbanks Climate Action

Coalition and has served on the Gwinch'in Steering Committee. Johnson's name is familiar to many since she is the creative producer and writer for the beloved PBS kids' show "Molly of Denali" which began in 2019 and continues still today. Children across the US identify with Molly, and in addition to cherishing her character and learning about Native traditions and songs, they learn about the Northern environment where Molly comes from. In this manner, activists like Johnson have used Hollywood, production, and television to amplify Indigenous voices in mainstream US culture and to highlight the unique role of the Arctic as a continued homeland for many Indigenous families, as well as non-human inhabitants that Native people have long lived alongside.

Publications have also contributed to public awareness on preserving the Arctic. Print publications, including *We Are the Arctic* (2016), feature photographs, biographies of local Gwich'in, environmentalists, and scholars who have pledged to protect the Arctic National Wildlife Refuge. This publication called for readers to make a pledge to the US Senate to protect the Arctic alongside 94,500 supporters.[46] Among the voices of Indigenous peoples centered in this publication, Native women are centered, including Lorraine Netro (Gwitchin First Nation, from Old Crow, Yukon, Canada), and Princess Daazhraii Johnson (Gwich'in Athabascan, from Fairbanks, Alaska). *We Are the Arctic* also features Robert Thompson (Iñupiat from Kaktovik) describing his support for protecting the Arctic. Thompson specifically mentioned his granddaughter as his reason to fight to protect the land for future generations.

Activism and art created to protect the Arctic from drilling in particular emerged within a wider political context where tensions exist between activists and leaders from certain Alaska Native Corporations who seek to profit from corporate extraction on their own Indigenous lands. The Native Corporations were established in 1971 after the passage of the Alaska Native Claims Settlement Act of 1971. While some corporations aim to center cultural values, including language revitalization and conservation projects, other corporations have created partnerships with oil companies and outside companies that have accelerated resource extraction. Tensions have also existed between Indigenous peoples and environmentalists who center a settler colonial perspective on environmental projects. For example, as a writer for *Tundra Times*, Paul Ongtooguk critiqued the showing of *The Last Great Wilderness*, a film about drilling in the Arctic as unfairly depicting Iñupiat corporate leaders who approved oil exploration. Ongtooguk remained cautious of environmentalists, particularly when environmentalists sought to terminate Indigenous whaling practices in northern Alaska during the 1970s, as discussed above.[47]

Like leadership in other governmental organizations, who is in charge determines the degree to which an Alaska Native Corporation prioritizes either land conservation efforts or accelerated resource extraction. One of the greater wedges resides between the Arctic Slope Regional Corporation, where leaders, including Tara Sweeney (Iñupiat), former Assistant Secretary of the BIA under the Trump Administration, continue to advocate for oil development. These calls for oil drilling are in direct opposition to the Gwich'in Athabascans who oppose drilling for oil and construction of a pipeline in the Arctic National Wildlife Refuge that would ultimately devastate the Porcupine caribou, the subsistence food that the Gwich'in rely upon. While there are still Native Corporations that endorse oil drilling, Iñupiat artists protest oil extraction.

While the media focuses attention on leaders of the Alaska Native Corporations who have political power and representation in the Western mainstream press, scholars, including Subhankar Banerjee, an environmental scholar and photographer, have chronicled local Indigenous communities who oppose Arctic oil drilling. Banerjee details, "Iñupiat communities across the North Slope are concerned about the offshore seismic testing and future oil exploration. They worry about the potential impact from offshore oil development on the marine life that they depend on for subsistence food-Bowhead and Beluga whales, seals, walrus, birds, and fish."[48]

Other artists and Indigenous activists have spoken out against Arctic oil drilling, including Allison Akootchook Warden (Iñupiat) from Kaktovik. Akootchook critiques oil extraction and the lofty promises made by oil companies. Akootchook is in the process of creating a traveling art installation titled, "Everybody Will Be a Millionaire!" highlighting the slogan that corporations promised to Indigenous communities on the North Slope about the reapings of oil drillings and how each person would monetarily benefit from a personal level.[49] Demonstrating the global nature of Indigenous issues and conservation efforts, this exhibit is traveling and is expected to tour the United Arab Emirates in addition to locally at the Anchorage Museum.

Within his art and published writing, artist Joseph "Joe" Senungetuk (Iñupiat) has propelled a narrative of Indigenous sovereignty forward while critiquing colonial structures of land decimation and resource extraction (Figure 5.1). Joe grew up in Wales in northwestern Alaska and he then attended elementary school in Nome. After attending Mt Edgecumbe Boarding School in southeast Alaska, Joe migrated farther south to attend the San Francisco Art Institute in the 1960s.[50] He lived in the San Francisco Bay Area for a period of time and then returned to Alaska and taught art classes at Sheldon Jackson College in Sitka, for Native students. As evidence of his continued activism and leadership in the arts, in recent years, Senungetuk has been the artist in residence at Alaska Pacific University.[51]

While primarily illuminating Iñuit history from an autobiographical standpoint, Senungetuk's book, *Give or Take a Century* (1971), includes explanations of his art that critique colonial politics and its environmental

Figure 5.1 Joe Senungetuk at the SEED Lab, Anchorage Museum. April 2022. Photo by the author.

impacts. Regarding his self-portrait created in 1970 (see Figure 5.2), Joe explained, "I assigned myself the job of a self-portrait and I put into the background various news articles from *Tundra Times*. That was a Native-owned Alaskan newspaper that I subscribed to, and I still have quite a few of those newspaper copies at home."[52] Joe's testimony and the headlines he chose to integrate in his self-portrait reveal the power of the Native press both socially and within his own life in representing meaning to political events. Some of these headlines included concerns about the impact of oil extraction impacting both the Iñuit and the land.

"There has been a great outcry from the oil company combine, that the proposed pipeline, which would destroy Alaska's environmental integrity, is needed in order to make jobs available for Native people."[53] This is how Senungetuk has decried oil companies. Senungetuk's art also depicts the challenges brought about by the discovery of oil in Alaska. One artistic representation Senungetuk created is titled, "Emergence of Resource- the Oil Spill Shadow" (Figure 5.3).

In this, Senungetuk explains his work is "meant to express the utter despair, felt by this writer in the mere contemplation of the construction of the Alaskan oil pipeline. It would destroy the land, the people, and the environment."[54] In an interview, Joe explained the impact of oil, "I usually think of an idea and when that art print was being made by me, I was

Figure 5.2 Joe Senungetuk's Self Portrait featured in *Give or Take a Century* (1971).

thinking about all of the concentrated efforts of the media and so on. All of the sudden being impacted by the oil discovery [...] and that was a huge difference as a way of reflecting on Native history. All of the sudden people were talking about the goings on of Iñupiaq people especially in Barrow and we began learning of their ways of transition to businesspeople." Here Senungetuk identified how the oil discovery upended the major Indigenous lifestyle and how many Native people converted to become Western-influenced businesspeople. Indeed, art and writing served as means for

Figure 5.3 Joe Senungetuk's "Emergence of Resource- the Oil Spill Shadow."

environmental justice activists like Senungetuk to share their message on environmental preservation.

Recently, the No on Pebble Mine constituted a grassroots movement that aligned with Indigenous activism to protect the land from corporate resource extraction in the 21st century. In driving around Anchorage or hiking in the Chugach Mountain range, one would see the bumper sticker "No on Pebble Mine" on cars and decorating hikers' water bottles (Figure 5.4). During the early millennium, these stickers circulated throughout Alaska. Their prevalence indicates grassroots efforts to terminate the proposed plan for Pebble Mine that fit the colonial pattern of corporate extraction. In the case of Pebble Mine, after discovering copper in 2001, a Canadian mining company acquired rights to mine the region.[55] Such a

Figure 5.4 Photo of No Pebble Mine bumper sticker (In author's personal collection—2022).

mining project would have decimated the local waterways, including the habitat of Alaska's salmon. The salmon represent livelihood as a subsistence food that the Yupiit have relied upon for thousands of years. To take away the salmon would be detrimental not only to the natural ecology but also to the Indigenous lifeways of the Yupiit.

A transnational approach to understanding water rights reveals pan-Indigenous movements across geographies to halt colonial resource extraction, as well as the role of Alaska in this broader trend. The No on Pebble Mine movement parallels other Indigenous environmental activist movements including the Water Protectors movement in 2016. This movement included the Standing Rock Indian Reservation protesting the Dakota Access Pipeline with the "water is sacred" movement and the "water is life" movement. American Studies scholar Nick Estes (Lower Brule Sioux Tribe) has contended that a wave of Indigenous activism centered on Standing Rock to protect the people, the land, and the water.[56] In addition to these movements in North America spanning from Alaska, through Canada, and the US, water protectors act throughout South America and across the globe. In Alaska, the efforts to protect Bristol Bay from mining at Pebble Mine continue to this day. Websites like savebristolbay.org have action plans and forms to submit online letters to Congress.

Even in mainstream US culture, popular books in children's literature highlight the water protectors. In the book, *We Are Water Protectors* (2020), author Carole Lindstrom (Ojibwe) highlights intergenerational Indigenous activism in securing sacred lands for future generations and protecting water from pipelines. Michaela Goade (Tlingit/Haida) illustrated this book with beautiful watercolors, landing the Caldecot medal in 2021. The partnership between Lindstrom and Goade reveals Native women's alliances in literary studies and environmental justice activism to advance Indigenous rights and social justice issues. The prize-winning nature of this book indicates that the issue resonates with a wider group of mainstream US audiences beyond just Native children. To have such a book receive national attention is significant for Native collaborators, but also for Native children who more readily see representation of their culture in classroom and library settings. Such a trend indicates that mainstream American culture has opened in some areas to the power of Indigenous-led activism to protect the environment. This change has been propelled by Native activists, artists, and cultural producers who have fought to defend Arctic ecologies of land, water, and animals.

Conclusion

Today, the Arctic environment that encompasses land and water is irreparably altering due to the climate crisis with melting permafrost and rising sea levels. In coastal Alaska, several Native homes are eroding into the ocean jeopardizing towns and forcing housing relocations. These changes directly impact Indigenous Arctic populations who now confront homelessness in their homelands. Rising sea levels and coastal erosion are altering the landscape and forcing migrations of entire Native villages. In Shishmaref, for example, where homes are disappearing into an eroding shoreline, the press has played a significant role in addressing this issue of housing instability. This town in addition to other Indigenous towns in Beringia, including Kivalina on Alaska's North Slope and Lorino in Russia, is at risk of erosion from sea ice no longer sheltering the shoreline of their towns.[57]

In coastal towns, residents are trying to preemptively arrange shoreline blocks to slow the speed of coastal erosion. In a comprehensive study about natural disasters in Alaska published in 1997, highlighting the importance of this issue since a majority of Alaskan residents live along the coast, scientists documented the imminent problem of rising sea levels and failed attempts in trying to build lasting seawalls in towns like Shishmaref for protection.[58] Other towns, including Kotzebue, have had some success with the seawalls.[59] In the Iñupiat town of Unalakleet, residents have constructed rock barriers positioned on the shoreline to prevent significant coastal erosion. It is estimated that eventually the town of Unalakleet will need to relocate to higher ground. According to oral histories, the town has already moved several times postcolonial contact due to epidemics and the impacts of Western colonization.

As Alaska Natives battle the impacts of climate change, on top of so many other environmental harms and catastrophes created or exacerbated by colonialism, they draw from and carry forward a long tradition of multi-faceted activism in pursuit of environmental justice in Alaska. Artists, advocates, authors, grassroots organizations, and the press continue to center Native issues and perspectives on preservation, which are inextricably intertwined with efforts to sustain forms of Indigenous sovereignty that have historically promoted environmental sustainability and justice. While the *Tundra Times* may no longer be in operation, Indigenous journalists such as Tripp J. Crouse (Ojibwe) and Alyssa London (Tlingit) continue to represent Alaska Native issues in media outlets such as Spruce Root, *Indian Country Today*, and MSNBC. In the worlds of arts and letters, Native people are continuing to create a legacy of activism through public-facing platforms that reach both local and national audiences. Both journalism and art have played a key role in mobilizing a collective Alaska Native consciousness and broader US public awareness about the environmental concerns in Alaska.

Similarly, Alaska Native environmental justice activism continues to present day with Indigenous-coordinated organizations that mobilize community empowerment to address the climate crisis from an Indigenous perspective. Native Movement, for example, represents an Indigenous grassroots movement in current day that organizes within the framework of an Indigenous worldview to support Indigenous decolonial movements, including ones that support Indigenous environmental justice.[60] According to their history statement, "Native Movement has provided leadership and support for grassroots-led projects that endeavor to ensure Indigenous Peoples' rights, the rights of Mother Earth, and the building of healthy & sustainable communities for all."[61] The Wilderness Society in Alaska is headed by Karlin Nageak Itchoak (Iñupiat). Part of the mission of The Wilderness Society is to advance Indigenous rights alongside conservation during climate change.[62] Other organizations include The Nature Conservancy (TNC), Alaska Chapter, whose Community Development Specialist, Crystal Yankawgé Nelson (Tlingit), argues that "'Conservation' is a Western concept, but when it's approached from a decolonized standpoint, Indigenous communities can create better social, environmental, and economic conditions for themselves. Thriving communities can then make better decisions about land use. This is why I'm focused on a community development approach to conservation."[63] Yankawgé articulates the shifting nature of conservation efforts that now work with Indigenous peoples, are facilitated by Indigenous leaders, and center conceptualizations and practices of environmental justice that support the sustainable ecological relationships furthered by Indigenous sovereignty.

Notes

1 "Alaskan Native Language Preservation and Advisory Council," https://www. commerce.alaska.gov/web/dcra/AKNativeLanguagePreservationAdvisory Council/Languages.aspx

2 Here are a few sources on defining Native sovereignty: Amy E. Den Ouden and Jean M. O'Brien (eds.), *Recognition, Sovereignty Struggles, & Indigenous Rights in the United States* (Chapel Hill: The University of North Carolina Press, 2013), 13–16; Lloyd L. Lee (ed.), *Navajo Sovereignty: Understandings and Visions of Dine People* (Tucson: The University of Arizona Press, 2017); David Wilkins and Heidi Kiiwetinepinesiik Stark, *American Indian Politics and the American Political System*, 4th ed. (Rowman & Littlefield, 2018, 2001), 1–21.

3 The Red Nation, *The Red Deal: Indigenous Action to Save Our Earth* (Brooklyn, NY: Common Notions, 2021), 5–6.

4 Finis Dunaway, *Defending the Arctic Refuge: A Photographer, an Indigenous Nation, and a Fight for Environmental Justice* (Chapel Hill: The University of North Carolina Press, 2021), 38–39.

5 Lawrence E. Davies, "Controversy Rages over Plans for Big Alaska Power Project," *The New York Times*, August 22, 1964, 25.

6 Claus-M. Naske and Herman E. Slotnick, *Alaska: A History*, 3rd ed. (Norman: University of Oklahoma Press, 2011), 288.

7 Ernest Gruening, *Many Battles: The Autobiography of Ernest Gruening* (New York: Liveright, 1973), 496.

8 Gruening, *Many. Battles*, 497.
9 Dunaway, *Defending the Arctic Refuge*, 38–40.
10 Personal interview with author with Alfred "Al" Wright, August 3, 2022, at his home in Fairbanks, Alaska.
11 Lawrence E. Davies, "Controversy Rages over Plans for Big Alaska Power Project, *The New York Times*, August 22, 1964, 25.
12 Ibid.
13 University of Michigan School of Natural Resources, *Rampart Dam and the Economic Development of Alaska* (Ann Arbor, MI: Rampart Dam-Alaska Economic Development Project, March 1966), 51.
14 University of Michigan School of Natural Resources, *Rampart Dam and the Economic Development of Alaska*, 53.
15 University of Michigan School of Natural Resources, *Rampart Dam and the Economic Development of Alaska*, 51.
16 Maria Shaa Tlaa Williams, "A Brief History of Native Solidarity," in *The Alaska Native Reader: History, Culture, Politics*, ed. Maria Shaa Tlaa Williams (Durham, NC: Duke University Press, 2009), 202–16, 210.
17 Tuzzy Consortium Library, "Tundra Times Photographic Record: Howard Rock." *Tuzzy Library*. https://tuzzy.org/TundraTimes/HowardRock, accessed January 31, 2022.
18 Lael Morgan, *Art and Eskimo Power: The Life and Times of Alaskan Howard Rock* (Fairbanks, AK: Epicenter Press, 1988).
19 Fred Paul, *Then Fight For It: The Largest Peaceful Redistribution of Wealth in the History of Mankind* (Trafford Publishing, 2007), 134–35.
20 Karl Francis, "Fear and Anger in Northwest Alaska," *Tundra Times* 31, no. 1 (October 15, 1992).
21 Maria Shaa Tlaa Williams, "A Brief History of Native Solidarity," 202–16, 210.
22 Dan O'Neill, *The Firecracker Boys: H-Bombs, Iñupiat Eskimos, and the Roots of the Environmental Movements* (New York: Basic Books, 1994).
23 *Anchorage Times*, June 19, 1961, p. 4. From: Ronald Lautaret, *Alaskan Historical Documents since 1867* (Jefferson: McFarland & Company, Inc., 1989), 133–34.
24 Arthur M. Piper, *Potential Effects of Project Chariot on Local Water Supplies*, https://pubs.usgs.gov/pp/0539/report.pdf
25 Piper, *Potential Effects of Project Chariot on Local Water Supplies*, 34.
26 Charles Etok Edwardsen Jr. "Etok" Opinion piece, "EM Radiation Test Opposed," *Tundra Times* XXXIII, no. 3 (December 1, 1993).
27 R.G. Fuller, J.B. Kirkwood, "Ecological Consequences of Nuclear Testing," in *The Environment of Amchitka Island, Alaska*, ed. Melvin L. Merritt et al. (Washington DC: Technical Information Center Energy Research and Development Administration, 1977), 646–47.
28 An image of Lisbourne's artistic depiction of death at Cape Thompson may be found in the Anchorage Museum.
29 Ed. Karen Brewster, *The Whales, They Give Themselves: Conversations with Harry Brower, Sr.* (Fairbanks: University of Alaska Press, 2004), 40.
30 Lautaret, *Alaskan Historical Documents since 1867*, 135–42.
31 Lautaret, *Alaskan Historical Documents since 1867*, 136.
32 Susan Hackley Johnson, *The Pribilof Islands: A Guide to St. Paul, Alaska* (Saint Paul Island, Alaska: Tanadgusix Corporation, 1978), 23.
33 Barbara Boyle Torrey, *Slaves of the Harvest* (Saint Paul Island, Alaska: TDX Corporation, 1978), 155–61.

34 Torrey, *Slaves of the Harvest*, 163–64.

35 Dunaway, *Defending the Arctic Refuge*, 140.

36 Bathsheba Demuth, "The Walrus and the Bureaucrat: Energy, Ecology, and Making the State in the Russian and American Arctic, 1870–1950," *The American Historical Review* 124, no. 2 (April 2019): 483–510.

37 Davis Hovey, "Native artisans worry ivory bans in other states could reverberate in Alaska," October 27, 2016, KNOM-Nome, https://www.ktoo.org/2016/10/27/native-artisans-worry-ivory-bans-states-reverberate-alaska/

38 Susie Silook, "The Anti-Depression *Uliimaaq*," *Alaska Native Writers, Storytellers & Orators: The Expanded Edition*, ed. Ronald Spatz, contributing editors Jeane Breinig, Patricia H. Partnow (Alaska Quarterly Review 2001), 245–48.

39 "Etsy Suspending Alaska Native Accounts for Ivory Works, SHI President Responds," February 7, 2018. https://www.kinyradio.com/news/news-of-the-north/etsy-suspending-alaska-native-accounts-for-ivory-works-shi-president-responds/

40 Dunaway, *Defending the Arctic Refuge*, 5–6.

41 Peter G. Wells, James N. Butler, and Jane Staveley Hughes (eds.), *Exxon Valdez Oil Spill: Fate and Effects in Alaskan Waters* (West Conshohocken, PA: American Society for Testing and Materials, 1995). Scientists studied the Prince William Sound and the Northern Gulf of Alaska.

42 Steve Pilkington, September 11, 1989, *Tundra Times* XXVI.

43 Margaret Bauman, "PWS Natives Still Fear Oil Contaminated Food," February 15, 1995, *Tundra Times* 34. Congress passed the 1971 Alaska Native Claims Settlement Act (ANCSA) establishing Alaska Native Corporations during the termination era. On ANCSA and the creation of Alaska Native Corporations see, Eve Tuck, "ANCSA as X-Mark: Surface and Subsurface Claims of the Alaska Native Claims Settlement Act" in *Transforming the University: Alaska Native Studies in the 21st Century*, ed. Beth Gigondidoy Leonard et al. (Maitland, FL: Mill City Press, 2014), 240–72; Shari M. Huhndorf and Roy M. Huhndorf, "Alaska Native Politics since the Alaska Native Claims Settlement Act," *The South Atlantic Quarterly* 110, no. 2 (Spring 2011): 385–401.

44 Oral history interview with Mike Webber, in Sharon Bushnell and Stan Jones, *Personal Stories from The Spill: Exxon Valdez Disaster* (Kenmore, WA: Epicenter Press, 2009), 228.

45 Jen Rose Smith, "Reclaiming Native knowledge through kelp farming in Cordova, Alaska," June 11, 2021, *Vogue* magazine, date accessed June 17, 2022. https://www.vogue.com/article/reclaiming-native-knowledges-through-kelp-farming-in-cordova-alaska

46 "Defending the Arctic Refuge from Oil Drilling," *Care2 Petitions*, https://www.thepetitionsite.com/takeaction/570/633/747/?TAP=1007&cid=causes_petition_postinfo

47 Dunaway, *Defending the Arctic Refuge*, 138–40.

48 Subhankar Banerjee "Terra Incognita: Communities and Resource Wars," in *The Alaska Native Reader: History, Culture, Politics* (Durham, NC: Duke University Press, 2009), 187.

49 "Everybody will be a millionaire," project proposal personal website of Allison Warden, date accessed June 28, 2022. https://www.allisonwarden.com/everybody-will-be-a-millionaire.html

50 Eric Bork, "Nature and Indigenous Alaskan Art with Joe and Martha Senungetuk," *Alaska Public Media*, December 7, 2021. https://www.alaskapublic.org/2021/12/07/nature-and-indigenous-alaskan-art-with-joe-and-martha-senungetuk/

51 Melissa Shaginoff, "The Re-Collected Images of Joseph Senungetuk," *Press Reader* website, September 15, 2019. https://www.pressreader.com/canada/inuit-art-quarterly/20190915/281569472428395

52 Personal interview with author with Joe Senungetuk, April 5, 2022, SEED Lab Anchorage Museum. To watch excerpts of Joe's interview on YouTube check out: https://www.youtube.com/watch?v=3rQKaqL0z68

53 Senungetuk, *Give or Take a Century*, 155.

54 Joseph Senungetuk, *Give or Take a Century: An Eskimo Chronicle* (San Francisco, CA: The Indian Historian Press, 1971).

55 Nicole Greenfield, "Alaska Natives Lead a Unified Resistance towards Pebble Mine," April 20, 2021, *Natural Resources Defense Council* website, https://www.nrdc.org/stories/alaska-natives-lead-unified-resistance-pebble-mine

56 Nick Estes, *Our History is the Future: Standing Rock versus the Dakota Access Pipeline, and the long tradition of Indigenous resistance* (New York: Verso Books, 2019).

57 Bathsheba Demuth, *Floating Coast: An Environmental History of the Bering Strait* (New York, NY: W.W. Norton & Company, 2019), 316.

58 Owen Mason et al., *Living with the Coast of Alaska* (Durham, NC: Duke University Press, 1997), 97–114.

59 Mason et al., *Living with the Coast of* Alaska, 106.

60 "Leadership development," *Native Movement* website, https://www.nativemovement.org/leadership-development

61 "Our History," *Native Movement* website, https://www.nativemovement.org/our-history, date accessed June 17, 2022.

62 "Karlin Nageak Itchoak," *The Wilderness*, https://www.wilderness.org/about-us/karlin-nageak-itchoak, accessed June 17, 2022.

63 "Q&A: Building Resiliency Through Community-Based Conservation," with Crystal Nelson, February 6, 2019, https://www.nature.org/en-us/about-us/where-we-work/priority-landscapes/emerald-edge/emerald-edge-stories/crystal-nelson-community-development-in-southeast-alaska/, accessed July 6, 2022.

Bibliography

Banerjee, Subhankar. "Terra Incognita: Communities and Resource Wars." In *The Alaska Native Reader: History, Culture, Politics*, edited by Maria Sháa Tláa Williams. Durham, NC: Duke University Press, 2009.

Bauman, Margaret. "PWS Natives Still Fear Oil Contaminated Food." *Tundra Times*, February 15, 1995.

Bork, Eric. "Nature and Indigenous Alaskan Art with Joe and Martha Senungetuk." *Alaska Public Media*, December 7, 2021. https://www.alaskapublic.org/2021/12/07/nature-and-indigenous-alaskan-art-with-joe-and-martha-senungetuk/

Brewster, Karen. *The Whales, They Give Themselves: Conversations with Harry Brower, Sr.* Fairbanks: University of Alaska Press, 2004.

Bushnell, Sharon, and Stan Jones. *Personal Stories from the Spill: Exxon Valdez Disaster*. Kenmore, WA: Epicenter Press, 2009.

"Cape Thompson!" *Anchorage Museum*. Accessed January 6, 2023. https://www.anchoragemuseum.org/exhibits/unsettled/gallery/cape-thompson/

Davies, Lawrence E. "Controversy Rages Over Plans for Big Alaska Power Project." *The New York Times*. August 22, 1964.

"Defend the Arctic Refuge From Oil Drilling." *Alaska Wilderness League and Patagonia*. Accessed January 6, 2023. https://www.thepetitionsite.com/takeaction/570/633/747/?TAP=1007&cid=causes_petition_postnfo

Demuth, Bathsheba. *Floating Coast: An Environmental History of the Bering Strait*. New York: W.W. Norton & Company, 2019.

_____. "The Walrus and the Bureaucrat: Energy, Ecology, and Making the State in the Russian and American Arctic, 1870–1950." *The American Historical Review* 124, no. 2 (April 2019): 483–510.

Den Ouden, Amy E., and Jean M. O'Brien. *Recognition, Sovereignty Struggles, & Indigenous Rights in the United States*. Chapel Hill: The University of North Carolina Press, 2013.

Dunaway, Finis. *Defending the Arctic Refuge: A Photographer, an Indigenous Nation, and a Fight for Environmental Justice*. Chapel Hill: The University of North Carolina Press, 2021.

Edwardsen, Etok Jr. "Etok", Charles. "EM Radiation Test Opposed." *Tundra Times*, December 1, 1993.

Estes, Nick. *Our History Is the Future*. London: Verso Books, 2019.

"Etsy Suspending Alaska Native Accounts for Ivory Works, SHI President Responds." *KINY Radio*, February 7, 2018. https://www.kinyradio.com/news/news-of-the-north/etsy-suspending-alaska-native-accounts-for-ivory-works-shi-president-responds/

Francis, Karl. "Fear and Anger in Northwest Alaska." *Tundra Times*, October 15, 1992.

Fuller, Robert G., and James B. Kirkwood. "Ecological Consequences of Nuclear Testing." In *The Environment of Amchitka Island, Alaska*, edited by Melvin L. Merritt et al., Technical Information Center Energy Research and Development Administration, 1977.

Greenfield, Nicole. "Alaska Natives Lead a Unified Resistance to the Pebble Mine." *The Natural Resources Defense Council*. April 20, 2021. https://www.nrdc.org/stories/alaska-natives-lead-unified-resistance-pebble-mine

Gruening, Ernest. *Many Battles: The Autobiography of Ernest Gruening*. New York: Liveright, 1973.

Hackley Johnson, Susan. *The Pribilof Islands: A Guide to St. Paul, Alaska*. Anchorage, AK: Tanadgusix Corporation, 1978.

Hovey, Davis. "Native Artisans Worry Ivory Bans in Other States Could Reverberate in Alaska." *KNOM-Nome*, October 27, 2016. https://www.ktoo.org/2016/10/27/native-artisans-worry-ivory-bans-states-reverberate-alaska/

Huhndorf, Shari M., and Roy M. Huhndorf. "Alaska Native Politics Since the Alaska Native Claims Settlement Act." *The South Atlantic Quarterly* 110, no. 2 (Spring 2011): 385–401.

Itchoak, Karlin Nageak, *The Wilderness Society*, Accessed January 6, 2023. https://www.wilderness.org/about-us/karlin-nageak-itchoak

Lautaret, Ronald. *Alaskan Historical Documents since 1867*. Jefferson, MO: McFarland & Company, Inc., 1989.

Lee, Lloyd. *Navajo Sovereignty: Understandings and Visions of Dine People.* Tucson: The University of Arizona Press, 2017.

Mason, Owen, William J. Neal, Orrin H. Pilkey, Jane Bullock, Ted Fathauer, Deborah Pilkey, and Douglas Swanston. *Living with the Coast of Alaska.* Durham, NC: Duke University Press, 1997.

Morgan, Lael. *Art and Eskimo Power: The Life and Times of Alaskan Howard Rock.* Fairbanks, AK: Epicenter Press, 1988.

Naske, Claus-M., and Herman E. Slotnick, *Alaska: A History.* Norman: University of Oklahoma Press, 2011.

Native Movement. Accessed January 6, 2022. https://www.nativemovement.org/leadership-development

Official Alaska State Website. "Alaska Native Language Preservation & Advisory Council." Accessed January 6, 2022. https://www.commerce.alaska.gov/web/dcra/AKNativeLanguagePreservationAdvisoryCouncil/Languages.aspx

O'Neill, Dan. *The Firecracker Boys: H-Bombs, Iñupiat Eskimos, and the Roots of the Environmental Movements.* New York: Basic Books, 1994.

"Our History," *Native Movement.* Accessed January 6, 2023. https://www.nativemovement.org/our-history

Paul, Fred. *Then Fight For It: The Largest Peaceful Redistribution of Wealth in the History of Mankind.* Bloomington, IN: Trafford Publishing, 2007.

Pilkington, Steve. "Cleanup Boom Threatens Village Sobriety Efforts." *Tundra Times* XXVI, September 11, 1989. Accessed July 12, 2022.

Piper, Arthur M. *Potential Effects of Project Chariot on Local Water Supplies.* Washington, DC: United State Government Printing Office, 1966. Accessed January 6, 2023. https://pubs.usgs.gov/pp/0539/report.pdf

"Q&A: Building Resiliency Through Community-Based Conservation." *The Nature Conservancy*, February 6, 2019. https://www.nature.org/en-us/about-us/where-we-work/priority-landscapes/emerald-edge/emerald-edge-stories/crystal-nelson-community-development-in-southeast-alaska/

Senungetuk, Joe. Interview by Holly Miowak Guise. April 5, 2022. SEED Lab, Anchorage Museum, Alaska.

Senungetuk, Joseph. *Give or Take a Century: An Eskimo Chronicle.* San Francisco, CA: The Indian Historian Press, 1971.

Shaginoff, Melissa. "The Re-Collected Images of Joseph Senungetuk." *Inuit Art*, September 15, 2019. https://www.pressreader.com/canada/inuit-art-quarterly/20190915/281569472428395

Silook, Susie, "The Anti-Depression *Uliimaaq.*" In *Alaska Native Writers, Storytellers & Orators: The Expanded Edition*, edited by Ronald Spatz, Jeane Breinig, Patricia H. Partnow, 245–48. Alaska Quarterly Review, 2001. Fairbanks, Alaska.

Smith, Jen Rose. "Reclaiming Native Knowledges through Kelp Farming in Cordova, Alaska." *Vogue*, June 11, 2021. https://www.vogue.com/article/reclaiming-native-knowledges-through-kelp-farming-in-cordova-alaska

The Red Nation. *The Red Deal: Indigenous Action to Save Our Earth.* Brooklyn, NY: Common Notions, 2021.

Torrey, Barbara Boyle. *Slaves of the Harvest.* Anchorage, AK: TDX Corporation, 1978.

Tuck, Eve. "ANCSA as X-Mark: Surface and Subsurface Claims of the Alaska Native Claims Settlement Act." In *Transforming the University: Alaska Native Studies in the 21st Century*, edited by Beth Gigondidoy Leonard et. al. Maitland, FL: City Press, 2014.

Tuzzy Library Database. "Tundra Times Photographic Record: Howard Rock." Accessed January 31, 2022. https://tuzzy.org/TundraTimes/HowardRock

University of Michigan School of Natural Resources, *Rampart Dam and the Economic Development of Alaska*. Ann Arbor, MI: Rampart Dam- Alaska Economic Development Project, March 1966.

Warden, Allison Akootchook. "Everybody Will Be a Millionaire!" *Allison Akootchook Warden*. Accessed January 6, 2023. https://www.allisonwarden.com/everybody-will-be-a-millionaire.html

Wells, Peter G., James N. Butler, and Jane Staveley Hughes. *Exxon Valdez Oil Spill: Fate and Effects in Alaskan Waters*. West Conshohocken, PA: American Society for Testing and Materials, 1995.

Wilkins, David, and Heidi Kiiwetinepinesiik Stark. *American Indian Politics and the American Political System*. 4th ed. Rowman & Littlefield, 2001, 2018. Lanham, MD.

Williams, Maria Sháa Tláa. "A Brief History of Native Solidarity." In *The Alaska Native Reader: History, Culture, Politics*, edited by Maria Sháa Tláa Williams, 202–16. Durham, NC: Duke University Press, 2009.

Wright, Alfred "Al." Interview by Holly Miowak Guise. August 3, 2022. Fairbanks, Alaska.

6 Indigenous Peoples in Canada and Beyond

The Inuit Circumpolar Council's Climate Change Advocacy Work

Lydia Schoeppner

Introduction: Climate Change and Indigenous Peoples in Canada

Section 35 of the Canadian Constitution Act of 1982 recognizes three different and distinct Indigenous groups: First Nations, Metis, and Inuit. Indigenous Peoples' health and well-being in Canada is impacted by a history of colonialism, discrimination, and systemic racism which resulted in forced displacement, the disruption and loss of Indigenous knowledges and languages, (intergenerational) trauma, and death.[1] Compared to its non-Indigenous population, Canada's 1.8 million Indigenous Peoples (5% of the population) are uniquely affected and more greatly impacted by climate change,[2] because of existing socio-economic and health-related disparities like shorter life spans, greater risk of food and water insecurity, less access to quality housing, poorer transportation systems, unstable infrastructure and buildings, costs for purchasing foods.[3] A 2022 report on *Health of Canadians in a Changing Climate* states:

> The changing climate will exacerbate the health and socio-economic inequities already experienced by First Nations, Inuit, and Métis peoples, including respiratory, cardiovascular, water- and food-borne, chronic and infectious diseases, as well as financial hardship and food insecurity. Natural hazards, coupled with unpredictable and extreme weather events, can result in temporary or long-term evacuations from traditional territories, in addition to greater risk of injury and death from accidents while out on the land. Infrastructure damage or instability due to climate change, particularly in Northern and remote locations, may restrict access to health systems and supplies. [...] Cross-cutting climate impacts will disrupt the livelihoods of First Nations, Inuit, and Métis peoples, families and communities, affecting their sense of identity and cultural continuity and compounding existing mental health issues. Indigenous knowledge systems and practices are key to First Nations, Inuit, and

DOI: 10.4324/9781003214380-9

Métis peoples' ability to observe, respond, and adapt to climate and environmental changes.[4]

The report also recognizes Indigenous knowledge systems and practices as equal to non-Indigenous science, and as increasingly recognized in climate change adaptation strategies. Furthermore, it recommends that Indigenous Peoples' "rights and responsibilities over their lands, natural resources, and ways of life" need to be "respected, protected and advanced through distinctions-based, Indigenous-led, climate change adaptation, policy, and research." It cautions that climate change threatens Indigenous Peoples' "ways of life, resilience, cultural cohesion, and opportunities for the transmission of Indigenous knowledges and land skills, particularly among youth."[5] Thus, for Indigenous Peoples, "environmental justice is not just about environmental destruction, but it is an aspect of colonization."[6]

Environmental (in-) Justice in Canada

The US Environmental Protection Agency defines environmental justice as "the fair treatment and meaningful involvement of all people regardless of race, color, national origin, or income with respect to the development, implementation and enforcement of environmental laws, regulations and policies. Fair treatment means no group of people should bear a disproportionate share of the negative environmental consequences resulting from industrial, governmental and commercial operations or policies." Meaningful involvement means inclusion of people in decision-making that might impact their health and environment, and that decision-making processes consider community concerns, connecting with the potentially affected and ensuring their involvement.[7] Agyeman et al. defined environmental injustice as disproportionate "exposure to the risks and externalities [...] borne by poor and racialized peoples and communities," while environmental well-being "means access to aesthetic and healthy environments rather than just the avoidance of deprivation or harm."[8] More comprehensively, environmental justice issues do involve not only documentable technical data and scientific assumptions about the world, but also fairness of treatment and the participatory ability of all marginalized peoples to be able to make substantive qualitative changes to the impositions of the larger society, especially those that adversely affect their rights and freedoms. [...] Environmental matters and justice are to a large extent about who gets to ask the questions, who gets to be heard (and listened to), and who benefits from how and if the questions are answered, researched, or considered relevant.[9]

Both sources emphasize the importance of fair treatment and inclusion in decision-making processes as the main elements of environmental justice.

Environmental injustice, then, can be identified through their absence. Others have pointed out that environmental injustice can be understood as environmental inequality where some experience higher environmental burdens and have unequal access to environmental resources.[10]

Environmental justice can be further nuanced. The term distributional justice describes the "demographic and geographic distribution of harmful industrial by-products and beneficial environmental goods."[11] Thus, climate change creates conditions of distributive injustice for Inuit, who experience the most severe effects of climate change, while industrial production sites are distant. Distributional justice with regard to climate change is a global issue that necessitates collective problem-solving and also a closer look at productive environmental justice that focuses on the "initial production of harm."[12] Procedural justice, on the other hand, focuses on the unequal participation in decision-making processes, illustrating how democratic and inclusive environmental justice mechanisms are. Social justice concentrates on recognition by looking at who is heard and on what concerns, what is power and who has power, and in what contexts the powerful operate. Environmental justice combines all these forms of justice (distributional, procedural, social) in one concept.[13]

Climate change "reflects and increases social inequality," with Indigenous Peoples facing the most severe (disproportionate) effects (which can be termed a result of "industrial, governmental and commercial operations") while meaningful involvement in climate change policy-making is lacking. Thus, global climate change is an example of environmental injustice for Indigenous Peoples.[14] In Canada and globally, Indigenous People contribute the least to the causes of climate change but find themselves at the forefront of its most severe impacts. Although the term environmental justice has not been commonly used by Indigenous groups in Canada, environmental justice movements have existed on this land for hundreds, if not thousands of years, as Indigenous Peoples have been stewards of the environment since time immemorial.[15] Importantly, for Indigenous Peoples, environmental protection goes beyond Western-based conceptions of power-relationships among and between peoples and existing colonial institutions. Rather, environmental justice is understood more broadly and holistically, and it consists of relations "among all beings of Creation."[16] Writing from a First Nations perspective, Deborah McGregor notes:

> Environmental justice includes our relationships with each other, including all plants and animals, the sun, the stars, the Creator, and so on. It is necessary to move beyond the human-centered approach to one of understanding, accepting, enacting, respecting, and honouring relationships with all of Creation. [...] Because of their intimate relationship with the land, any injustice to Aboriginal people is an

environmental injustice to the extent that it impairs the ability of Aboriginal people to fulfill their responsibilities to Creation. Conversely, any injustice to the environment that impedes the ability of Creation to fulfil its duties to Aboriginal people is an injustice to Aboriginal people. Of course, this is true of all people: we cannot survive without an environment that fulfils our needs for survival.[17]

All three Indigenous groups in Canada have noted that climate change threatens their connections with the land and environment. To realize environmental justice, First Nations, Metis, and Inuit leaders have called for urgent action to respect the environment, reduce pollution, and adapt to changes. The Pan-Canadian Framework (PCF) on Clean Growth and Climate Change was adopted by Canada's First Ministers in 2016. It contains over 50 action items intended to help Canada meet its Paris Agreement greenhouse gas emissions reduction target for 2030. To realize this target, the framework focuses on four aspects: "pricing carbon pollution," emissions reduction, resilience and adaptation, and "clean technology, innovation, and jobs."[18] The framework also involves three distinct bilateral tables for collaboration with the country's three Indigenous groups' national organizations: the Assembly of First Nations (AFN), the Metis Nation, and the Inuit Tapiriit Kanatami (ITK). These platforms were established to provide opportunities for senior federal officials and Indigenous representatives to discuss and cooperate on climate change action, based on recognition and partnership.[19] The PCF's ongoing work with Indigenous groups in Canada will be briefly described in the next three sections below.

First Nations

Over 1 million individuals in Canada identify as First Nations.[20] In 2019, the AFN – established as the First Nations' collective voice in 1982 – declared a climate emergency, recognizing climate change's impact on the waters, lands, peoples, and animals and committed to work locally, nationally, and internationally to maintain global warming below 1.5 degrees Celsius. It called for the protection of First Nations' rights and instructed the AFN to develop a First Nations climate strategy.[21] At its first national climate gathering in March 2020, the AFN emphasized the dangers stemming from climate change: dramatic rise in temperatures, "extinction of a million species," and overall "catastrophic consequences" if no action is taken. It also noticed an alarming warming trend in Canada by 2.3 degrees Celsius and that the country was "not projected to meet its 2030 targets," while a growing number of First Nations suggested "rapid decarbonization" to stop the speed of climate change. The AFN further pointed out interconnections between the climate crisis and First Nations' daily experiences, emphasizing climate change as an amplifier of colonization (e.g. in mental health and well-being, food and water insecurity, housing, poverty, erosion of culture, rights, and access to land) and that addressing the climate crisis also means working towards self-determination and reconciliation. Based on the intricate interconnections between the people and the land, the AFN recommended solutions grounded in First Nations' "law, knowledge, language, and governance" that are unique for each First Nation but simultaneously represent a commonly shared conception of "natural, spiritual, and environmental law." It further noted that First Nations' "traditional government and knowledge must stand equally with western systems." A "transformational shift" in Canada's and the world's approach is needed to tackle climate change, the AFN argued, and it supported the creation of an AFN National Climate Strategy.[22]

At the First Nations-Canada Joint Committee on Climate Action (JCCA), AFN representatives met with federal officials to discuss climate change approaches. The JCCA was created to help First Nations improve access to federal programmes, while promoting self-determination in "emerging clean-growth opportunities." The third and most recent report (2021) states that "First Nations leaders, organizations, and communities have reinforced the need for Canada to take an ambitious and holistic approach to reducing carbon pollution, adapting to the impacts of climate change, and improving the ways in which the natural environment is respected and protected."[23] The report suggests focusing on emissions reduction and income inequality.

Another noteworthy development in First Nations dealing with climate change started in the Yukon Territory in 2020, when First Nations Youth held their first-ever Climate Action Gathering in Whitehorse. Leaders, knowledge-keepers, community members, and youth from northern British Columbia (BC), the Yukon, and the Northwest Territories gathered to address climate change through "talking circles, collaborative art, presentations, song, and ceremony." The historic gathering culminated in the signing of the Climate Change Emergency Declaration which endorsed the development of a youth-led Yukon First Nations Climate Vision and Action Plan. One result of this was the creation of a 20-month Yukon First Nations Climate Action Fellowship for 14 Yukon and northern BC youth to work with Yukon First Nations on the creation of a Climate Vision and Action Plan to guide governments, industry, and communities in their responses to climate change "with spirit and actions that reflect a Yukon First Nations worldview."[24] Acknowledging the need to engage with climate change as "whole people," the fellows' approach included all four quadrants of the medicine wheel: in addition to the mental and physical approaches (that tend to focus on carbon dioxide and physical impacts of climate change), spiritual and emotional approaches also need to receive attention. The fellows argue that dominant approaches aim at treating the symptoms, while neglecting the root causes of climate change, pointing out that "quick fixes" like solar panels and e-cars are insufficient. Instead, climate action happens when humans re-connect with spirit, self, each other, and earth – through practicing, experiencing, and learning. This involves learning language, storytelling, song, art, humour, ceremony, and the harvesting of medicines.[25]

Metis

Over 624,000 Canadians identify as Metis, a distinct people that emerged in the late 18th century as the offspring of relations between local Indigenous individuals and immigrant fur traders.[26] Metis traditional ways of life (e.g. fishing, harvesting, hunting) were and continue to be negatively impacted by climate change, and Metis women wrote about climate change causing "ecogrief, ecoparalysis, solastalgia (existential distress caused by climate change) and eco-anxiety."[27] Since 2017, the Metis Nation-Canada Joint Table on Clean Growth and Climate Change has been meeting to enhance relationship-building, information-sharing and collaboration in policy-making. The Metis Nation identified the following priorities for the collaborative project: capacity building; collecting of Metis traditional knowledge; need for research and data collection for Metis policy; creating training and education opportunities on climate change; finding solutions to climate change that center around environmental stewardship and nature; issues related to "emergency management and disaster-risk mitigation," health, transportation, renewable energy, and energy-efficient renovations.[28]

A 2020 report prepared for the Metis National Council notes the following climate change risks for Metis in Canada: forest fires, flooding, extreme heat and drought, vector-borne diseases, invasive species, glacial retreat, sea-level rise, ocean acidification, hypoxia, and it notes the need for more information on climate change impacts on the health of Metis. The Metis National Council is instructed to work with researchers and federal agencies at all levels to create a "national Metis and climate change data collection strategy and research agenda."[29] The report recommends improvements in identifying Metis realities in data sets, and to develop Metis health and well-being indicators; establishing a fund to help Metis develop adaptation and mitigation plans (including youth training, food and traditional medicine programmes, support for knowledge translation from Elders, and working with provincial and federal agencies to report and monitor health risks); local research on climate change; targeted funding for Metis students and research that includes Metis knowledge; creating an Expert Group on climate change, a National Metis Youth Advisory Council on Climate Change, and also to develop the position of a Climate Change and Health Specialist to coordinate Metis climate change work and to function as a liaison in the PCF. The report recommends creating an intersectional and inclusive national Metis climate change strategy, similar to the ITK's Climate Change Strategy.[30]

Inuit

About 70,500 individuals in Canada identify as Inuit, the smallest of the three Indigenous groups.[31] However, their homeland – Inuit Nunangat, located in the circumpolar north – covers about 35% of Canada's land-mass and half of the country's coastline,[32] and this area experiences the most extreme impacts of climate change. The ITK was created in 1971 as the national voice of Inuit in Canada. In 2016, it published the "Inuit Priorities for Canada's Climate Strategy—A Canadian Inuit Vision For Our Common Future in Our Homelands" to provide the government with recommendations, for example inclusion of the Inuit voice in climate policies, assistance for Inuit food security, long-term funding for Inuit-led climate change research, lowering of carbon emissions and dealing with infrastructure deficits that are worsened by climate change, ensuring Inuit free, prior, and informed consent for all economic development projects. The vision paper also outlined how climate change serves to exacerbate existing social risks (to food security, the built environment, physical and mental health, travel safety, family economics/self-sufficiency, culture/traditional knowledge, and education).

The Inuit-Canada Table on Clean Growth and Climate Change was created in 2017 to provide a forum for representatives from ITK, Regional Land Claims Organizations, and federal officials to discuss and advance joint climate priorities. Since then, the ITK has shifted its focus to its National Inuit Climate Change Strategy (NICCS) to help advance Inuit-determined actions to strengthen the sustainability and resilience of Inuit Nunangat in the face of a rapidly changing climate and landscape. The NICCS was published in 2019, recognizing the need to develop Inuit regional strategies while also engaging in international climate change policy-making.[33] The Canadian government provided $1 million to the strategy in the same year to help "advance (...) Inuit-determined actions to strengthen the sustainability and resilience of Inuit Nunangat in the face of a rapidly changing climate and landscape." The money was meant to help advance Inuit agency in decision-making, food security, wellness and environmental health; building of climate-resilient infrastructure, renovation, adaptation strategies, and support for "Inuit energy independence."[34]

This brief overview of the more recent experiences of Canada's Indigenous Peoples with climate change illustrates the importance of Indigenous non-governmental advocacy organizations to make their voice heard by decision-makers. Given that the Arctic is currently warming about four times faster than the rest of the world (in some areas up to seven times faster),[35] the north is not just a "hot spot" for climate change, but also the "health barometer" of the planet. For this reason, the rest of this chapter focuses on the Inuit's concept of environmental justice and its application in the face of climate change conflicts through the Inuit Circumpolar Council – in Canada and beyond.

The Creation of the Inuit Circumpolar Council

As a result of colonial demarcation in the Arctic, Inuit Nunaat today spans across four countries: Canada, the US (Alaska), Greenland (politically associated with Denmark), and Russia (Chukotka). The Inuit Circumpolar Council (ICC) – an international Indigenous Peoples organization – represents Inuit from all four nation states. The ICC's creation in the 1970s was a result of several cumulative developments. The totalizing impacts and goals of colonial endeavours in the north forced Inuit to assimilate into a Western-type, imported, mainstream society, resulting in loss of culture, identity, and even life.[36] These experiences of rapid change, including an increasing loss of language, values, knowledge, and customs, generated a sense of urgency to develop an Inuit advocacy organization. The 1960s and 1970s mark a time when Inuit found themselves in dependency on external forces holding power, but it is also a period when the first generation of Inuit children who attended day- or residential schools came to maturity, and "dependency turned to a constructive defiance."[37] It was a time when the international decolonization movement of Indigenous Peoples gained momentum and led to political empowerment, when Inuit stood up against their oppressors to "administer their own affairs."[38] With the "Generation 68 [...] a new form of Inuit diplomacy, domestically and internationally, was developed."[39]

Another strong momentum came from Alaskan Inuit who recently had gone through complex and challenging processes of establishing self-empowerment structures. This had become necessary in light of the discovery of a massive oil reserve containing about 25 billion barrels of oil in Prudhoe Bay, Alaska, in 1968, that turned out to be the largest conventional oil field in the US and one of the 20 largest oil fields worldwide.[40] The beginning of large-scale oil exploration "was the very dramatic development which prompted the Inuit to see the need to cooperate internationally in order to counteract the presence of the global oil industry."[41] In Alaska, the pressure for oil extraction in Prudhoe Bay (and the planned construction of infrastructure to transport the oil) led to land claims discussions, and Alaskan Inuit successfully realized important milestones towards increased regional self-determination through the creation of the Alaska Native Claims Settlement Act in 1971 and the North Slope Borough in 1972.[42]

As a result of these concerns, the first Inuit Circumpolar Conference (ICC), a pan-Arctic Inuit gathering, took place in June 1977 in Barrow, Alaska. The transnational, consensus-based, non-profit, non-governmental, Indigenous Peoples' organization recognizes the shared homeland and culture of Inuit across national borders. It "discuss[es], represent[s], lobby[s], and protect[s Inuit] interests on the international level,"[43] based on representation from the four national chapters in Alaska, Canada, Greenland, and Russia (since 1989).[44] The quadrennial General Assemblies (Gas) not only offer opportunities to gather, network, exchange information, celebrate Inuit culture, and discuss and release a new declaration on the organization's positions and achievement but also inform the ICC's future work.[45] The ICC – renamed Inuit Circumpolar Council in 2006 – is an important international NGO representing about 180,000 Inuit.[46]

Inuit Connections with the Environment and Inuit Conflict Resolution

Indigenous and non-Indigenous ontologies and epistemologies differ in many ways, and this includes conceptions of the environment. Thus, it is important to approach "environment" and "justice" from an Inuit-informed perspective. "Environmentalism" is a newer term to describe an old practice of "caring for and about the world we live in."[47] Environmental justice has been practiced by Inuit without explicitly labelling it as such.

The discovery of oil and related environmental concerns caused a heightened urgency for environmental protection among Inuit in the 1960s. However, Inuit have always been strong advocates for respecting and protecting the environment, which, in turn, provides resources essential for survival. The land is even part of the creation process: according to Inuit beliefs, a tiny part of the land helps shape the foetus in a mother's womb.[48] The late Mario Aupilaarjuk shared the following about interconnections between the land and Inuit:

> The living person and the land are actually tied up together because without one the other doesn't survive and vice versa. You have to protect the land in order to receive from the land. If you start mistreating the land, then it won't support you. [...] In order to survive from the land, you have to protect it. The land is so important for us to survive and live on; that's why we treat it as part of ourselves.[49]

Interactions between humans and the land are mutually beneficial: if the land is respected (and, thus, protected), it will provide for humans who, in turn, will continue protecting the land. Treating the land respectfully means, for example, having good thoughts, keeping campsites clean, and giving gifts to the land. On the other hand, disrespectful treatment of the land could have catastrophic consequences (e.g. bad weather or lack of hunting success).[50] Guidelines for environmental protection are provided by Inuit Qaujimajatuqangit (IQ), a "holistic knowledge concept [...] specific for the Inuit, with emphasis on both social and environmental aspects and most importantly human-environment interrelations, including values on appropriate behaviour and good governance."[51] One of IQ's basic laws of relationship, *maligait*,[52] instructs Inuit to respect all living things and to continually plan and prepare for a better future.[53] One of the communal laws or IQ principles (*Inuit piqujangit*) reminds Inuit of their responsibility for environmental stewardship (*avatimik kamattiarniq*), which means respecting the environment, animals, and the land.[54]

The ICC's climate change work is often informed and shaped by Inuit pre-contact conflict resolution mechanisms that help create mutually beneficial alliances and exchanges[55]: collaborating on a long-term basis towards a common goal (collective persistence), a close analysis of power and resulting best strategic actions (political realism), recognizing and responding quickly to changed circumstances (adaptability), and the use of indirect criticism and reluctance to corner an opponent (avoiding win-lose confrontations).[56]

Climate change is a global problem with regard to its causes and effects. Due to the severity and omnipresence of climate change impacts in the Arctic and in Inuit lives, addressing environmental and ecological threats caused by climate change constitutes one of the focal points in the ICC's international work.[57] The strong connections of Inuit with their land and environment are severely challenged by the primary and secondary effects of anthropogenic climate change, constituting an essential threat to Inuit survival, threatening Inuit economies, culture, knowledge, food security, rights, health, and (mental) well-being. The following analysis is a subjective attempt to describe Inuit environmentalism as it is reflected through the ICC's climate change advocacy work.

Early Observations of Environmental Damage

In its early years, the ICC was mainly concerned about potential damages to the environment caused by resource extraction and oil spills. Thus, its declarations in 1977 and 1980 called for the protection and sustainable use of the environment, information exchange on harms caused by oil and other pollutants, and the provision of government funding for cleanup operations and compensation payments. Additionally, environmental impact assessments and Inuit inclusion in all decision-making processes are mentioned as important prerequisites for resource extraction projects.[58] At its third GA in Iqaluit in 1983, the ICC issued the Circumpolar Environmental Principles. In addition to the previously mentioned requirements, this document highlights the need for recognition and inclusion of Inuit expertise on the Arctic environment in scientific databases and management regulations for the environment. Moreover, the ICC called for international co-operation to properly research, protect, and manage the circumpolar environment, and it suggests oil spill response standards and support for alternative energy systems.[59]

At the next GA in 1986 in Kotzebue, the ICC adopted its Inuit Regional Conservation Strategy (IRCS) – a policy paper outlining a strategic approach to regional conservation and environmental development in the Arctic.[60] The 96-page document was the first conservation strategy focusing on a region, and also the first by an Indigenous People.[61] It was internationally praised for its "outstanding efforts to protect and enhance the Arctic environment,"[62] and it won the UN Environmental Programme's (UNEP) Global 500 Award in 1988 as "an example of the world conservation strategy in action."[63] In 1996, it received the Environmental Award of the Nordic Council of Ministers.[64] According to the IRCS, the Arctic is a "sink for atmospheric pollution"[65] by majority societies (in which Inuit are a minority). Although climate change was not directly mentioned as a threat, the document noted negative implications of non-renewable resource development and pollution that can damage the "habitats of harvested species" and could increase the "risk to other ecological processes."[66] At this point in time, pollution was understood as caused by "oil spills, [...] from tanker accidents, oil well blowouts, [...] discharges from vessels at sea, and accidental discharges from oil platforms, pipelines, terminals and storage facilities."[67]

The IRCS illustrated some early observations of environmental change in the Arctic – pieces of a larger picture that would gradually come together in the future. The document focused on environmental protection of the Arctic from harmful impacts originating externally[68] The document cautioned to develop "hydrocarbons[69] [...] with great care, because of the potentially destructive impact on harvested resources needed for subsistence."[70] Similarly, dangers stemming from bioaccumulation of mercury – a heavy metal – in the food chain were mentioned, but the document only acknowledged it's discharging in the air (dust) and water (tailing ponds) as a result of mining activities.[71] Later it would become known that melting permafrost also releases large volumes of mercury into the atmosphere. Mercury can enter the food chain and bio-accumulate in humans where it can cause serious health problems.[72] Furthermore, the IRCS recognized a growing threat by hydrocarbons, air pollution, and changes in climate that decimate animal numbers, and the need for sustainable development and environmental protection were clearly on the ICC's radar.

At the same GA in Kotzebue, the ICC acknowledged the environmental vulnerability of the Inuit lands, in relation to the transportation of resources like oil, gas, and minerals.[73] In resolution 17 on Arctic Haze,[74] the declaration mentioned the negative impacts of changes to the climate in the Arctic for the first time, explaining that Arctic air pollution was caused by the northward circulation of externally caused industrial pollution. It recognized the need for scientific research to learn more about the regional and global implications of pollution in the Arctic, recommending an "Arctic wide systematic sampling of air" and calling upon governments, science, and the industry to cooperate in "solving the general problem of air pollution."[75]

A Stronger Focus on Climate Change and Its Various Impacts (1989)

The ICC's GA in Sisimiut in July 1989 was overshadowed by the Exxon Valdez accident three months earlier. On its way from Alaska to California, the supertanker had hit a reef in Prince William Sound, Alaska, releasing 11 million gallons of black Prudhoe Bay crude oil into the water,[76] killing a large number of local marine-based animals.[77] For the ICC, this accident was a testament of increased shipping activity in Arctic waters – made possible by global warming that caused the melting of Arctic sea ice, which, in turn, allowed marine traffic to travel in and through the Arctic more easily and for longer periods of time. Thus, the ICC's 1989 declaration stated that "a global warming trend has begun and that the greatest impacts will be felt in the Arctic region." The oil spill is referenced twice, highlighting dangers stemming from shipping accidents, and noting shortcomings in safety assurances, governmental and industrial response failures, and their efforts to minimize the effects of the Exxon Valdez accident. Based on these realities, the ICC feared that similar events in the future would not be adequately dealt with, calling for new and adequate shipping and vessel construction standards, full consultation and participation in industrial resource exploitation endeavours in Arctic offshore waters, and requesting governments to provide "proof of safe operations, oil spill clean-up capability, and non-interference with wildlife resources and the lifestyle of the Inuit people." Furthermore, the declaration suggested Inuit receive more benefits from industrial resource exploitation, pointed out the need for a review and reform of existing response mechanisms and environmental protection frameworks as well as the need for legislation for fair compensation for loss of wildlife and habitat due to environmental disaster. With regard to the oil spill, the ICC requested full reports on cleanup efforts and short- and long-term environmental impacts.

Shaping International Mechanisms (1990–2001)

In 1992, the UN Conference on Environment and Development (UNCED) took place in Rio de Janeiro. At this conference, the Declaration on Environment and Development, and Agenda 21 were created, and the Convention on Biological Diversity (CBD) and the UN Framework Convention on Climate Change (UNFCCC) were developed. Since then, the nations of the world gather annually at the Conference of the Parties (COP) to discuss climate change. At COP 3 in 1997, the Kyoto Protocol was signed. This was the first international treaty for cutting greenhouse gas emissions that set binding emission reduction targets. The ICC participated at UNCED in 1992, and it has been active within UNFCCC since its inception, continuing to highlight the dangerous effects of climate change in the Arctic, and advocating to have the Arctic included in UNFCCC as a specific region affected by climate change.[78]

Another important milestone in the ICC's climate change work was the creation of the Arctic Council (AC) in 1996. Although Indigenous participants still remain non-voting partners, the Permanent Participant category is praised by many as a unique mechanism for Indigenous Peoples in international politics, because it provides stronger participatory rights than regular observer status, allowing for relationship-building with high-ranking politicians, direct negotiation with Arctic states, and tabling own proposals.[79] Indigenous Peoples' direct and substantial participation in Arctic environmental cooperation in the AC and its various Working Groups was especially important for the ICC – to reflect and accommodate their special connections with the region.[80]

Persistent Organic Pollutants (POPs) and the Stockholm Convention (1980s and 1990s)

In the late 1980s, more scientific knowledge became available about the enhancing function of climate change in the transport of persistent organic pollutants (POPs) and toxic particles and chemicals to the Arctic where they can accumulate and cause great harm.[81] For Inuit, these consequences are physically and emotionally challenging:

> The emotions we now feel – shock, panic, rage, grief, despair – as we discover that the food which for generations has nourished us and keeps us whole physically and spiritually is now poisoning us. […] We go out on the land to hunt, fish, trap and gather. The environment

is our supermarket. [...] As we put our babies to our breasts we feed them a noxious chemical cocktail that foreshadows neurological disorders, cancer, kidney failure, reproductive dysfunction, etc. This is truly worrying.[82]

In its 1989 declaration, the ICC asked governments to better protect the Arctic environment, to set and follow minimum international standards for the "use and dispersal of chemical toxins and contaminants," to conduct scientific studies on the impacts of chemical contaminants in the Arctic, and to formulate an international agreement between nation states to reduce and eliminate toxic contaminants in the Arctic. Also, the ICC called for more research on the food chain and on health problems of circumpolar residents caused by the "bioaccumulation of toxic materials," connecting climate change for the first time to the issue of physical health and food security.[83] In 1997, the ICC participated in official negotiations to help determine harmful chemicals that would be recognized in UNEP's protocol on POPs.

The ICC's 1998 declaration focused strongly on POPs, advising the Executive Council to treat circumpolar POP contamination as the highest priority, to advocate internationally for the elimination of POPs in the Arctic, and to communicate with regional and local Inuit organizations on related issues. Increased financial support is recommended for the AC's Arctic Monitoring and Assessment Programme (AMAP), as it focuses on measuring pollutants and the impact of climate change on Arctic ecosystems and human health. Furthermore, emitting countries should develop a global treaty to "eliminate POPs of concern" in the Arctic. The ICC is also instructed to cooperate with Indigenous groups and other NGOs and environmental organizations in the International POPs Elimination Network (IPEN) to jointly develop a global treaty and to educate key international organizations about the importance of "transboundary contaminants in the Arctic."[84]

The Stockholm Convention, adopted in 2001 and entered into force in 2004, is a legally binding document that aims at the elimination or restriction of the use and production of several POPs.[85] The ICC contributed to this global treaty through its membership in a national delegation.[86] During the first negotiations in Montreal, the ICC participated as a member of the Canadian Arctic Indigenous Peoples against POPs delegation (CAIPAP), with Sheila Watt-Cloutier as the group's spokesperson.[87] As ICC International Vice President, she simultaneously represented the interests of Inuit of Greenland, the USA, and Russia.[88] While CAIPAP usually operated from the back room, Watt-Cloutier – a "gifted public speaker able to convey technical information to a large audience (...) from the heart" – intervened at "strategically important moments."[89]

At the third session in Geneva, a conflict developed around the planned prohibition of dichloro-diphenyl-trichloroethane (DDT), a POP that was harmful in the eyes of Inuit, but that was also an important agent to control malaria in other areas of the world. The ICC "respond(ed) in a way that would be appropriate and helpful to the bigger picture:" Inuit would "not be party to any agreement that threatens others," suggesting that the convention should continue to aim at a complete elimination of POPs while simultaneously ensuring alternatives for DDT in the developing world.[90] This tactic is an example of Inuit pre-contact diplomacy creating win-win scenarios, co-operation, and emphasizing collective wisdom. Watt-Cloutier concluded that the ICC's ability to speak "reasonably, wisely, and professionally" and always from the "high moral ground"[91] and its preference for "politics of influence" over "politics of protest"[92] helped develop a successful partnership with the Canadian government and bridged a wide gulf between the developing and the developed world. The creation process of the Stockholm Convention serves as an example of the ICC's contribution to a global treaty through its work in a national delegation.[93]

The influence exerted by Indigenous Peoples in the Stockholm Convention negotiations was "out of proportion to their numbers"[94] and Inuit made a point of representing the voice of the land:

> Inuit are few in numbers and don't constitute a major lobby group. [...] But we are back now and wish to speak out on behalf of the land that has sustained us for hundreds of generations. We are the land and the land is us. We cannot stand by, waiting for slow moving governments to step in and make everything right, rather we must try to effect what change we can.[95]

Advocacy for Existing International Co-Operation and Frameworks (2002/2003)

In its 2002 declaration, the first in the new millennium, Canadian Inuk Sheila Watt-Cloutier, who had served as ICC Canada's president from 1995 to 2002, was elected as NGO's new International Chair, leading it into an era of intense, creative, and persistent climate change advocacy. The declaration strongly promoted the need to keep the Arctic environment safe from POPs, climate change, heavy metals, and unsustainable development by calling upon governments to enact national and "multi-lateral agreements to reduce or eliminate environmental damage and resulting human health problems in the Arctic." Governments are called upon to ratify the Kyoto Protocol, 1998, POPs Protocol to the UN-ECE-LRTAP, and the recent Stockholm Convention. Furthermore, the ICC is instructed to partner with governments and NGOs to "develop global initiatives to combat climate change in general, and an Arctic climate change program in particular" and to represent and protect Inuit interests in international for a like the World Summit on Sustainable Development (WSSD).[96]

The ICC has been active in all the key processes involving Indigenous Peoples in the UN.[97] It regularly participates at the Permanent Forum on Indigenous Issues' (UNPFII's) annual two-week sessions in New York, and (in cooperation with the Sami peoples) it coordinates a united voice for both Arctic Indigenous groups at UNFCCC's COP meetings.[98]

Climate Change as a Human Rights and Environmental Justice Issue (2005–2007)

Given that the circumpolar north is the "health barometer of the planet," Inuit understood their role as advocates for the health of the globe and its people.[99] To reach the hearts and minds of policymakers more effectively, it begun to focus more on the protection of the *people* in the environment. Watt-Cloutier, in cooperation with representatives from Earthjustice and the Center for International Environmental Law (CIEL), compiled a human rights petition arguing that the US failure of controlling greenhouse gas emissions damaged Inuit livelihoods and constituted a violation of Inuit social, economic, and cultural rights (which had already been recognized internationally). The ICC Canada supported the petition that was submitted on behalf of the Inuit in the US and Canada to the Inter-American Commission on Human Rights (IACHR) in 2005.[100] Importantly, the document was endorsed by 63 testimonials from Inuit hunters in Alaska and Canada who had shared their eyewitness accounts and expertise – the document's "lifeblood."[101]

The petition argued that climate change violated the right to "practice and enjoy their culture," and that we were being denied the right to use and enjoy our traditional lands, as the land was either changing or becoming inaccessible. The fact that we were unable to hunt as before for food and for hides and skins for clothing and that the loss of ice and snow was damaging our snow machines, our sleds, and our other tools was a violation of our right to personal property. The Western store-bought diet we were being forced to adopt, the accidents caused by melting ice and snow, and our increasing exposure to UV radiation [...] meant that our rights to health and life were being severely constrained.

[...] Our fundamental right to residence and movement was being violated as our homes were damaged and the land upon which many of our communities were built was being eroded by melting permafrost. [...] Inuit's fundamental right to their own means of subsistence was being denied as climate change was hurting almost every aspect of our hunting culture: the quantity and quality of wildlife, the length of the hunting season, methods of traveling and the ability of our elders to pass on traditional knowledge.[102]

Understanding environmental damage caused by climate change as a violation of human rights was a new development at this point, as governments had previously perceived climate change only as environmental and economic problems.[103] The petition was an example of Inuit seeking environmental/climate justice using legal avenues: it was the first "international legal action on climate change, [that] opened the door to the recognition of collective rights for Indigenous Peoples and firmly established the link between climate change and human rights within the mainstream global discourse."[104] Although the IACHR was ultimately unable to rule on the issue, the petition was a strategic and powerful way to make an Indigenous voice heard, exposing deficiencies in international law, opening up new space for courts as a means to enforce action, and portraying climate change as a national responsibility.[105] Its line of argumentation was soon picked up by others, for example in the Human Rights Council, and the Arctic Athabaskan Council submitted a similar petition to the IACHR in 2013.

The 2015 Paris Agreement was the first international climate change agreement that specifically acknowledged the connections between human rights and the environment. Two years later, the IACHR also recognized this connection, now permitting "cross-border human rights claims related to transboundary environmental damage," which was one of the reasons the Inuit petition was rejected.[106] In a historic move, the UN passed an adoption in July 2022 that declares access to a clean, healthy, and sustainable environment a universal human right with 161 votes in favour and eight abstentions. The UN paper states that the effects of climate change "interfere with the enjoyment of this right – and that environmental damage has negative implications, both direct and indirect, for the effective enjoyment of all human rights." Although not legally binding, the UN recognition is an important step towards holding governments accountable for Arctic environmental protection.[107] "Today, environmental justice and human rights movements are merging together as a global force for social change and democratization."[108]

Over a decade prior to these international developments, Watt-Cloutier already illustrated the petition's global benefits, describing it as

A means of *inviting* the United States to *talk with us* [...]. Our petition aims to *change hearts, minds*, and the climate change policy of the Government of the United States. [...] What we are doing is *for the world.* [...] Our petition is a *'gift'* from Inuit hunters and elders to the world. It is an *act of generosity* from an ancient culture deeply tied to the natural environment [...]. Let us work together to protect the Arctic so that we may save the planet.[109]

The accusation of the US at an international human rights body was expressed as a benevolent invitation, and the petition was described as a generous gift to the world – a win-win scenario for everyone. It is also noteworthy that US politics were criticized by a Canadian Inuk leader and the Canadian branch of the ICC, a face-saving strategy for Inuit in the US.[110] This is Inuit pre-contact diplomacy in action.

The Many Strong Voices Project and the Arctic Wisdom Event (2005) and Concern for Open Polar Waters and "Human-Induced Climate Change" (2006)

In 2005, the ICC co-created a new international forum, the Many Strong Voices (MSV) programme, as one of three NGOs representing Arctic Indigenous Peoples.[111] Developed at COP 11, the platform allowed for collaboration of non-government and government organizations to raise awareness about the impact of climate change in the Arctic and in Small Island Developing States (SIDS) in the Caribbean, South Pacific, and Indian Ocean.[112] MSV is a platform for Indigenous Peoples in both regions to share their climate change experiences, to highlight the connection between melting ice and rising sea levels, and to share best practices for relocation.[113] MSV helped draft passages for official UNFCCC documents, and through its observer status at the Intergovernmental Panel on Climate Change (IPCC), it supports efforts by its Arctic and SIDS members.

MSV's Portraits of Resilience – an exhibition of photographs and stories collected by youth in the Arctic and in the Seychelles, the Marshall Islands, Fiji, Kiribati, Tuvalu, and Samoa – toured various museums around the world, it was showcased at various COP events, and it was featured by UNEP.[114] MSV also held a global consultation with affected communities and peoples at COP 19 in Warsaw in 2013.[115] MSV is an example of Inuit political realism and adaptability, as it is a creative, innovative, and inclusive mechanism that focuses co-operation and relationship-building with Indigenous Peoples, nation states, and the UN.

In the same year, the ICC organized the "Arctic Wisdom" event in Iqaluit – another platform for sharing information about cultural, scientific, and political issues related to climate change in the Arctic. On Earth Day, about 1000 Inuit, joined by Hollywood celebrities Salma Hayek and Jake Gyllenhaal, stood together on the snow to form a "dramatic human aerial art image" communicating the message "Arctic Warning: Listen" and the outline of a drum dancer. The photo, taken from an airplane, raised public and media interest and it quickly drew the world's attention on the human face of climate change in the Arctic.[116]

The 2006 declaration is again deeply concerned about climate change impacts in the Arctic. Due to the "fragility of the Arctic's biological diversity," industrial activity and other developments are "a much greater concern than elsewhere."[117] The ICC is instructed to help eliminate POPs through regional and national participation in the development of "plans to implement the global Stockholm Convention on POPs."[118] For the first time, an ICC declaration used the phrase "human-induced climate change," and the NGO is becoming more cognizant of the impacts of climate change on Inuit culture, society, and health. Also, this declaration discussed for the first time the connection between climate change and resulting open polar waters, calling for the study of the environmental, socio-economic, and cultural impacts of the opening of the Arctic Ocean on Inuit communities.[119]

"No Longer in a Period of Climate Change but in Climate Crisis"[120] – The Indigenous Peoples' Global Summit on Climate Change and the Anchorage Declaration (2009)

To address the impact of climate change on Indigenous Peoples worldwide, the ICC organized and hosted the Indigenous Peoples' Global Summit on Climate Change in April 2009. This was the first climate change meeting that concentrated completely on Indigenous Peoples' realities.[121] Almost 500 Indigenous representatives met in Anchorage for discussion, information-sharing, relationship-building, and to create alliances for long-term cooperation. The event also allowed for networking between Indigenous and non-Indigenous participants (including UN agencies, the World Bank, NGOs, donors, and representatives from the private sector). The Anchorage Declaration – the summit's final declaration – reflects the global Indigenous Peoples' common standpoint on climate change and environmental justice. It stated alarmingly that "Mother Earth is no longer in a period of climate change but in climate crisis," impacting Indigenous Peoples' cultures, health and well-being, rights, food systems, infrastructure, and survival. The declaration demanded an "immediate end to the destruction and desecration of the elements of life," highlighting the crucial role of Indigenous Peoples in the defence and healing of Mother Earth. It upheld the importance of the United Nations Declaration on the Rights of Indigenous Peoples (UNDRIP) in all decision-making processes related to climate change and demanded the reflection of these principles in UNFCCC. Reiterating the "urgent need for collective action," Indigenous Peoples offered to share their innovations, practices, and Traditional Knowledge with regard to climate change if their "fundamental rights as intergenerational guardians of this knowledge are fully recognized and respected."[122]

The declaration's 14 calls for action encompassed recommendations for the upcoming COP 15 conference[123]: Nation states were asked to decrease their dependency on fossil fuels, to recognize Indigenous Peoples' knowledge and practices in developing approaches to deal with climate change, and to abandon "false solutions to climate change" (like nuclear energy or "clean coal") that have negative impacts on Indigenous Peoples' rights and environments.[124] It also directed states to respect and implement Indigenous Peoples' human rights (including collective rights, the right to mobility and protection from forced removal, and for voluntary isolation). Nation states were also urged to address the rights, conditions, and status of climate change migrants, and to "return and restore lands, territories, waters, forests, oceans, sea ice and sacred sites that have been taken from Indigenous Peoples."[125]

The summit's participants agreed to use the many existing international bodies and directed the IPCC, the Millennium Ecosystem Assessment, and other institutions to support Indigenous Peoples' climate change assessments. While Indigenous communities were encouraged to engage in information exchange, the declaration also called upon UNFCCC to provide more formal structures that include Indigenous participation.[126] Furthermore, the declaration urged that Indigenous "local mitigation and adaptation measures," be documented and made public, and it called for Indigenous Peoples' involvement in all climate change processes, including the promotion and implementation of UNDRIP, Indigenous engagement in UNFCCC, and capacity-building of Indigenous Peoples' climate change mitigation and adaptation and self-governance.[127] In December 2009, the summit's recommendations were presented to COP 15 in Copenhagen.[128]

The ICC's Regional Report

At the Anchorage summit, the ICC submitted a regional report, pointing out the specific climate-induced changes and challenges faced specifically by Inuit, supported by quotes from local hunters.[129] The report recommended finding a balance between "old and new ways" in the future[130]: that ways of dealing with climate change need to include teaching youth the necessary skills; cooperating with researchers, politicians, and business leaders; researching the impact of climate change; making research results available to decision-makers and translating results into Indigenous languages; learning about new economic and commercial opportunities and new hunting techniques; finding financial resources; and identifying barriers to adaptation.[131] Responses to climate change, the report suggested, need to include adaptation strategies while simultaneously preserving cultural elements, requiring a combination of scientific and experience-based knowledge. It stated that Inuit are open to new partnerships in the private and public sector and with civil society, but such partnerships need to include consultation, knowledge-sharing, and relational accountability.[132]

Highlighting the importance of education and awareness-raising, the report discussed the need for Inuit learning centres to help Inuit address climate change issues. Continued international co-operation and climate change impact avoidance were considered major benchmarks for a successful post-Kyoto process. Inuit inclusion and consultation in Arctic research and policy-making, long-term climate change policies, and Inuit self-determination as laid out in the ICC's Inuit Declaration on Arctic Sovereignty were considered important elements for dealing with climate change.[133] The ICC's regional report was a fine example of Inuit peacemaking as it reflected typical Inuit conflict resolution approaches like relationship-building, international co-operation, skills and knowledge acquisition, adaptation, and information-sharing to create a more inclusive and participatory environment and to prepare for the future.

Climate Change as a Food Security Issue (2010)

In the 2010 Nuuk declaration, climate change was listed as a threat to Inuit food security, urging the NGO to link Inuit food security to all its health- and climate change-related work.[134] In addition to understanding climate change as a violation of human rights, the ICC made an important point here, moving further towards a more holistic approach to tackle climate change. The document also discussed Inuit adaptability and pragmatism in the face of climate change, stating that "Inuit have a history of finding resources within their communities and elsewhere to adapt and meet challenges created by change."[135] The ICC was also instructed to find ways to adapt to the "new Arctic reality," for example by including Inuit communities in a suggested $20 billion International Climate Change Adaptation Fund.[136] Furthermore, the ICC was called to continue its regional, national, and international efforts to reduce emissions of contaminants that bio-accumulate in the Arctic and to "engage in activities that advance and strengthen the provisions of international instruments" like the Stockholm Convention, the International Agreement on Mercury Pollution (under negotiation at the time), and the 1972 Convention on the Prevention of Marine Pollution by Dumping of Wastes and Other Matter.[137]

The Inuit Arctic Policy (1983, 1992, 2010)

The Inuit Arctic Policy is a comprehensive document that addresses Inuit concerns to protect the northern environment.[138] The ICC released a total of three Arctic Policies in 1983, 1992, and 2010. Over the years, it has become an important position paper with recommendations for concrete action related to rights; peace; security; environment; social, cultural, and economic issues; education; and research. While not mentioning climate change, the first two papers took notice of noise, biological, and chemical pollution in the Arctic.[139] The 2010 Inuit Arctic Policy focused substantially on climate change and environmental protection. It helped shape the concept of environmental justice from a legal angle, stating the "right to a safe and healthy environment" as an "emerging human right," and that "Inuit have the right and responsibility to ensure the integrity of the Arctic environment and its resources, as a continuing source of life, livelihood and well-being for present and future generations."[140] It also suggested the creation of "laws and enforcement procedures" to protect the Arctic environment, and it supported research into the development of an Arctic Environmental Bill of Rights, including

- the right to a reasonable level of environmental quality based on Indigenous rights, individual and collective rights of Inuit to lands and resources, and the right to be compensated for damage to their culture and to the Arctic environment and its resources, upon which they depend;
- rights for individuals and peoples to appear in courts of law in order to prevent environmental offenses affecting the Arctic;
- rights of access to information on a timely basis;
- the right to participate in the setting of environmental standards by state governments or agencies [...];
- provision of legislative standards [...], so that the actions of agencies mandated to protect the Arctic environment may be reviewed by the courts.[141]

These elements indicate that the ICC understands environmental justice as distributional, procedural, and social justice, emphasizing the importance of financial compensation, participation, and recognition of Inuit as power-holders. Furthermore, the policy supported regional, national, and international punishment of crimes against the environment.[142]

The policy reflects Inuit pre-contact conflict resolution mechanisms of inclusivity, skill acquisition, knowledge-sharing, communication, relationship-building, caring for others, co-operation, and working towards a common goal. The ICC spelled out that the "health of the Arctic environment" was the foundation for Inuit livelihood, but also for the Earth's well-being. It made concrete recommendations for a post-2012 agreement to "stabilize greenhouse gas (GHG) concentrations" and it called for recognition of Inuit traditional knowledge by the UN, the IPCC, researchers, and decision-makers. Global leaders (especially of G20 countries) are urged to commit to significant funding and to "use an International Climate Change Adaptation Fund [...] to help Indigenous Peoples [...] adapt to the inevitable changes and to accelerate technology transfer."[143] Moreover, the document recommended that rules should be created "for international marine coordination in the Arctic on safety, emissions, and infrastructure," which should involve closely working with the United Nations Convention on the Law of the Sea (UNCLOS) and the International Maritime Organization (IMO), advocacy for the ratification of UNCLOS, and greater respect for existing international human, cultural, and Indigenous rights documents.[144]

The document also emphasized the need for a holistic and inclusive approach[145]; and that development projects should respect the natural environment and be preceded by impact assessment procedures based on Inuit free, prior, and informed consent. Furthermore, it recommended creating international monitoring agencies to measure national environmental protection performances, developing binding procedures to help resolve transnational environmental conflicts, and creating uniform Arctic-wide impact assessment procedures.[146] Initiatives to protect the Arctic environment were also recognized as economic opportunities for Inuit.[147]

The Artic Policy also emphasized the dangerous impacts of climate change on Arctic freshwater resources through acid snow and rain, POPs, and mercury pollution, and it highlighted Inuit collective and individual subsistence rights over Arctic land and marine areas, including renewable resources like hydro-power.[148] The ICC is instructed to lobby governments to enact national legislation and to implement and promote international agreements to limit or eliminate "harmful environmental damage and resulting human health problems in the Arctic." Sea ice and polynyas (areas of year-round open water that are important for marine life and Inuit culture and subsistence economy) should be protected "as a habitat and platform for marine mammals" and recognized as important to Inuit for harvesting, travelling, camping, and for recreational reasons.[149] Intensified Arctic marine traffic brings a multitude of resulting new challenges for Inuit, including vessel noise, potential oil spills and oil pollution, ship tracks, and their impact on marine mammal migration and Inuit subsistence economy. The policy also instructed the ICC to study the cultural, environmental, and socio-economic impacts of the opening of the Arctic Ocean, the Northwest Passage, and other waterways on Inuit communities. More research – with direct Inuit participation in all stages and including traditional Inuit knowledge – was also required on pollution caused by marine transportation.[150]

The policy paper also clarified that the right to health includes the "protection against external risks likely to endanger health," such as environmental health risks in the Arctic, and it warned of serious individual and collective impacts on Inuit should environmental contaminants not be better controlled or contained to prevent illnesses.[151] Moreover, Arctic nation states and other developed countries were called to end the release of toxic chemicals into the oceans[152] and updated safety features were recommended to stabilize airport runways threatened by melting permafrost.

Adaptation, and Recognition of Indigenous Peoples in COP (2014–2015)

The ICC's 2014 Declaration recognized the importance of food security, Inuit well-being, and environmental stewardship, and it mandated the ICC to continue highlighting the human impact of climate change on Inuit, to urge for international co-operation to mitigate climate change, to "develop adaptation strategies," and to promote future renewable resource use. The ICC is called to continue its existing work in the CBD, IUCN, and CITES, and it is instructed to strengthen international instruments like the Stockholm and Minamata Convention. Acknowledging access to clean water as a human right, the document mandated the ICC to lobby for "regulatory policies that protect Arctic freshwater systems from unsustainable pollution and depletion."[153] Generally, this declaration continues to develop a more holistic understanding of climate change, illustrating its impact on nutrition and food security, advocating for "improved access to sufficient Inuit traditional food sources."[154]

In 2015, world leaders met in Paris for the next COP conference. The ICC and the Saami Council represented the Arctic Indigenous Peoples in the International Indigenous Peoples Forum on Climate Change (IIPFCC), the Indigenous Peoples' caucus at UNFCCC processes, established in 2008.[155] At the conference, the ICC delegation participated in negotiations to have Indigenous Peoples' rights recognized in the Paris Agreement,[156] resulting in the recognition of Indigenous Peoples' climate change experiences, knowledge, and technologies, and COP's commitment to establishing a new forum that would later be called the Local Communities and Indigenous Peoples Platform.[157]

In a position paper that was distributed to all national delegations at COP 2015, the ICC advocated for financial support for projects that enhance climate adaptation and mitigation, lobbied for recognition of Indigenous knowledge and for understanding climate change as a human rights issue, and it called for a threshold of 1.5 degrees Celsius of global temperature increase compared to pre-industrial levels (in opposition to the more generally accepted increase of 2 degrees).[158] Together with the governments of Nunavut and Greenland, the ICC also issued a joint statement on climate change, calling for the protection of Inuit food security, and noting everyone's equal rights to development.[159] Furthermore, the ICC suggested that the UN establishes a Green Climate Fund (funded by developed nations), and it called for support for renewable energy creation.

Climate Change as an Opportunity (2018)

In the 2018 Utqiagvik Declaration, climate change continued to holistically interwoven with other issues, directing the ICC to continue its advocacy in existing international fora (e.g. UNEP and WHO), to closely monitor the implementation of the 2030 Sustainable Development Goals (SDG),[160] and to ensure that Inuit are informed.[161] While the document acknowledged opportunities stemming from climate change, it mainly discussed its negative repercussions: loss of multi-year sea-ice and melting permafrost (enhancing the cycling of contaminants to the Arctic), Arctic warming (requiring adjustments of food storing and hunting strategies), changes in coverage and movement of sea ice and animal migration patterns, more extreme storms, and the arrival of new species in Inuit Nunaat. Furthermore, the document pointed out that Inuit food security and sovereignty necessitate holistic and multi-agency approaches, innovation, Inuit management authority over their living resources, as well as research, advocacy, and multi-level education about "Inuit food security priorities" – which are connected to Inuit self-governance. Also, the ICC is directed to lobby for the enforcement of the IMO's Polar Code[162] and other international regulations, to "advance emergency response" and to "phase out heavy fuel oil" (HFO).[163]

The declaration also explained that existing human rights instruments confirm Inuit rights to self-determination, including the right "to govern wildlife management," but that existing international treaties and trade bans prevent Inuit from exercising their "rights to use Arctic living resources" which, in turn, has negative repercussions for Inuit culture, economies, and health. To protect Inuit rights to hunting, fishing, and gathering, the ICC focused on human rights instruments like UNDRIP.[164] The ICC supports various bodies (e.g. the Circumpolar Inuit Wildlife Committee, the Circumpolar Inuit Wildlife Network, AC, CAFF, IUCN, and CBD) in its work on food sovereignty, Inuit rights and wildlife management. The declaration encourages the ICC to engage in the creation of an International Union Conservation Nature Indigenous Peoples Organization, another new international space for Indigenous Peoples.[165]

The declaration focused most strongly on climate change in its section on the environment where it is acknowledged that the Arctic is undergoing "profound, rapid and unpredictable change," and that Inui are engaged in multiple ways to eliminate the causes of climate change, for example in research, knowledge-sharing, adapting, and negotiating international and bilateral agreements. Additionally, the declaration acknowledged the ICC's trailblazing role as an environmental advocate when it was the first to case for "the precautionary principle."[166] In concrete terms, the declaration suggests the inclusion of Inuit views and enhancing Inuit self-determination through participation in Arctic research efforts and in high-level ministerial interactions; participating in UNFCCC's Local Communities and Indigenous Peoples Platform; enhancing information exchange and knowledge-sharing across Inuit Nunaat regarding climate resilience, and innovation and adaptation strategies.

Challenges due to Melting Sea Ice (2017–2019)

The effects of climate change allow for eased commercial and touristic access to Arctic waters. While tourists mostly understand the negative effects of climate change in the Arctic, the very melting of ice constitutes one of the reasons sparking their explorative curiosity, and it allows for expeditions to encroach further into Inuit Nunangat. The 2010 Inuit Arctic Policy already noticed these changes and recommended significant Inuit involvement in Arctic tourism. The ICC's 2017 Pikialasorsuaq report more broadly voices concerns about increased shipping, tourism, resource development, and commercial fishing in the area.[167] As a result of global warming, the ice bridge surrounding the North Water Polynya (a year-round ice-free area — see Figure 6.1) is becoming unstable – a problematic development, given the polynya's extreme biological productivity and importance for millions of animals – and Inuit hunters.[168] The ICC's important work in the Pikialasorsuaq region was mentioned in the IPCC's 2019 Special Report on the Ocean and Cryosphere in a Changing Climate.

"Resolve Is Not Action" (2021)

The language of the ICC's position paper delivered at COP 26 in Glasgow, Scotland, on October 2021 is noticeably sharper than previously. Noting that, while Inuit have been witnessing climate change in the Artic for over 30 years ("we were among the first to sound the alarm") and have contributed negligible emissions, Inuit suffer from the most severe effects of climate change while governments are distracted by questions around equity. This is, essentially, a statement about environmental injustice. Reminding governments that "the Arctic as we have known it for thousands of years

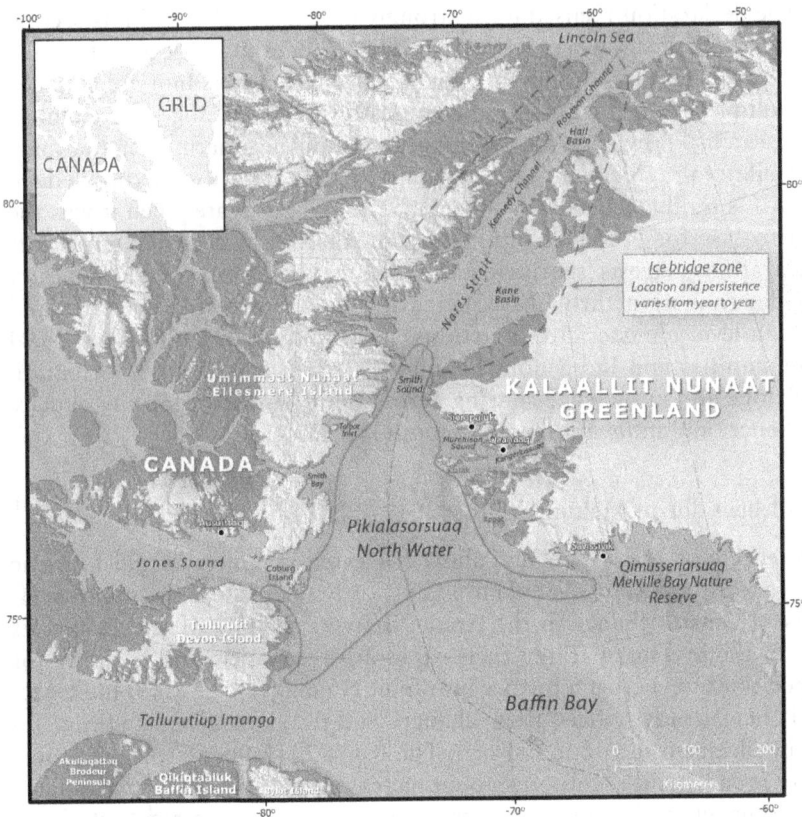

Figure 6.1 Map of the North Water Polynya (Pikialasorsuaq). Courtesy of Oceans North (https://www.oceansnorth.org/en/where-we-work/north-water-polynya/).

is slipping away [...] our home is becoming unrecognizable," the paper confronts governments with the dire truth that "resolve is not action." The ICC notes that the Arctic is warming multiple times faster than the world, urging that "the global community MUST (sic) act now to work with us," challenging nation states to view the world more holistically by connecting climate change with ocean governance, human rights, and food security.

In addition to describing climate change effects in the Arctic ("changes in weather and ice patterns, changes in distribution and abundance of wildlife, new species, stressed infrastructure, and the cultural and social impacts of these changes"), the publication also contextualizes these changes globally (floods, droughts, irreversible changes that are now also felt in other regions) to emphasize the urgency to act. In very clear language, the ICC paper calls on global leaders to genuinely act to "make unprecedented and massive efforts to cap global temperature rise," valuing and including Indigenous knowledge and leadership in climate action and governance, and the protection of oceans and cryosphere as critical ecosystems. This means financial support for Indigenous Peoples, investment in "Inuit-driven renewable energy development," inclusion of Inuit in Arctic shipping discussions at the IMO, and enabling Indigenous/Inuit participation in "the Intergovernmental Conference on an internationally legally binding instrument [...] (on) marine biological diversity" under UNCLOS.[169] The ICC has spent decades, and a huge amount of energy, time and resources for its climate change advocacy work, and even the Inuit's patience is wearing off in the face of these dire realities and world leaders' refusal or reluctance to develop constructive policies and make binding commitments. The frustration and disappointment can be sensed in this paper.

Trailblazing New Involvement in IPCC and IMO (2021/2022)

In March 2021 the ICC, as the first Indigenous Peoples Organization (IPO), is recognized with formal observer status at IPCC. It participates for the first time in this new role at IPCC's 55th session in February 2022. During the approval session for the Summary for Policymakers document on the Working Group 2, ICC International chair Dalee Sambo Dorough formally intervenes to emphasize the importance of the right of self-determination for Inuit to further nuance existing formulations around recognition of Indigenous Peoples' inherent rights. In its press release, the ICC called her intervention a "historic moment in IPCC history and a significant step forward in enhancing Indigenous engagement in the IPCC process." The ICC also participates in IPCC's approval sessions as part of the Canadian delegation, and it applauds the Canadian government's support of a Heavy Fuel Oil (HFO) ban, noting that "no effective way to deal with an HFO spill in the Arctic" is in place. This is in line with the ICC's 2018 GA declaration's recommendation to phase out HFO to protect the environment.[170]

In the fall of 2021, the ICC was the first Indigenous Organization receiving provisional consultative status at the IMO, allowing the NGO to share expertise about climate change, the economic importance of shipping for Arctic communities, and "sustainable use of Arctic waters."[171] The ICC has started advocating for the reduction of black carbon – a short-lived climate forcer produced from the burning of HFO – and decarbonization of global shipping to protect the Arctic marine environment and Inuit food security.[172] A major milestone was realized when the IMO subcommittee on ship noise pollution adopted the ICC's suggestion to "explicitly include

[...] engagement and utilization of Indigenous Knowledge" in its future work.[173] Due to an increase in Arctic shipping and Inuit dependency on the maritime environment, this work is crucial for the ICC.

New Challenges and Directions: The 2022 ICC Declaration

The 2022 General Assembly was overshadowed by the COVID-19 pandemic and the Russian war in Ukraine as well as the subsequent pause of the AC's work. The ICC is concerned about the halt of the forum's efforts to protect the northern environment, and the declaration noticeably emphasizes the need for unity and peace in the Arctic with good governance and Arctic security as top priority areas, noting that the Arctic "shall continue forever to be used exclusively for peaceful and environmentally safe purposes." This declaration also stands out as the most inclusive of all of ICC's declarations, focusing more intentionally on the ICC as a "safe and inclusive space where Inuit of all ages, gender, sexual orientation, and degree of cultural knowledge and language are welcome."

Criticizing that rule-making entities governing marine areas – like UN-CLOS – have been created without Inuit input, the ICC clearly demands that "this must not happen any longer." Rather, such institutions "must stem from Inuit and cannot be governed without us." Furthermore, the declaration states that "the Arctic must be protected through partnership with Inuit." Using the term "must" in these sentences conveys the sense of urgency felt among Inuit, indicating an observable change in the ICC's approach compared to the past when the NGO tended to focus more on making recommendations rather than on demands. The world is called to act immediately to "address the inequity of climate change impacts by respecting our inherent right of self-determination in decision-making processes in the Arctic," to cap global warming to 1.5 degrees Celsius, and to ensure equitable participation climate change mechanisms. Again, the declaration points out the Arctic's role in global biodiversity, temperature regulation and well-being, and it remains committed to continuing its existing participation in international fora to combat climate change and ensure the inclusion of Inuit leadership, knowledge, and participation in the creation of new instruments.

The ICC's 2022 Circumpolar Inuit Protocols for Equitable and Ethical Engagement instruct researchers and decision-makers to ensure the inclusion of Inuit knowledge in climate change research, knowledge generation, and policy-making.[174] Moreover, the document calls for financial support for climate change adaptation and mitigation strategies led and defined by Inuit, and it advocates for climate-resilient infrastructure, and for reducing and preventing black carbon in the Arctic through decarbonizing efforts in shipping and reducing local diesel consumption. While the declaration is in favour of renewable energy sources, it simultaneously clarifies that "we will not in any way bear the cost of transitioning to safe alternative fuel sources."[175]

Conclusion

As national minorities, Inuit disproportionately experience the negative impacts of climate change, while missing out on the benefits generated by industrial activities that others get to experience. Climate change creates conditions of environmental inequality for Inuit – environmental injustice – and the ICC's advocacy aims at realizing environmental justice for Inuit. Importantly, the ICC has urged for distributional justice, for example through the establishment of green funds to distribute financial gains more fairly, while its tireless work for fair and meaningful inclusion and participation of Inuit in decision-making processes are examples of efforts to realize procedural justice. The ICC's advocacy for recognition of Inuit epistemologies, practices, and knowledges as well as its work to Indigenize the international political system is aimed at creating a socially just reality for Inuit. Like other Indigenous groups, Inuit have argued that environmental justice processes need to go beyond taking into account Indigenous knowledges, as "contributing to a conversation that has until recently been a monochrome monologue with an occasional dash of multicultural phrase or clause for colour."[176] Rather, full inclusion and participation of Indigenous Peoples in environmental justice mechanisms need to be aimed at, for a truly inclusive approach that can help change the culture of environmental law.

Inuit have been urging for decades for environmental protection and justice: fair treatment and respect for Indigenous rights and self-determination and inclusion and meaningful involvement of Inuit knowledge in global climate change debates and policy-making processes. Environmental protection is a right and responsibility of Inuit – in the Arctic and globally. The ICC's climate change advocacy is informed by and grounded in Inuit environmental knowledge, and its practices reflect not only the importance of Inuit environmental stewardship, but also a number of other foundational IQ values like sharing, helping others, being resourceful and open-minded, developing new skills, respect, and concern for others.

The ICC's approach is also a reflection of typical Inuit conflict resolution mechanisms that are known to be pragmatic, future-oriented, creative, and innovative; they aim at collaboration and reciprocity; and they involve patience, sharing, and respect for all living things. The ICC's interest in collaborating with nation states and other Indigenous groups reflect the preference for cooperation and relationship-building over confrontation, and it also mirrors Inuit pre-contact practices of political realism and adaptability that instruct Inuit to study their context, recognize changed circumstances, and react constructively to create win-win scenarios.[177] As an IPO, the ICC's climate change advocacy aims at creating space for Indigenous Peoples and their knowledges in international fora, for example as Permanent Participants in the AC, in UNPFII or UNFCCC's Local Communities and Indigenous Peoples Platform.

The ICC's international climate change work has illustrated this conflict's holistic and interconnected nature. This includes discussions around food security, rights, self-determination, health, economic processes, wildlife management, sustainable development, shipping, and tourism, to name a few. Environmental health is synonymous with cultural, ecological, and physical health, and the continuation of cultural practices is intimately linked to a healthy and functioning ecosystem. These forms of sustainability are requited for productive justice. For Inuit, environmental protection and justice are connected to ecological, social, economic, and cultural sustainability, and solutions require both mitigation and adaptation strategies.[178]

In many ways, the ICC is a trailblazer in shaping international climate change work. It was the first transnational forum in the Arctic, and it inspired the creation of the AC, a platform that brings together Arctic politicians and Indigenous groups to collaborate on environmental protection. The ICC was also instrumental in the creation of the Permanent Participant mechanism in the AC that serves to strengthen Arctic Indigenous voices. Moreover, the ICC was the first to help shape climate change as a human rights issue, it organized the first global climate change conference for Indigenous Peoples, and it was the first IPO to obtain IMO consultative status and to be recognized as an observer at IPCC. All these milestones are moments of historic significance, because they help enhance Inuit engagement in international climate change mechanisms, to make the Inuit voice heard where international guidelines and policies are shaped, and to Indigenize the international system.

Landmark ICC policy papers inform and advice governments and international organizations on Arctic realities and suggest ways to move forward: the award-winning IRCS in 1986 – the first regional conservation strategy in the world, and the first one by an Indigenous People, three Arctic Policy Papers (1983, 1992, and 2010), the Inuit petition at IACHR (2005), and the Pikialasorsuaq Report (2017). In addition to its own papers, the ICC also supports international guidelines that are important milestones in addressing global climate change: for example CBD, the Kyoto Protocol, the Paris Agreement, and the Minamata Convention. The NGO has been instrumental in the shaping of global climate change regulations like the Stockholm Convention and COP conferences. Time and again, it advocated for the creation of and commitment to equitable and enforceable international rules to reduce global warming and to limit its effect, and the ICC continues to shape international regulations on POPs, mercury and other heavy metals, hydrocarbons, and HFO to protect the fragile Arctic environment. To strengthen Inuit and Indigenous rights, the ICC submitted a petition to the IACHR, contributed to and holds up UNDRIP, and it supports the creation of an Arctic Environmental Bill of Rights (Arctic Policy 2010). It continuously works to educate a global audience about Inuit rights and how climate change effects violate Inuit social, economic, cultural, Indigenous Peoples', and human rights.

Safe shipping, one of the main concerns that led to the creation of the ICC in the 1970s, remains prominently on its climate change agenda. The Pikialasorsuaq Report (2017) and the ICC's stronger ties with the IMO and UNCLOS suggest that the NGO is increasingly focusing on the protection and Inuit transboundary use of open Arctic waters, an interesting area of national and international jurisdiction where melting sea ice has led to increased competition for control and exploration, and, as a result, heightened urgency for conflict resolution.[179] For the ICC, these developments also include concerns around ethical tourism and research in Inuit Nunaat. Increasing tensions between North America and Russia also pose new military and environmental threats in the Arctic.

Since its creation, the NGO has successfully participated in strengthening and supporting existing mechanisms and in the delivery of new guidelines, structures, ground-breaking conventions, and fora to address climate change, and it has added a human face to the global climate change debate.[180] "The ICC has been an important player in international affairs with regard to environmental and human rights issues,"[181] and its advocacy, perseverance, professionalism, patience, and diplomacy are respected and relied upon internationally. Through the ICC, Inuit have placed themselves at the vanguard of worldwide environmental protection,[182] and the ICC is one of the most visible and important IPOs in the international arena. Its advocacy, position papers, and the processes it has initiated and impacted have substantially shaped our understanding of climate change and environmental justice. Although the ICC represents a relatively small number of people, its environmental stewardship, insights, and decade-long climate change advocacy are wake-up calls and, indeed, "a gift from Inuit hunters and elders to the world."[183]

Notes

1 Public Health Agency of Canada, "From Risk to Resilience: An Equity Approach to COVID-19: Chief Public Health Officer of Canada's Report on the State of Public Health in Canada 2020" (Ottawa: Government of Canada, 2021), https://www.canada.ca/en/public-health/corporate/publications/

chief-public-health-officer-reports-state-public-health-canada/from-risk-resilience-equity-approach-covid-19.html

2 Statistics Canada, "Demographic Characteristics and Indigenous Groups," November 11, 2022, https://www150.statcan.gc.ca/n1/en/subjects/indigenous_peoples/demographic_characteristics_and_indigenous_groups

3 Health Canada, "Health of Canadians in a Changing Climate. Advancing Our Knowledge for Action" (Health Canada, 2022), 46–47.

4 National Collaborating Centre for Indigenous Health, "Climate Change and Indigenous Peoples' Health in Canada," Health of Canadians in a Changing Climate: Advancing Our Knowledge for Action (Health Canada, 2022), 57, https://changingclimate.ca/health-in-a-changing-climate/chapter/2-0/

5 National Collaborating Centre for Indigenous Health, 57–58.

6 Deborah McGregor, "Honouring Our Relations: An Anishnaabe Perspective on Environmental Justice," in *Speaking for Ourselves. Environmental Justice in Canada*, ed. Julian Agyeman et al. (Vancouver: UBC Press, 2009), 28.

7 OP United States Environmental Protection Agency, "Learn About Environmental Justice," Overviews and Factsheets, 2021, https://www.epa.gov/environmentaljustice/learn-about-environmental-justice

8 Randolph Haluza-DeLay et al., "Speaking for Ourselves, Speaking Together: Environmental Justice in Canada," in *Speaking for Ourselves. Environmental Justice in Canada*, ed. Julian Agyeman et al. (Vancouver: UBC Press, 2009), 5.

9 Haluza-DeLay et al., 9.

10 Sarah Fleisher Trainor et al., "Environmental Injustice in the Canadian Far North: Persistant Organic Pollutants and Arctic Climate Impacts," in *Speaking for Ourselves. Environmental Justice in Canada*, ed. Julian Agyeman et al. (Vancouver: UBC Press, 2009), 145. The authors reference Pellow (2000) here.

11 Fleisher Trainor et al., 153.

12 Fleisher Trainor et al., 153.

13 Haluza-DeLay et al., "Speaking for Ourselves, Speaking Together: Environmental Justice in Canada," 10.

14 Sarah F. Trainor et al., "Arctic Climate Impacts: Environmental Injustice in Canada and the United States," *Local Environment* 12, no. 6 (December 1, 2007): 628, 636, https://doi.org/10.1080/13549830701657414

15 Randolph Haluza-DeLay et al., 2, 4.

16 McGregor, "Honouring Our Relations: An Anishnaabe Perspective on Environmental Justice," 27.

17 McGregor, 33, 40.

18 Government of Canada, "Pan-Canadian Framework on Clean Growth and Climate Change Second Annual Report: Section 1," August 8, 2019, https://www.canada.ca/en/environment-climate-change/services/climate-change/pan-canadian-framework-reports/second-annual-report/section-1.html

19 "Joint Committee on Climate Action Annual Report Highlights First Nations Leadership in Addressing Climate Change," news releases, Government of Canada, August 13, 2021, https://www.canada.ca/en/environment-climate-change/news/2021/08/joint-committee-on-climate-action-annual-report-highlights-first-nations-leadership-in-addressing-climate-change.html; "About AFN," Assembly of First Nations, accessed August 2, 2022, https://www.afn.ca/about-afn/.

20 Statistics Canada, "Demographic Characteristics and Indigenous Groups," November 11, 2022.

21 Assembly of First Nations, "Declaring a First Nations Climate Emergency," July 2019.

22 Assembly of First Nations, "Framing a First Nations Climate Lens" (White-horse, March 3, 2020), https://www.afn.ca/wp-content/uploads/2019/08/19-05-Declaring-a-First-Nations-Climate-Emergency.pdf

23 "Joint Committee on Climate Action Annual Report Highlights First Nations Leadership in Addressing Climate Change."

24 "Origin and Purpose," Yukon First Nations Climate Action Fellowship, accessed August 2, 2022, https://www.yfnclimate.ca/origin-and-purpose

25 "Reconnection Vision and Action Plan," Yukon First Nations Climate Action Fellowship, 2021, https://www.yfnclimate.ca/yfnrvap

26 Statistics Canada, "Demographic Characteristics and Indigenous Groups," November 11, 2022.

27 Women of the Metis Nation, "Negative Impacts of Climate Change on Culture and Cultural Rights," accessed August 3, 2022.

28 "Canada's Partnership with Indigenous Peoples on Climate," Government of Canada, January 26, 2022.

29 JF Consulting, "Metis Nation Climate Change and Health Vulnerability Assessment" (Metis National Council, 2020), 91.

30 JF Consulting, 92–94.

31 Statistics Canada, "Demographic Characteristics and Indigenous Groups," November 11, 2022.

32 "Statistics on Indigenous Peoples," Statistics Canada, 2022.

33 Inuit Tapiriit Kanatami, "National Inuit Climate Change Strategy" (Ottawa: Inuit Tapiriit Kanatami, 2019); "Canada's Partnership with Indigenous Peoples on Climate."

34 "Canada's Partnership with Indigenous Peoples on Climate."

35 Mika Rantanen et al., "The Arctic Has Warmed Nearly Four Times Faster than the Globe since 1979," *Communications Earth & Environment* 3, no. 1 (August 11, 2022): 1–10.

36 Peter Kulchyski, "Colonization of the Arctic," in *Encyclopedia of the Arctic*, ed. Mark Nuttall, vol. 1 (New York: Routledge, 2005), 405–11.

37 Frances Abele and Thierry Rodon, "Inuit Diplomacy in the Global Era: The Strengths of Multilateral Internationalism," *Canadian Foreign Policy* 13, no. 3 (2007): 51.

38 Aqqaluk Lynge, *Inuit. The Story of the Inuit Circumpolar Conference* (Nuuk, 1993), 87.

39 Abele and Rodon, "Inuit Diplomacy in the Global Era," 51.

40 W. Dallam Masterson and Albert G. Holba, "North Alaska Super Basin: Petroleum Systems of the Central Alaskan North Slope, United States," *AAPG Bulletin* 105, no. 6 (June 2021): 1234.

41 Lynge, *Inuit*, 14.

42 Eben Hopson, "Inupiat under Four Flags: Planning the First Inuit Circumpolar Conference" (North Slope Borough, AK, 1976), 12.

43 Inuit Circumpolar Conference, "First General Assembly. Summary and Resolutions" (Barrow, June 13, 1977), sec. 77.01.

44 Inuit Circumpolar Council, "Charter & Bylaws," sec. 1.8, accessed October 16, 2018, http://www.inuitcircumpolar.com/icc-charter.html

45 Mads Faegteborg, "Inuit Circumpolar Conference, an Indigenous Organization as an Instrument of Change," in *Nordic Arctic Research on Contemporary Arctic Problems*, ed. Lise Lyck (Aarhus: Aarhus Universitetsforlag, 1992), 243–44.

46 "About ICC | Inuit Circumpolar Council Canada," July 2022, https://www.inuitcircumpolar.com/about-icc/

47 Haluza-DeLay et al., "Speaking for Ourselves, Speaking Together: Environmental Justice in Canada," 5.

48 Christopher G. Trott, "Social Relations among Inuit: Tuqłuraqtuq and Ilagiit," in *The Inuit World*, ed. Pamela Stern (London: Routledge, 2021), 288–303.

49 John Bennett and Susan Diana Mary Rowley, *Uqalurait: An Oral History of Nunavut* (Montreal: McGill-Queen's University Press, 2004), 118.

50 Ibid., 120; Mark Nuttall, "Locality, Identity and Memory in South Greenland," *Études/Inuit/Studies* 25, no. 1/2 (2001): 53–72.

51 Parnuna Egede Dahl and Pelle Tejsner, "Review and Mapping of Indigenous Knowledge Concepts in the Arctic," in *Inuit Worlds*, ed. Pamela Stern, 1st ed. (London: Routledge, 2021), 239.

52 Nunavut Department of Education, "Inuit Qaujimajatuqangit Education Framework for Nunavut Curriculum" (Iqaluit: Nunavut Department of Education, 2007), 27–36, https://www.gov.nu.ca/sites/default/files/files/Inuit%20Qaujimajatuqangit%20ENG.pdf

53 Shirley Tagalik, "Inuit Qaujimajatuqangit: The Role of Indigenous Knowledge in Supporting Wellness in Inuit Communities in Nunavut" (Prince George: National Collaborating Centre for Aboriginal Health, 2009–2010), 1.

54 Alexina Kublu, Frederic Laugrand, and Jarich Oosten, "Interviewing the Elders," in *Interviewing Inuit Elders*, ed. Frederic Laugrand and Jarich Oosten, vol. 1 (Iqaluit: Nunavut Arctic College, 1999), 1–12.

55 Bennett and Rowley, *Uqalurait*, 127.

56 Abele and Rodon, "Inuit Diplomacy in the Global Era," 45–63.

57 Aqqaluk Lynge, *Inuit*, 107; Lydia Schoeppner, "The Inuit Circumpolar Council – Agent of Peacemaking for Inuit in Nunavut and Greenland" (Doctoral Dissertation, Winnipeg, University of Manitoba, 2020).

58 Hopson, "Inupiat under Four Flags: Planning the First Inuit Circumpolar Conference," 28; Inuit Circumpolar Conference, "First General Assembly. Summary and Resolutions," sec. 77–06; Inuit Circumpolar Conference, "Resolutions 1980" (Nuuk, 1980), secs. 7–80, 18–80.

59 Inuit Circumpolar Conference, "Resolutions 1983" (Iqaluit, 1983), secs. 21–23.

60 Faegteborg, "Inuit Circumpolar Conference, an Indigenous Organization as an Instrument of Change," 245; Inuit Circumpolar Conference, "Resolutions 1986" (Kotzebue, 1986), sec. 18; Inuit Circumpolar Conference, "Towards an Inuit Regional Conservation Strategy," (Kotzebue: Inuit Circumpolar Conference, 1986), 5, 85.

61 Inuit Circumpolar Conference, "Towards an Inuit Regional Conservation Strategy," V.

62 Inuit Circumpolar Conference, "Resolutions 1989" (Sisimiut, 1989), sec. 3; UN Environmental Programme, "Inuit Regional Conservation Strategy for the Arctic" (Kotzebue: ICC General Assembly, 1988), http://www.global500.org/index.php/thelaureates/online-directory/item/631-inuit-regional-conservation-strategy-for-the-arctic

63 UN Environmental Programme, "Inuit Regional Conservation Strategy for the Arctic."

64 Mads Faegteborg, "Inuit Circumpolar Conference (ICC)," in *Encyclopedia of the Arctic*, ed. Mark Nuttall, vol. 2 (New York: Routledge, 2005), 1001; Nordic Co-operation, "Nordic Council of Ministers," Nordic cooperation, accessed January 21, 2019, https://www.norden.org/en/nordic-council-ministers

65 Inuit Circumpolar Conference, "Towards an Inuit Regional Conservation Strategy," 5.

66 Inuit Circumpolar Conference, V, VI.

67 Inuit Circumpolar Conference, 68.

68 Inuit Circumpolar Conference, VII.

69 The combustion of fossil fuels (coal, gas, oil) creates large amounts of carbon dioxide (CO_2), a hydrocarbon, whose atmospheric concentration is the main cause of climate change.

70 Inuit Circumpolar Conference, "Towards an Inuit Regional Conservation Strategy," 5.

71 Inuit Circumpolar Conference, 68–69.

72 Chris Mooney, "The Arctic Is Full of Toxic Mercury, and Climate Change Is Going to Release It," *Washington Post*, February 5, 2018; World Health Organization, "Mercury and Health," accessed July 8, 2022, https://www.who.int/news-room/fact-sheets/detail/mercury-and-health

73 Inuit Circumpolar Conference, "Resolutions 1986," sec. 86–14.

74 Peter Hough, *International Politics of the Arctic Coming in from the Cold* (London: Routledge, 2013), 54–55.

75 Inuit Circumpolar Conference, "Resolutions 1986," sec. 17; International Union for Conservation of Nature, "IUCN – A Brief History," IUCN, 2018, https://www.iucn.org/about/iucn-brief-history; Anja Miller, "Membership Information," December 19, 2018.

76 Hough, *International Politics of the Arctic Coming in from the Cold*, 55; Alan Taylor, "The Exxon Valdez Oil Spill: 25 Years Ago Today," *The Atlantic*, March 24, 2014, https://www.theatlantic.com/photo/2014/03/the-exxon-valdez-oil-spill-25-years-ago-today/100703/; Alaska Oil and Gas Association, "History1960's," 2019, https://www.aoga.org/industry-history/

77 Liesel Ashley Ritchie, "Individual Stress, Collective Trauma, and Social Capital in the Wake of the Exxon Valdez Oil Spill," *Sociological Inquiry* 82, no. 2 (2012): 187–211.

78 Sheila Watt-Cloutier, *The Right to Be Cold: One Woman's Story of Protecting Her Culture, the Arctic and the Whole Planet* (Toronto: Allen Lane, 2015), 194–95, 205; United Nations, "Earth Summit," 1997, http://www.un.org/geninfo/bp/enviro.html; United Nations, "United Nations Framework Convention on Climate Change," 1992, para. 2; United Nations Framework Convention on Climate Change, "What Is the Kyoto Protocol?," 2019, https://unfccc.int/process-and-meetings/the-kyoto-protocol/what-is-the-kyoto-protocol/what-is-the-kyoto-protocol

79 Leena Heinämäki, "Rethinking the Status of Indigenous Peoples in International Environmental Decision-Making: The Role of Arctic Indigenous Peoples and the Challenge of Climate Change," in *Climate Governance in the Arctic*, ed. Timo Koivurova, E. Carina H. Keskitalo, and Nigel Bankes, vol. 50, Environment and Policy (New York: Springer, 2009), 209–10, 248–49, 253.

80 Leena Heinämäki, "Rethinking the Status of Indigenous Peoples," 248; Mary Simon, "Proposed Objectives for an Arctic Sustainable and Equitable Development Strategy," Protecting the Arctic environment (Report on the Yellowknife Preparatory Meeting, Yellowknife, NWT, April 18–23, 1990) (Ottawa, 1990), 183; "The Arctic Council Turns 25," Arctic Council, 2021, 25, https://arctic-council.org/about/timeline/25/

81 Pernilla Carlsson et al., "Polychlorinated Biphenyls (PCBs) as Sentinels for the Elucidation of Arctic Environmental Change Processes: A Comprehensive Review Combined with ArcRisk Project Results," *Environmental Science and Pollution Research* 25, no. 23 (2018): 22499–528. In the 1980s several

studies confirmed that POPs are transported to the Arctic through atmospheric and oceanic pathways. Due to their strong bioaccumulation potential, POPs have been found in high concentration in the tissues of Arctic marine mammals which constitute the major food source for Inuit. Consequently, Inuit who consumed a lot of country food showed elevated POP levels (Ibid.). Tair Teran, Lara Lamon, and Antonio Marcomini, "Climate Change Effects on POPs' Environmental Behaviour: A Scientific Perspective for Future Regulatory Actions," *Atmospheric Pollution Research* 3, no. 4 (2012): 466–76, https://doi.org/10.5094/APR.2012.054

82 Watt-Cloutier (1999): The Need for a Global Treaty on Persistent Organic Pollutants, as cited in Terry Fenge, "POPs and Inuit: Influencing the Global Agenda," in *Northern Lights against POPs. Combatting Toxic Threats in the Arctic*, ed. David Leonard Downie and Terry Fenge (Montreal: McGill-Queen's University Press, 2003), 192–213.
83 Inuit Circumpolar Conference, "Resolutions 1989," secs. 12, 14.
84 Inuit Circumpolar Conference, "Resolutions 1998" sec. 6.
85 Watt-Cloutier, *The Right to Be Cold,* 178.
86 Heinämäki, "Rethinking the Status of Indigenous Peoples," 251.
87 Watt-Cloutier, *The Right to Be Cold*, 155.
88 Sheila Watt-Cloutier, "The Inuit Journey towards a POPs-Free World," in *Northern Lights against POPs. Combatting Toxic Threats in the Arctic*, ed. David Leonard Downie and Terry Fenge (Montreal: McGill-Queen's University Press, 2003), 258.
89 Fenge, "POPs and Inuit: Influencing the Global Agenda," 202.
90 Watt-Cloutier, "The Inuit Journey towards a POPs-Free World," 259–60; Watt-Cloutier, *The Right to Be Cold*, 163.
91 Watt-Cloutier, "The Inuit Journey towards a POPs-Free World," 264–65.
92 Ibid., 267.
93 Heinämäki, "Rethinking the Status of Indigenous Peoples," 251.
94 Fenge, "POPs and Inuit: Influencing the Global Agenda," 210.
95 Sheila Watt-Cloutier, quoted in: Fenge, 203.
96 Inuit Circumpolar Conference, "Kuujjuaq Declaration" (Kuujjuaq, 2002), sec. 7.
97 Jens Dahl, *The Indigenous Space and Marginalized Peoples in the United Nations* (New York: Palgrave Macmillan, 2012), 89.
98 United Nations, "UNPFII Members," accessed January 13, 2019, https://www.un.org/development/desa/indigenouspeoples/about-us/members.html; Abele and Rodon, "Inuit Diplomacy in the Global Era," 56.
99 Watt-Cloutier, *The Right to Be Cold*, 205.
100 Ibid., 223–24.
101 Ibid., 237.
102 Sheila Watt-Cloutier, *The Right to Be Cold*, 236–37.
103 Agnieszka Szpak, "Arctic Athabaskan Council's Petition to the Inter-American Commission on Human Rights and Climate Change—Business as Usual or a Breakthrough?," *Climatic Change* 162, no. 3 (October 1, 2020): 1587, https://doi.org/10.1007/s10584-020-02826-y
104 "Sheila Watt-Cloutier," The Right Livelihood Award, 2016, https://www.rightlivelihoodaward.org/laureates/sheila-watt-cloutier/
105 E. Carina H. Keskitalo, Timo Koivurova, and Nigel Bankes, "Conclusions on Climate Governance in the Arctic," in *Climate Governance in the Arctic*, ed. Timo Koivurova, E. Carina H. Keskitalo, and Nigel Bankes, vol. 50, Environment and Policy (New York: Springer, 2009), 439; Heinämäki, "Rethinking

the Status of Indigenous Peoples," 208–9, 218–19; Joyeeta Gupta, "A History of International Climate Change Policy," *Wiley Interdisciplinary Reviews: Climate Change* 1, no. 5 (2010): 648; Fleisher Trainor et al., "Environmental Injustice in the Canadian Far North," 152.

106 Sheila Watt-Cloutier, *The Right to Be Cold*, 237–38, 240 The same argument was reflected in the Male Declaration on the Human Dimension of Global Climate Change (2007), UN Human Rights Council Resolution 7/23 (2008), a study by the UN Human Rights Commissioner (2009), at following COP conferences, in an Oxfam study (2008), and communities applied the argumentation in their lawsuits against energy companies. Years later, the IACHR acknowledged the connection in another case (Ibid., 255–57).

107 "UN General Assembly Declares Access to Clean and Healthy Environment a Universal Human Right," UN News, July 28, 2022, https://news.un.org/en/story/2022/07/1123482; UN General Assembly, "The Human Right to a Clean, Healthy and Sustainable Environment " (UN, August 1, 2022), https://digitallibrary.un.org/record/3983329; Inuit Circumpolar Council, "International Day of the World's Indigenous Peoples: Inuit Welcome the Historic UN General Assembly Recognition of a Clean, Healthy, Sustainable Environment as a Human Right" (Inuit Circumpolar Council, August 9, 2022), https://iccalaska.org/wp-icc/wp-content/uploads/2022/08/Indigenous-Peoples-Day-2022_ENG.pdf

108 Paul Mohai, David Pellow, and J. Timmons Roberts, "Environmental Justice," *Annual Review of Environment and Resources* 34 (2009): 425, https://doi.org/10.1146/annurev-environ-082508-094348

109 Quoted in: Abele and Rodon, "Inuit Diplomacy in the Global Era," 58. Emphasis added.

110 Abele and Rodon, "Inuit Diplomacy in the Global Era."

111 The other two NGOs are the Aleut International Association and the Arctic Athabascan Council.

112 Many Strong Voices, "About the Many Strong Voices Programme," 2014, https://www.manystrongvoices.org/documents/MSVBook2014.pdf

113 Watt-Cloutier, *The Right to Be Cold*, 243–4.

114 Many Strong Voices, "About the Many Strong Voices Programme"; Many Strong Voices, "Global Engagement in Climate Negotiations & IPCC," 2014, https://www.manystrongvoices.org/documents/MSVBook2014.pdf; Many Strong Voices, "Portraits of Resilience," 2014, https://www.manystrongvoices.org/documents/MSVBook2014.pdf

115 Many Strong Voices, "Climate Change and Community-Based Relocation," 2014, https://www.ciel.org/wp-content/uploads/2014/11/COP19_CB_Relocation_7Feb2014.pdf

116 Watt-Cloutier, *The Right to Be Cold*, 245; "Hollywood's Hayek, Gyllenhaal Join Inuit for Arctic Circle Event on Earth Day," *The New York Times*, April 23, 2005, sec. Arts, https://www.nytimes.com/2005/04/23/arts/hollywoods-hayek-gyllenhaal-join-inuit-for-arctic-circle-event-on-earth.html; the story, including the aerial photo, can be found here: https://www.spokesman.com/stories/2005/apr/23/global-warming-activists-go-north/

117 Inuit Circumpolar Conference, "Utqiagvik Declaration" (Utqiavik, 2006), 1.

118 Inuit Circumpolar Conference, sec. 11.

119 Inuit Circumpolar Conference, secs. 21, 26.

120 Indigenous Peoples' Global Summit on Climate Change, "Report of the Indigenous Peoples' Global Summit on Climate Change" (Anchorage: Indigenous

Peoples' Global Summit on Climate Change, 2009), 6, http://www.un.org/ga/president/63/letters/globalsummitoncc.pdf

121 Indigenous Peoples' Global Summit on Climate Change, 9, 13.

122 Indigenous Peoples' Global Summit on Climate Change, 6, 8.

123 The signatories of the Anchorage Declaration recommend in their first call for action that the developed countries meeting at COP15 "support a binding emissions reduction target" for developed countries "of at least 45% below 1990 levels by 2020 and at least 95% by 2050."

124 Indigenous Peoples' Global Summit on Climate Change, "Report of the Indigenous Peoples' Global Summit on Climate Change," 6, 7.

125 Indigenous Peoples' Global Summit on Climate Change, 8.

126 Indigenous Peoples' Global Summit on Climate Change, 7.

127 Indigenous Peoples' Global Summit on Climate Change, 12–13.

128 IISD, "UN Backed Indigenous Peoples' Global Summit on Climate Change Ends with Declaration Calling for Fossil Fuel Phase-Out," April 24, 2009, https://unfccc.int/resource/docs/2009/smsn/ngo/168.pdf; Patricia Cochran, "Displaced by Climate Change," https://www.alaskapublic.org/wp-content/uploads/2017/07/Displaced-by-Climate-Change-Patricia-Cochran-Slides-6.6.2017.pdf

129 Indigenous Peoples' Global Summit on Climate Change, "Report of the Indigenous Peoples' Global Summit on Climate Change," secs. 57–58.

130 Indigenous Peoples' Global Summit on Climate Change, 9–11.

131 Indigenous Peoples' Global Summit on Climate Change, sec. 59.

132 Indigenous Peoples' Global Summit on Climate Change, secs. 7, 69–70.

133 Indigenous Peoples' Global Summit on Climate Change, secs. 63, 68.

134 Inuit Circumpolar Council, "Nuuk Declaration – Inoqatigiinneq, Sharing Life" (Nuuk, 2010), secs. 33–38.

135 Inuit Circumpolar Council, 2.

136 Adaptation Fund, "Adaptation Fund," Adaptation Fund, accessed July 16, 2019, https://www.adaptation-fund.org/about/; the Fund was established under the Kyoto Protocol and helps developing countries and vulnerable communities adapt to climate change.

137 Inuit Circumpolar Council, "Nuuk Declaration – Inoqatigiinneq, Sharing Life," sec. 40.

138 Inuit Circumpolar Conference, "Principles and Elements for a Comprehensive Arctic Policy" (Montreal: Centre for Northern Studies and Research, McGill University, 1992), 2.

139 Inuit Circumpolar Conference, "Resolutions 1983," sec. 83.30.2.

140 Inuit Circumpolar Council, "Inuit Arctic Policy," 2010, sec. 3, pg. 26, points 1, 2.

141 Inuit Circumpolar Council, sec. 3, pg. 28, points 12 and 13.

142 Inuit Circumpolar Council, sec. 3, pg. 28, point 14.

143 Inuit Circumpolar Council, sec. III.7.

144 UN Universal Declaration of Human Rights (1948); International Covenants of 1966 on Economic, Social and Cultural Rights (ICESCR) and on Civil and Political Rights (ICCPR); International Labour Organization (ILO) Convention on Indigenous and Tribal Peoples, 1989 (No. 169), UNDRIP, 2007.

145 Ibid., sec. 3, pg. 27, 28, point 3, 7.

146 Ibid., sec. 3, pg. 29, points 15 and 16.

147 Ibid., sec. 3, pg. 27, points 8 to 11.

148 Ibid., sec. 3, pg. 29, points 1 and 2.

149 Ibid., sec. 3, pg. 38–45.
150 Ibid., sec. 3, pg. 45–48.
151 Ibid., sec. 4, pg. 52, sections 10 and 11.
152 Inuit Circumpolar Council, sec. 5, pg. 88, point 8. The ICC criticizes the delivery of toxic products like PCBs and chlordane (pesticide) to "less developed countries that are not in a position to properly store and dispose of toxic substances." Due to their bioaccumulation in the Arctic Ocean, marine mammals, and eventually Inuit, will be exposed to these dangerous chemicals through their food chain, creating serious health problems among Inuit and possibly leading to the extinction of some marine mammals (Ibid.).
153 Inuit Circumpolar Council, "Kitigaaryuit Declaration – Ukiuqta'qtumi Hivuniptingnun, One Arctic, One Future" (Inuvik, 2014), secs. 10–12, 14, 15.
154 Inuit Circumpolar Council, secs. 36, 37.
155 "The Arctic," International Indigenous Peoples Forum on Climate Change, accessed January 20, 2019, http://www.iipfcc.org/the-arctic/; "Who Are We?," International Indigenous Peoples Forum on Climate Change, accessed January 20, 2019, http://www.iipfcc.org/who-are-we/
156 United Nations, "Paris Agreement" (Paris: United Nations, 2015), https://unfccc.int/sites/default/files/english_paris_agreement.pdf; the Paris Agreement directs nation states to consider the rights of Indigenous Peoples in their climate change actions (preamble) and that adaptation actions should take into consideration the knowledge of Indigenous Peoples (article 7.5) (Ibid.).
157 UN, "Report of the Conference of the Parties on Its Twenty-First Session, Held in Paris from 30 November to 11 December 2015," 2016, sec. 135.
158 ICC, "Chair's Message. Inuit Voices Informing Action," January 2016, https://www.inuitcircumpolar.com/wp-content/uploads/chairs_message_-_january_2016_.pdf; ICC, "Progress Made in Paris: ICC Urges Action Now" (Ottawa, Dez 2015), https://www.inuitcircumpolar.com/press-releases/progress-made-in-paris-icc-urges-action-now/; Nunatsiaq Online, "Climate Change Linked to Human Rights, Inuit Leader Says," December 4, 2015, https://nunatsiaq.com/stories/article/65674climate_change_linked_to_human_rights_inuit_leader_says/
159 Government of Greenland, Government of Nunavut, and ICC, "Joint Statement of Inuit Circumpolar Council, the Government of Greenland and the Government of Nunavut on Climate Change," December 8, 2015, http://www.inuitcircumpolar.com/uploads/3/0/5/4/30542564/joint_statement_on_climate_change unfcc_cop21-eng.pdf; this refers to the 1986 Declaration on the Right to Development, an "inalienable human right." UN General Assembly, "Declaration on the Right to Development," 1986, https://www.ohchr.org/en/professionalinterest/pages/righttodevelopment.aspx
160 UN Sustainable Development Goals, "Sustainable Development Goals," accessed December 5, 2018, https://sustainabledevelopment.un.org/?menu=1300; United Nations, "Transforming Our World: The 2030 Agenda for Sustainable Development" (New York: United Nations, 2015), https://sdgs.un.org/2030agenda
161 Inuit Circumpolar Council, "Utqiagvik Declaration – The Arctic We Want" (Utqiagvik, 2018), secs. 8, 9.
162 International Maritime Organization, "International Code for Ships Operating in Polar Waters (Polar Code)," n.d., http://www.imo.org/en/MediaCentre/HotTopics/polar/Documents/POLAR%20CODE%20TEXT%20AS%20ADOPTED.pdf; International Maritime Organization, "Shipping in Polar

Waters. Adoption of an International Code of Safety for Ships Operating in Polar Waters (Polar Code)," accessed December 6, 2018, http://www.imo. org/en/MediaCentre/HotTopics/polar/Pages/default.aspx

163 Inuit Circumpolar Council, "Utqiagvik Declaration – The Arctic We Want," 3,4; Inuit Circumpolar Council, secs. 16–18.

164 Ibid., 7–8.

165 "In 2016, IUCN Members made the landmark decision to alter its membership structure for the first time in 60 years by creating an Indigenous Peoples Organisation category – making IUCN the first intergovernmental organisation to recognise and include Indigenous Peoples' Organisations as a distinct membership constituency." IUCN, "Indigenous Peoples Launch Self-Determined Agenda at IUCN World Conservation Congress," IUCN, September 3, 2021, https://www.iucn.org/news/governance-and-rights/202109/indigenous-peoples-launch-self-determined-agenda-iucn-world-conservation-congress-4

166 The "precautionary principle" refers to being cautious in advance to prevent harm from happening in a situation where further scientific insight (that could impact the decision) is not yet available. In this case, the ICC refers to its own early warnings of the impacts of climate change (and the ICC's recommended actions) in the absence of concrete knowledge.

167 Inuit Circumpolar Council, "People of the Ice Bridge: The Future of the Pikialasorsuaq. Report of the Pikialasorsuaq Commission," 2017, http://pikialasorsuaq.org/en/Resources/Reports

168 UN, "The Arctic Local Communities and Indigenous Peoples Platform," Local Communities and Indigenous Peoples Platform Web Portal, accessed July 18, 2022, https://lcipp.unfccc.int/about-lcipp/un-indigenous-sociocultural-regions/arctic

169 Dalee Sambo Dorough, "Inuit Call for the Tools Needed to Protect the Arctic" (ICC, October 28, 2021), https://www.inuitcircumpolar.com/project/inuit-call-for-the-tools-needed-to-protect-the-arctic/

170 Inuit Circumpolar Council, "Inuit Urge Protection and Preparation Now – To Avoid Paying the Ultimate Cost" (Inuit Circumpolar Council, February 18, 2020), https://www.inuitcircumpolar.com/news/inuit-urge-protection-and-preparation-now-to-avoid-paying-the-ultimate-cost/; Inuit Circumpolar Council, "Inuit Circumpolar Council, First Indigenous Peoples Organization to Formally Participate as an Observer at the IPCC" (Inuit Circumpolar Council, March 1, 2022), https://www.inuitcircumpolar.com/news/inuit-circumpolar-council-first-indigenous-peoples-organization-to-formally-participate-as-an-observer-at-the-ipcc/

171 Inuit Circumpolar Council, "ICC Presents Case for Consultative Status to IMO Council" (Inuit Circumpolar Council, March 1, 2021), https://www.inuitcircumpolar.com/news/icc-presents-case-for-consultative-status-to-imo-council/

172 Inuit Circumpolar Council, "Inuit Voices to Be Heard at IMO on Critical Shipping Issues | Inuit Circumpolar Council Canada," accessed June 7, 2022, https://www.inuitcircumpolar.com/news/inuit-voices-to-be-heard-at-imo-on-critical-shipping-issues/; Inuit Circumpolar Council, "Inuit and Small Islands Aligned on Reducing Black Carbon and the Decarbonization of the Global Shipping Fleet to Protect Vulnerable Communities | Inuit Circumpolar Council Canada" (Inuit Circumpolar Council, December 7, 2021), https://www.inuitcircumpolar.com/news/inuit-and-small-islands-aligned-on-reducing-black-carbon-and-the-decarbonization-of-the-global-shipping-fleet-to-protect-vulnerable-communities/

173 Inuit Circumpolar Council, "A Victory for Inuit at IMO – Indigenous Knowledge to Be Included to Deal with Underwater Ship Noise Pollution" (Inuit Circumpolar Council, January 24, 2022), https://www.inuitcircumpolar.com/news/a-victory-for-inuit-at-imo-indigenous-knowledge-to-be-included-to-deal-with-underwater-ship-noise-pollution/

174 Inuit Circumpolar Council, "Circumpolar Inuit Protocols for Equitable and Ethical Engagement" (Inuit Circumpolar Council, 2022), https://secureserver cdn.net/45.40.145.201/hh3.0e7.myftpupload.com/wp-content/uploads/EEE-Protocols-LR-WEB.pdf

175 Inuit Circumpolar Council, "2022 Inuit Circumpolar Council Declaration" (Inuit Circumpolar Council, July 2022), https://iccalaska.org/wp-icc/wp-content/uploads/2022/07/2022ICC-DECLARATION-1.pdf

176 Haluza-DeLay et al., "Speaking for Ourselves, Speaking Together: Environmental Justice in Canada," 10.

177 Abele and Rodon, "Inuit Diplomacy in the Global Era"; Frances Abele and Thierry Rodon, "Coming in from the Cold. Inuit Diplomacy and Global Citizenship," in *Indigenous Diplomacies*, ed. J. Marshall Beier (New York: Palgrave Macmillan, 2009), 115–85; Gary N. Wilson, "Inuit Diplomacy in the Circumpolar North," *Canadian Foreign Policy Journal* 13, no. 3 (January 1, 2007): 65–80, https://doi.org/10.1080/11926422.2007.9673443

178 Fleisher Trainor et al., "Environmental Injustice in the Canadian Far North," 152, 157.

179 Lydia Schoeppner, "The Role of International Institutions and Organizations in Sovereignty Conflicts in the Arctic," *Journal for Peace and Justice Studies* 24, no. 1 (2014): 50–86.

180 Sheila Watt-Cloutier, "It's Time to Humanize Climate Change Issues, Says Sheila Watt-Cloutier," *National Observer*, October 19, 2018, https://www.nationalobserver.com/2018/10/19/opinion/its-time-humanize-climate-change-issues-says-sheila-watt-cloutier

181 Dahl, *The Indigenous Space and Marginalized Peoples in the United Nations*, 93.

182 Mark Nuttall, "Inuit," in *Encyclopedia of the Arctic*, ed. Mark Nuttall, vol. 2 (New York: Routledge, 2005), 997.

183 Quoted in: Abele and Rodon, "Inuit Diplomacy in the Global Era," 58. Emphasis added.

Bibliography

Abele, Frances, and Thierry Rodon. "Inuit Diplomacy in the Global Era: The Strengths of Multilateral Internationalism." *Canadian Foreign Policy* 13, no. 3 (2007): 45–63.

_____. "Coming in from the Cold. Inuit Diplomacy and Global Citizenship." In *Indigenous Diplomacies*, edited by J. Marshall Beier, 115–85. New York: Palgrave Macmillan, 2009.

"About ICC | Inuit Circumpolar Council Canada," July 2022. https://www.inuitcircumpolar.com/about-icc/.

Adaptation Fund. "Adaptation Fund." Adaptation Fund. Accessed July 16, 2019. https://www.adaptation-fund.org/about/

Alaska Oil and Gas Association. "History – 1960's," 2019. https://www.aoga.org/industry/history-1960s.

Arctic Council. "The Arctic Council Turns 25," 2021. https://arctic-council.org/about/timeline/25/

Assembly of First Nations. "Declaring a First Nations Climate Emergency," July 2019. https://www.afn.ca/wp-content/uploads/2019/08/19-05-Declaring-a-First-Nations-Climate-Emergency.pdf

Assembly of First Nations. "Framing a First Nations Climate Lens." Whitehorse, March 3, 2020. https://www.afn.ca/wp-content/uploads/2019/08/19-05-Declaring-a-First-Nations-Climate-Emergency.pdf

_____. "About AFN." Accessed August 2, 2022. https://www.afn.ca/about-afn/

Bennett, John, and Susan Diana Mary Rowley. *Uqalurait: An Oral History of Nunavut*. Montreal: McGill-Queen's University Press, 2004.

Carlsson, Pernilla, Knut Breivik, Eva Brorström-Lundén, Ian Cousins, Jesper Christensen, Joan O. Grimalt, Crispin Halsall, et al. "Polychlorinated Biphenyls (PCBs) as Sentinels for the Elucidation of Arctic Environmental Change Processes: A Comprehensive Review Combined with ArcRisk Project Results." *Environmental Science and Pollution Research* 25, no. 23 (2018): 22499–528. https://doi.org/10.1007/s11356-018-2625-7

Cochran, Patricia. "Displaced by Climate Change," 2017. https://www.alaskapublic.org/wp-content/uploads/2017/07/Displaced-by-Climate-Change-Patricia-Cochran-Slides-6.6.2017.pdf

Dahl, Jens. *The Indigenous Space and Marginalized Peoples in the United Nations*. New York: Palgrave Macmillan, 2012.

Egede Dahl, Parnuna, and Pelle Tejsner. "Review and Mapping of Indigenous Knowledge Concepts in the Arctic." In *Inuit Worlds*, edited by Pamela Stern, 1st ed., 233–48. London: Routledge, 2021.

Faegteborg, Mads. "Inuit Circumpolar Conference, an Indigenous Organization as an Instrument of Change." In *Nordic Arctic Research on Contemporary Arctic Problems*, edited by Lise Lyck, 243–49. Aarhus: Aarhus Universitetsforlag, 1992.

_____. "Inuit Circumpolar Conference (ICC)." In *Encyclopedia of the Arctic*, edited by Nuttall, Mark, 2:999–1002. New York: Routledge, 2005.

Fenge, Terry. "POPs and Inuit: Influencing the Global Agenda." In *Northern Lights against POPs. Combatting Toxic Threats in the Arctic*, edited by David Leonard Downie and Terry Fenge, 192–213. Montreal: McGill-Queen's University Press, 2003.

Government of Canada. "Pan-Canadian Framework on Clean Growth and Climate Change Second Annual Report: Section 1," August 8, 2019. https://www.canada.ca/en/environment-climate-change/services/climate-change/pan-canadian-framework-reports/second-annual-report/section-1.html

_____. "Joint Committee on Climate Action Annual Report Highlights First Nations Leadership in Addressing Climate Change." News releases, August 13, 2021. https://www.canada.ca/en/environment-climate-change/news/2021/08/joint-committee-on-climate-action-annual-report-highlights-first-nations-leadership-in-addressing-climate-change.html

_____. "Canada's Partnership with Indigenous Peoples on Climate," January 26, 2022. https://www.canada.ca/en/environment-climate-change/services/climate-change/indigenous-partnership.html

Government of Greenland, Government of Nunavut, and ICC. "Joint Statement of Inuit Circumpolar Council, the Government of Greenland and the Government of Nunavut on Climate Change," December 8, 2015. http://www.inuitcircumpolar.com/uploads/3/0/5/4/30542564/joint_statement_on_climate_change_unfcc_cop21-eng.pdf

Gupta, Joyeeta. "A History of International Climate Change Policy." *Wiley Interdisciplinary Reviews: Climate Change* 1, no. 5 (2010): 636–53. https://doi.org/10.1002/wcc.67.

Haluza-DeLay, Randolph, Pat O'Riley, Peter Cole, and Julian Agyeman. "Introduction. Speaking for Ourselves, Speaking Together: Environmental Justice in Canada." In *Speaking for Ourselves. Environmental Justice in Canada*, edited by Julian Agyeman, Peter Cole, Randolph Haluza-DeLay and Pat O'Riley, 1–26. Vancouver: UBC Press, 2009.

Health Canada. "Health of Canadians in a Changing Climate. Advancing Our Knowledge for Action." Health Canada, 2022.

Heinämäki, Leena. "Rethinking the Status of Indigenous Peoples in International Environmental Decision-Making: The Role of Arctic Indigenous Peoples and the Challenge of Climate Change." In *Climate Governance in the Arctic*, edited by Timo Koivurova, E. Carina H. Keskitalo, and Nigel Bankes, 50:207–62. Environment and Policy. New York: Springer, 2009. http://ebenhopson.com/request-for-lilly-endowment-grant-support-first-international-inuit-conference/

Hopson, Eben. "Inupiat under Four Flags: Planning the First Inuit Circumpolar Conference." North Slope Borough, AK, 1976.

Hough, Peter. *International Politics of the Arctic Coming in from the Cold*. London: Routledge, 2013.

ICC. "Progress Made in Paris: ICC Urges Action Now." Ottawa, Dec 2015. http://www.inuitcircumpolar.com/uploads/3/0/5/4/30542564/press_release_progress_made_in_paris_icc_urges_action_now.pdf

———. "Chair's Message. Inuit Voices Informing Action," January 2016. http://www.inuitcircumpolar.com/uploads/3/0/5/4/30542564/chairs_message_-_january_2016_.pdf

IISD. "UN Backed Indigenous Peoples' Global Summit on Climate Change Ends with Declaration Calling for Fossil Fuel Phase-Out," April 24, 2009. https://unfccc.int/resource/docs/2009/smsn/ngo/168.pdf

Indigenous Peoples' Global Summit on Climate Change. "Report of the Indigenous Peoples' Global Summit on Climate Change." Anchorage: Indigenous Peoples' Global Summit on Climate Change, 2009. http://www.un.org/ga/president/63/letters/globalsummitoncc.pdf

International Indigenous Peoples Forum on Climate Change. "The Arctic." Accessed January 20, 2019. http://www.iipfcc.org/the-arctic/

———. "Who Are We?" Accessed January 20, 2019. http://www.iipfcc.org/who-are-we/

International Maritime Organization. "International Code for Ships Operating in Polar Waters (Polar Code)," n.d. https://www.imo.org/en/ourwork/safety/pages/polar-code.aspx

———. "Shipping in Polar Waters. Adoption of an International Code of Safety for Ships Operating in Polar Waters (Polar Code)." Accessed December 6, 2018. http://www.imo.org/en/MediaCentre/HotTopics/polar/Pages/default.aspx.

International Union for Conservation of Nature. *IUCN – A Brief History*. IUCN, 2018. https://www.iucn.org/about/iucn-brief-history

Inuit Circumpolar Conference. "First General Assembly. Summary and Resolutions." Barrow, June 13, 1977.

_____. "Resolutions 1980." Nuuk, 1980.

_____. "Resolutions 1983." Iqaluit, 1983.

_____. "Resolutions 1986." Kotzebue, 1986.

_____. "Towards an Inuit Regional Conservation Strategy." Kotzebue: Inuit Circumpolar Conference, 1986.

_____. "Resolutions 1989." Sisimiut, 1989.

_____. "Principles and Elements for a Comprehensive Arctic Policy." Montreal: Centre for Northern Studies and Research, McGill University, 1992.

_____. "Resolutions 1998." Nuuk, 1998.

_____. "Kuujjuaq Declaration." Kuujjuaq, 2002.

_____. "Utqiagvik Declaration." Utqiavik, 2006.

Inuit Circumpolar Council. "Inuit Arctic Policy," 2010.

_____. "Nuuk Declaration – Inoqatigiinneq, Sharing Life." Nuuk, 2010.

_____. "Kitigaaryuit Declaration – Ukiuqta'qtumi Hivuniptingnun, One Arctic, One Future." Inuvik, 2014.

_____. "People of the Ice Bridge: The Future of the Pikialasorsuaq. Report of the Pikialasorsuaq Commission," 2017. http://pikialasorsuaq.org/en/Resources/Reports

_____. "Charter & Bylaws." Accessed October 16, 2018. http://www.inuitcircum polar.com/icc-charter.html

_____. "Utqiagvik Declaration – The Arctic We Want." Utqiagvik, 2018.

_____. "Inuit Urge Protection and Preparation Now – To Avoid Paying the Ultimate Cost." Inuit Circumpolar Council, February 18, 2020. https://www.inuitcircum polar.com/news/inuit-urge-protection-and-preparation-now-to-avoid-paying-the-ultimate-cost/

_____. "ICC Presents Case for Consultative Status to IMO Council." INuit Circumpolar Council, March 1, 2021. https://www.inuitcircumpolar.com/news/icc-presents-case-for-consultative-status-to-imo-council/

_____. "Inuit and Small Islands Aligned on Reducing Black Carbon and the Decarbonization of the Global Shipping Fleet to Protect Vulnerable Communities | Inuit Circumpolar Council Canada." Inuit Circumpolar Council, December 7, 2021. https://www.inuitcircumpolar.com/news/inuit-and-small-islands-aligned-on-reducing-black-carbon-and-the-decarbonization-of-the-global-shipping-fleet-to-protect-vulnerable-communities/

_____. "A Victory for Inuit at IMO – Indigenous Knowledge to Be Included to Deal with Underwater Ship Noise Pollution." Inuit Circumpolar Council, January 24, 2022. https://www.inuitcircumpolar.com/news/a-victory-for-inuit-at-imo-indigenous-knowledge-to-be-included-to-deal-with-underwater-ship-noise-pollution/.

_____. "Inuit Circumpolar Council, First Indigenous Peoples Organization to Formally Participate as an Observer at the IPCC." Inuit Circumpolar Council, March 1, 2022. https://www.inuitcircumpolar.com/news/inuit-circumpolar-council-first-indigenous-peoples-organization-to-formally-participate-as-an-observer-at-the-ipcc/

_____. "Inuit Voices to Be Heard at IMO on Critical Shipping Issues | Inuit Circumpolar Council Canada." Accessed June 7, 2022. https://www.inuitcircumpolar. com/news/inuit-voices-to-be-heard-at-imo-on-critical-shipping-issues/

_____. "2022 Inuit Circumpolar Council Declaration." Inuit Circumpolar Council, July 2022. https://iccalaska.org/wp-icc/wp-content/uploads/2022/07/2022ICC-DECLARATION-1.pdf

_____. "Circumpolar Inuit Protocols for Equitable and Ethical Engagement." Inuit Circumpolar Council, 2022. https://secureservercdn.net/45.40.145.201/ hh3.0e7.myftpupload.com/wp-content/uploads/EEE-Protocols-LR-WEB.pdf

_____. "International Day of the World's Indigenous Peoples: Inuit Welcome the Historic UN General Assembly Recognition of a Clean, Healthy, Sustainable Environment as a Human Right." Inuit Circumpolar Council, August 9, 2022. https://iccalaska.org/wp-icc/wp-content/uploads/2022/08/Indigenous-Peoples-Day-2022_ENG.pdf

Inuit Tapiriit Kanatami. "National Representational Organization for Inuit in Canada." Accessed October 14, 2018. https://www.itk.ca/

_____. *National Inuit Climate Change Strategy*. Ottawa: Inuit Tapiriit Kanatami, 2019.

IUCN. "Indigenous Peoples Launch Self-Determined Agenda at IUCN World Conservation Congress." IUCN, September 3, 2021. https://www.iucn.org/news/ governance-and-rights/202109/indigenous-peoples-launch-self-determined-agenda-iucn-world-conservation-congress-4

JF Consulting. "Metis Nation Climate Change and Health Vulnerability Assessment." Metis National Council, 2020.

Keskitalo, E. Carina H., Timo Koivurova, and Nigel Bankes. "Conclusions on Climate Governance in the Arctic." In *Climate Governance in the Arctic*, edited by Timo Koivurova, E. Carina H. Keskitalo, and Nigel Bankes, 50:429–43. Environment and Policy. New York: Springer, 2009.

Kublu, Alexina, Frederic Laugrand, and Jarich Oosten. "Interviewing the Elders." In *Interviewing Inuit Elders*, edited by Frederic. Laugrand and Jarich Oosten, 1:1–12. Iqaluit: Nunavut Arctic College, 1999.

Kulchyski, Peter. "Colonization of the Arctic." In *Encyclopedia of the Arctic*, edited by Mark Nuttall, 1:405–11.. New York: Routledge, 2005.

Lynge, Aqqaluk. *Inuit. The Story of the Inuit Circumpolar Conference*. Nuuk, 1993.

Many Strong Voices. "About the Many Strong Voices Programme," 2014. http:// www.manystrongvoices.org/about.aspx?id=5068

_____. "Climate Change and Community-Based Relocation," 2014. http://www. manystrongvoices.org/activities.aspx?id=5155

_____. "Global Engagement in Climate Negotiations & IPCC," 2014. http:// www.manystrongvoices.org/activities.aspx?id=5153

_____. "Portraits of Resilience," 2014. https://www.manystrongvoices.org/documents/ MSVBook2014.pdf

Masterson, W. Dallam, and Albert G. Holba. "North Alaska Super Basin: Petroleum Systems of the Central Alaskan North Slope, United States." *AAPG Bulletin* 105, no. 6 (June 2021): 1233–91. https://doi.org/10.1306/01282120057

McGregor, Deborah. "Honouring Our Relations: An Anishnaabe Perspective on Environmental Justice." In *Speaking for Ourselves. Environmental Justice in*

Canada, edited by Julian Agyeman, Peter Cole, Randolph Haluza-DeLay, and Pat O'Riley, 27–41. Vancouver: UBC Press, 2009.

Miller, Anja. "Membership Information," December 19, 2018.

Mohai, Paul, David Pellow, and J. Timmons Roberts. "Environmental Justice." *Annual Review of Environment and Resources* 34 (2009): 405–30. https://doi.org/10.1146/annurev-environ-082508-094348

Mooney, Chris. "The Arctic Is Full of Toxic Mercury, and Climate Change Is Going to Release It." *Washington Post*, February 5, 2018. https://www.washingtonpost.com/news/energy-environment/wp/2018/02/05/the-arctic-is-full-of-toxic-mercury-and-climate-change-is-going-to-release-it/

National Collaborating Centre for Indigenous Health. "Climate Change and Indigenous Peoples' Health in Canada." Health of Canadians in a Changing Climate: Advancing Our Knowledge for Action. Health Canada, 2022. https://changingclimate.ca/health-in-a-changing-climate/chapter/2-0/

Nordic Co-operation. "Nordic Council of Ministers." Nordic cooperation. Accessed January 21, 2019. https://www.norden.org/en/nordic-council-ministers

Nunatsiaq Online. "Climate Change Linked to Human Rights, Inuit Leader Says," December 4, 2015. https://nunatsiaq.com/stories/article/65674climate_change_linked_to_human_rights_inuit_leader_says/

Nunavut Department of Education. *Inuit Qaujimajatuqangit Education Framework for Nunavut Curriculum*. Iqaluit: Nunavut Department of Education, 2007. https://www.gov.nu.ca/sites/default/files/files/Inuit%20Qaujimajatuqangit%20ENG.pdf

Nuttall, Mark. "Locality, Identity and Memory in South Greenland." *Études/Inuit/Studies* 25, no. 1/2 (2001): 53–72.

_____. "Inuit." In *Encyclopedia of the Arctic*, edited by Mark Nuttall, 2:990–97. New York: Routledge, 2005.

Public Health Agency of Canada. "From Risk to Resilience: An Equity Approach to COVID-19: Chief Public Health Officer of Canada's Report on the State of Public Health in Canada 2020." Ottawa: Government of Canada, 2021. https://www.canada.ca/en/public-health/corporate/publications/chief-public-health-officer-reports-state-public-health-canada/from-risk-resilience-equity-approach-covid-19.html

Rantanen, Mika, Alexey Yu Karpechko, Antti Lipponen, Kalle Nordling, Otto Hyvärinen, Kimmo Ruosteenoja, Timo Vihma, and Ari Laaksonen. "The Arctic Has Warmed Nearly Four Times Faster than the Globe Since 1979." *Communications Earth & Environment* 3, no. 1 (August 11, 2022): 1–10. https://doi.org/10.1038/s43247-022-00498-3

Ritchie, Liesel Ashley. "Individual Stress, Collective Trauma, and Social Capital in the Wake of the Exxon Valdez Oil Spill*." *Sociological Inquiry* 82, no. 2 (2012): 187–211. https://doi.org/10.1111/j.1475-682X.2012.00416.x

Sambo Dorough, Dalee. "Inuit Call for the Tools Needed to Protect the Arctic." ICC, October 28, 2021. https://www.inuitcircumpolar.com/project/inuit-call-for-the-tools-needed-to-protect-the-arctic/

Schoeppner, Lydia. "The Role of International Institutions and Organizations in Sovereignty Conflicts in the Arctic." *Journal for Peace and Justice Studies* 24, no. 1 (2014): 50–86. https://doi.org/10.5840/peacejustice20142413

_____. "The Inuit Circumpolar Council – Agent of Peacemaking for Inuit in Nunavut and Greenland." Doctoral Dissertation, University of Manitoba, 2020.

Simon, Mary. "Proposed Objectives for an Arctic Sustainable and Equitable Development Strategy." Protecting the Arctic environment (Report on the Yellowknife Preparatory Meeting, Yellowknife, NWT, April 18-23, 1990). Ottawa, 1990.

Statistics Canada. "Demographic Characteristics and Indigenous Groups," November 11, 2022. https://www150.statcan.gc.ca/n1/en/subjects/indigenous_peoples/demographic_characteristics_and_indigenous_groups

_____. "Statistics on Indigenous Peoples," 2022. https://www.statcan.gc.ca/en/subjects start/indigenous_peoples

Szpak, Agnieszka. "Arctic Athabaskan Council's Petition to the Inter-American Commission on Human Rights and Climate Change—Business as Usual or a Breakthrough?" *Climatic Change* 162, no. 3 (October 1, 2020): 1575–93. https://doi.org/10.1007/s10584-020-02826-y

Tagalik, Shirley. *Inuit Qaujimajatuqangit: The Role of Indigenous Knowledge in Supporting Wellness in Inuit Communities in Nunavut*. Prince George: National Collaborating Centre for Aboriginal Health, 2009–2010. https://www.ccnsa-nccah.ca/docs/health/FS-InuitQaujimajatuqangitWellnessNunavut-Tagalik-EN.pdf

Taylor, Alan. "The Exxon Valdez Oil Spill: 25 Years Ago Today." The Atlantic, March 24, 2014. https://www.theatlantic.com/photo/2014/03/the-exxon-valdez-oil-spill-25-years-ago-today/100703/

Teran, Tair, Lara Lamon, and Antonio Marcomini. "Climate Change Effects on POPs' Environmental Behaviour: A Scientific Perspective for Future Regulatory Actions." *Atmospheric Pollution Research* 3, no. 4 (2012): 466–76. https://doi.org/10.5094/APR.2012.054

The New York Times. "Hollywood's Hayek, Gyllenhaal Join Inuit for Arctic Circle Event on Earth Day," April 23, 2005, sec. Arts. https://www.nytimes.com/2005/04/23/arts/hollywoods-hayek-gyllenhaal-join-inuit-for-arctic-circle-event-on-earth.html

The Right Livelihood Award. "Sheila Watt-Cloutier," 2016. https://www.right livelihoodaward.org/laureates/sheila-watt-cloutier/

Trainor, Sarah F., F. Stuart Chapin, Henry P. Huntington, David C. Natcher, and Gary Kofinas. "Arctic Climate Impacts: Environmental Injustice in Canada and the United States." *Local Environment* 12, no. 6 (December 1, 2007): 627–43. https://doi.org/10.1080/13549830701657414

Trainor, Sarah F., Anna Godduhn, Lawrence K. Duffy, F. Stuart Chapin III, David C. Natcher, Gary Kofinas, and Henry P. Huntington. "Environmental Injustice in the Canadian Far North: Persistant Organic Pollutants and Arctic Climate Impacts." In *Speaking for Ourselves. Environmental Justice in Canada*, edited by Julian Agyeman, Peter Cole, Randolph Haluza-DeLay, and Pat O'Riley, 144–62. Vancouver: UBC Press, 2009.

Trott, Christopher G. "Social Relations among Inuit: Tuqłuraqtuq and Ilagiit." In *The Inuit World*, edited by Pamela Stern, 288–303. London: Routledge, 2021.

UN. "Report of the Conference of the Parties on Its Twenty-First Session, Held in Paris from 30 November to 11 December 2015," 2016.

_____. "The Arctic | Local Communities and Indigenous Peoples Platform." Local Communities and Indigenous Peoples Platform Web Portal. Accessed July 18, 2022. https://lcipp.unfccc.int/about-lcipp/un-indigenous-sociocultural-regions/arctic

UN Environmental Programme. *Inuit Regional Conservation Strategy for the Arctic.* Kotzebue: ICC General Assembly, 1988. http://www.global500.org/index.php/thelaureates/online-directory/item/631-inuit-regional-conservation-strategy-for-the-arctic

UN General Assembly. "Declaration on the Right to Development," 1986. https://www.ohchr.org/en/professionalinterest/pages/righttodevelopment.aspx

_____. "The Human Right to a Clean, Healthy and Sustainable Environment." UN, August 1, 2022. https://digitallibrary.un.org/record/3983329

United Nations. "United Nations Framework Convention on Climate Change," 1992.

United Nations. "Earth Summit," 1997. http://www.un.org/geninfo/bp/enviro.html

_____. *Paris Agreement.* Paris: United Nations, 2015. https://unfccc.int/sites/default/files/english_paris_agreement.pdf

_____. *Transforming Our World: The 2030 Agenda for Sustainable Development.* New York: United Nations, 2015.

_____. "UNPFII Members." Accessed January 13, 2019. https://www.un.org/development/desa/indigenouspeoples/about-us/members.html

United Nations Framework Convention on Climate Change. "What Is the Kyoto Protocol?," 2019. https://unfccc.int/process-and-meetings/the-kyoto-protocol/what-is-the-kyoto-protocol/what-is-the-kyoto-protocol

United States Environmental Protection Agency, OP. "Learn About Environmental Justice." Overviews and Factsheets, 2021. https://www.epa.gov/environmental justice/learn-about-environmental-justice

UN News. "UN General Assembly Declares Access to Clean and Healthy Environment a Universal Human Right," July 28, 2022. https://news.un.org/en/story/2022/07/1123482

UN Sustainable Development Goals. "Sustainable Development Goals." Accessed December 5, 2018. https://sustainabledevelopment.un.org/?menu=1300

Watt-Cloutier, Sheila. "The Inuit Journey Towards a POPs-Free World." In *Northern Lights Against POPs. Combatting Toxic Threats in the Arctic*, edited by Downie, David Leonard, and Terry Fenge, 256–67. Montreal: McGill-Queen's University Press, 2003.

_____. *The Right to Be Cold: One Woman's Story of Protecting Her Culture, the Arctic and the Whole Planet*, edited by Downie, David Leonard, and Terry Fenge. Toronto: Allen Lane, 2015.

_____. "It's Time to Humanize Climate Change Issues, Says Sheila Watt-Cloutier." National Observer, October 19, 2018. https://www.nationalobserver.com/2018/10/19/opinion/its-time-humanize-climate-change-issues-says-sheila-watt-cloutier

Wilson, Gary N. "Inuit Diplomacy in the Circumpolar North." *Canadian Foreign Policy Journal* 13, no. 3 (January 1, 2007): 65–80. https://doi.org/10.1080/11926422.2007.9673443

Women of the Metis Nation. Negative impacts of climate change on culture and cultural rights. Accessed August 3, 2022. https://www.ohchr.org/sites/default/files/Documents/Issues/CulturalRights/Call_ClimateChange/Women-of-the-metis-nation.docx

World Health Organization. "Mercury and Health." Accessed July 8, 2022. https://www.who.int/news-room/fact-sheets/detail/mercury-and-health

Yukon First Nations Climate Action Fellowship. "Reconnection Vision and Action Plan," 2021. https://www.yfnclimate.ca/yfnrvap

_____. "Origin and Purpose." Accessed August 2, 2022. https://www.yfnclimate.ca/origin-and-purpose

7 Ecocide, Ethnic Rights and Extractivism

Struggles for Environmental Justice in Mexico

Alessandro Morosin

> We live in difficult times. Defenders of their rivers and lands have been killed throughout our nation.
>
> Manuel Antonio Ruiz, Indigenous Zapotec educator
> San Francisco Ixhuatan, Oaxaca
> Addressing a 2016 community meeting about how Special Economic Zones will lead to dispossession of territories and natural resources

Overview: Generations of Social and Ecological War

Mexico has been experiencing social and ecological war for five centuries, and Mexico continues to occupy a strategic position in the world-economy. Since the Spanish conquest of the Aztec empire in 1519–1521 by the ruthless Hernán Cortés and his *conquistadores*, and the later Mexican-American War in which the U.S. seized half of Mexico's territory in 1848 (principally in order to expand America's slavery-based economy), Mexican society and its natural resources have undergone immense colonial and neocolonial processes.

More recently, in the past few generations, *both* foreign investment *and* a growing section of Mexican business conglomerates have reaped benefits and consolidated control of assets, land, infrastructure and natural resources that were previously under public or communal management. Since the 1990s, the Mexican government has opened the country to markets and capital to an unprecedented degree. This latest arrival of privatization and transnational finance capital has clashed with myriad traditional cultural norms and values; it has damaged and ruined the livelihoods of many people (while enriching a very small privileged elite who enjoys political connections); and it has wiped out or threatened key natural spaces and public health. Meanwhile, thousands of community efforts, activist organizations, movements, uprisings and social conflicts have come forward *in response to* environmental violence as one salient form of injustice in our era.[1]

DOI: 10.4324/9781003214380-10

One Mexican ethnobiologist defines environmental violence as "historically structured asymmetrical power relations that are reproduced and maintained to foster capital accumulation."[2] Proceeding from this important understanding, and looking at the history of Mexico, we can discern that colonialism laid this foundation through inequalities of wealth and power that are perpetuated today throughout cities, towns and regions.

Given Mexico's historical position within the world-economy, its natural resources have helped other imperialist nations, banks and corporations "develop." But this has impoverished the country's economy while also degrading its ecology/biodiversity. Mexico's bountiful food production, energy, forests, water and other raw materials that are today so coveted by globalized markets tend to be located in rural areas that are populated by Indigenous ethnic groups. This is pointed out by Nemer Narchi, a trained oceanographer who has been conducting research as an environmental anthropologist in Mexico's coastal and marine communities since the year 2000: "Everyone, not only the poor and marginalized, is a victim of environmental violence. Conversely, because of our consumption practices, we are all partners in the generation and administration of violence toward the environment."[3]

Struggles for environmental justice are one organized reaction to this larger social and ecological war. Environmental justice in Mexico spans many years in a large and diverse multicultural nation. This chapter cannot offer any full summary or review of the voluminous literature on Mexico's social movements and environmental conflicts. Instead, it highlights the interrelations among ecocide, ethnic rights and extractivism. By focusing on several recent and ongoing examples of why and how Indigenous people as well as Mestizos (mixed-race Mexicans) have fought back against exploitation of their communal resources and landscapes, we can better appreciate why social justice and ecological considerations are both essential to any genuinely sustainable future.

The title of this chapter names three key terms. First, what is meant by *ecocide*? Simply put, ecocide refers to destruction of the environment "to such an extent that peaceful enjoyment of a part of the planet will be substantially diminished."[4]Many cultures have retained some notion of ecocide throughout history based on their own environmental ethics and traditions, but it is only in recent years that scientists, criminologists and even some government officials have developed their thinking on ecocide, in tandem with the worsening global ecological crisis.[5]As of November 2020, international lawyers began to formally define ecocide as a crime that is punishable under the International Criminal Court. This growing awareness of ecocide is bringing about "a turning-point in how the relationship between humans and the natural world is understood."[6]

Native scholars have always maintained that colonialism equals eco-cide. On May 4, 1493, Pope Alejandro VI established the Bula *Noverunt Universi* which divided the lands of the "New World" between Spain and Portugal. As Indigenous communities and civilizations were separated from what were once communally owned lands and tools, lands became privately and individually appropriated for the first time, by Spanish royal decree.[7] This signals a longstanding connection between colonial violence and the destruction of the natural environment which humans belong to.

This builds a bridge to the second key term, ***ethnic rights***. Ever since Spanish colonialism, there has been a constant and steady agitation for rights, recognition and revolution in what is today Mexico. Oppressed ethnic groups in Mexico, often descendants from the land's original/pre-Hispanic inhabitants, have expressed their demands for environmental justice in countless forms.

In present-day Mexico, the 62 groups of Indigenous peoples (population of 12.7 million) represent 10.5 percent of the country's 130 million inhabitants.[8] In six states of Mexico, the rights of Indigenous communities over land use take the form of social properties, where lands are collectively owned (either in *ejidos* or *comunidades*) and recognized as such by the Mexican Constitution. These states are Chiapas, Veracruz, Yucatán, Oaxaca, Hidalgo and San Luis Potosí. Francisco López Barcenas (a historian, lawyer and activist of Indigenous Mixtec descent from the state of Oaxaca) emphasizes how ethnic rights and culture go beyond mere demographics: Indigenous people live on 22.9 percent of the country's *socially owned properties*. Based on their disproportionate membership in agrarian communities where property is socially owned, they legally possess and manage 28 percent of the country's forests and 50 percent of its jungles. Many of these communities have culturally rich forms of perceiving *life* (past, present and future; as well as human and non-human).[9]

As this chapter will illustrate, the Mexican government regularly takes the side of transnational corporations to further privatize these Indigenous and/or socially owned lands through coercive "legal" processes and/or armed violence. For this reason, the past two decades have witnessed not just a resurgence of publicly expressed Indigenous discontent, but an increasingly assertive uprising of Indigenous people in defense of traditional collective forms of managing the water, forests, landscapes and ancestral knowledge. The fight against ecocide and the battle for ethnic rights overlap to a large extent in contemporary Mexico: defeating the privatization of Indigenous lands also entails protecting pluri-ethnic languages, cultures and forms of communal governance. This has been increasingly recognized by scholars in recent decades and seems to have always been grasped by the movements themselves.

Finally, how can we name the enemy who is waging social and ecological war? How do contemporary environmental justice activists in Mexico name the system or the economic forces against which they struggle? Here we can introduce the term *extractivism*. Eduardo Gudynas, a prolific intellectual from Uruguay, defines extractivism as "activities which remove great quantities of natural resources that are not then processed (or are done so in a limited fashion) and that leave a country as exports."[10] Dominant transnational actors in Latin America over the past two decades or so have revived "the extraction of minerals, hydrocarbons, and agrifoods."[11] Many movements in Mexico understand "extractivism" as an economic model that plunders and removes natural resources, while ignoring the interests of the broad population or leaving them even worse off. Like other nations in the Americas, Mexico has seen many social movements intensify actions at the local level to stop these kinds of projects. This includes movements against hydroelectric dams, wind farms (or so-called wind factories), gold mines, mega-tourism, special economic zones (SEZs) and large agribusiness plantations, to name a few. Some of these Mexico-based environmental justice campaigns have also gained notoriety at the national and international level.

We will examine four cases of environmental justice struggles in Northern, Central and Southern Mexico. These experiences attest to how people have exposed and resisted ecocide, ethnic rights and extractivism from the local level, with national reverberations.

Wixárika People Challenge Extractivism in Wirikuta (Central-Western Mexico)

One of Mexico's oldest surviving pre-Colombian cultures has attracted international notoriety and solidarity by challenging extractivist interests and extensive government plans couched as development in the state of San Luis Potosí. The Wixárika people are known to the outside world as Huicholes. Out of respect for their quest for self-determination, I refer to them by their self-definition as *Wixárika*, or *Wixaritari* plural (rather than "Huicholes"). This ancient group with a population of over 40,000 inhabits communities across parts of the Sierra Madre Occidental (neighboring states of Jalisco, Nayarit, Durango and Zacatecas). Their ancestral territory, or *kiekari*, spans the desert of San Luis Potosí. Among their pantheon of non-human landscapes, this section focuses on the anti-mining struggle and chemically dependent agribusiness.

The town of Real de Catorce in San Luis Potosí was established by the Spanish as a colonial mining center in the 1670s. Wixárika communities in the mountains are located some 250 miles west of Real de Catorce. They consider the town to be part of a larger region with sacred significance to which members make annual pilgrimages from October to May. This entails ceremonial migrations across desert steppes to a mountain known as Cerro Quemado (or *Leunaxü* in Wixárika language), collecting *peyote* cactus to perform rituals. The United Nations proclaimed this a UNESCO heritage site in 2000, highlighting its endemic flora and fauna. But co-existing with this narrative of biodiversity and conservation is a more recent language of cultural survival and cultural patrimony that the traditionally reclusive Wixaritari have proven quite skillful in fomenting throughout their contentious negotiations with corporate and government forces.

From 2010 through 2015, the Wixaritari and their allies led a relatively successful non-violent campaign that resulted in a state sanctioned moratorium on mining activity. There were some compromises made by Canadian mining company First Majestic Silver and its Mexican partner Real Bonanza. This also resulted in a presidential promise that no sacred lands would be affected by mining activity. At the same time, support for the mining project tends to remain consolidated among non-Indigenous locals of Real de Catorce. Aside from mining, the cultural autonomy of Wirikuta as well as its sensitive ecology remains threatened to this day by a confluence of other projects which include agro-industrial activity, peyote cactus hunting and drug-trafficking. Due to the corresponding de-forestation, monocropping is equally "extractive" as mining, even though mining toxifies landscapes to an even greater degree. Below, I draw from scholarly and journalistic sources, as well as Wixaritari publications, to parse through the evolution of this major social and environmental justice campaign and to survey some overall lessons from this case.

The land that this Indigenous group recognizes as Wirikuta exists in about 540 square miles of the San Luis Potosí desert ecosystem. There are footpaths, mountains, slopes and springs. The Wixaritari consider the underground aquifers to be "veins" that connect the desert to temples in their mountain communities to the west. In an effort to ensure good harvests plus individual and collective well-being, they wish for their ceremonies to remain unimpeded. In 2010, the Canadian mining company First Majestic Silver was granted 22 concessions for silver mining on 15,000 acres in the state of San Luis Potosí. This proposed mining project lies adjacent to Wirikuta's sacred lands, including Cerro Quemado as well as a mountain of solar emergence that the Wixaritari call *Reu'unaxi*.

Since the imposition of neoliberal economic policies in the early 1990s, large-scale mining projects in Mexico have provoked new alliances and forms of protest (although to date, no national anti-mine constituency has been able to significantly roll back, cancel or renegotiate the thousands of mining projects in Mexico as a whole). The movement that grew out of protecting Wirikuta is significant for being one of the first successful anti-mining campaigns. In a pattern, that has been shared and replicated by other environmental mobilizations, different segments of the Mexican population supported the Wixárika people to help foment a multi-generational environmental front. They demonstrated opposition to First Majestic Silver and to the federal government's granting of exploration permits to this company for open-pit gold mining. Soon thereafter, a much broader constituency of support emerged and was given organized expression, which in turn inspired to greater unity of action among the Wixaritari. In this particular case, the coalition of Indigenous people and diverse allies known as the Frente en Defensa de Wirikuta (or FDW) began to receive a great deal of publicity.

On October 27, 2010, Wixaritari community members organized a mass march down La Reforma Boulevard in Mexico City. This announced the national political presence of the heretofore more withdrawn and insular lives of the Wixaritari. With this coordinated mass action in Mexico City, the community and its aspirations became better known to the larger public in Mexico and the world. Then on February 6, 2011, representatives of more than 20 separate *peregrinos* (pilgrims) staged a common ceremony at the top of the tall peak of Cerro Quemado/*Reu'unaxi*. Wixaritari have come from time immemorial to leave corn, feathers, water from springs and other sacred objects to this site. But with the granting of mining concessions on this land by the Mexican government, the separate pilgrimages traditionally conducted by different Wixárika families had become *one single event*. This gathering of 600 people also served as a declaration to defend this desert from mining and other invasive activities: "Dozens of members of the press, academia, and the public invited by the Huichols" and by the newly formed FDW organization attended the event, witnessing shamans chanting to hundreds of tribe members.[12] At this meeting and since then, Wixaritari have produced and inspired a voluminous diffusion of media which frame their ancestral knowledge as an essential legacy not just for Mexicans, but for humanity and the world. An environmental justice movement was being inaugurated with Mexico's most ancient Indigenous group at its center, in the valley where they consider the world to have first originated. Journalist and researcher Dawn Paley captured the stakes of the conflict in Wirikuta by commenting "[i]magine drilling for oil under the Vatican, or bulldozing Eden to make room for a golf course."[13]

But it should be noted that the town of Real de Catorce itself has never been monolithic. As with many sites of socio-environmental conflicts, "the people" here are not so unified, but culturally and economically divided. What was once a more civil relation between *catorcenos and huicholes* (non-Indigenous locals and Wixaritari) became transformed. In this highland town of Real de Catorce and its satellites located 8 hours from the San Luis Potosí state capital, First Majestic Silver as well as the FDW and pro-Wirikuta rights activists have attempted to sway locals and win support from the townspeople. Water is also a fundamental point of contention. Around this time, there were groupings of scholars, native activists and other locals who were meeting to discuss recent studies that had been done on the area's precarious water resources. In conversations with some residents who were in favor of the mine, scholars warned that installing the mine would deplete Real de Catorce's water sources by 2030.

The historic mining industry in this town has waned throughout the 20th century while tourism has also declined to a trickle. Many of the job-seeking townspeople are loyal to the mining industry—hopeful for its return. For example, Wirikuta activists have joined forces with environmentalists and conservationists to propose certain alternative projects that would meet the pro-mine townspeople's demand for employment. But tensions increased when the federal government became involved without first contacting the Wixaritari. Locals in Catorce seemed to have a strong preference for mining over these "alternative projects." Mexico's federal conservation agency (CONANP) in 2012 proposed a "reserve" area that would limit (though not ban) mining within a certain spatial area. Upon receiving this news, some Mestizos in Catorce suspected the Wixaritari to have fallen under the supposed manipulation of the government and conservationists who (supposedly) wanted to appropriate the territory for themselves. This echoed a time-honored anti-Indigenous trope, heard throughout North American history and also expressed by recent Brazilian President Jair Bolsonaro, that "Indians" are selfishly occupying vast lands that they have no moral or legal right to control. These pro-mine locals in Catorce hoisted posters in the streets with slogans such as "No to the biosphere reserve" and "Mining goes hand in hand with sustainable development and the root of Real de Catorce is mining."[14]

Since this area has already been the site of 200 years of metal mining, the landscape was toxified in many regards before First Majestic Silver even existed. Again, this colonial history is latched onto by locals who make the argument that mining is linked with local identity. Activists in defense of Wirikuta are operating in this context in order to prevent the even more drastic damage that would be wrought by today's chemically intensive methods of extracting minerals. Salvador Chava Contreras, an anthropologist who has been working in the area and collecting interviews since 2014, has underscored this point: "It is worth mentioning that people are protecting a place that has essentially already been ruined by colonial mining, and if mining continues, it will make the place completely uninhabitable by humans and animals."[15]

Mining companies around the world engage in extensive public relations maneuvers to paint their industry as sustainable, natural and necessary for the development (or revival) of rural commerce. This sentiment has a material basis in some locales, even if it can be argued that pro-mine views are ultimately misguided. It appeals to people who are "trapped" in now-declining communities that have active memories of being mining towns. But ecological crises emerging in the 21st century, including in the desert of central Mexico, are often downplayed in much of the pro-mining discourse, or misunderstood by its proponents. The questions and objections pertaining to drought, climate change and toxicity have been expounded on by scientists and environmental activists for decades. For its part, the FDW staked out rhetorical and scientific claims that were picked up by some national and international media, such as this one:

> The methods and chemical substances implied in mining activities bring destruction to the regions where they are employed. The quantity of water that this company [First Majestic Silver] aims to utilize to obtain the mineral is irrational ... we are talking about a desert. The economic renumeration promised to the region's inhabitants is miserable compared to the profits that the company would exploit from natural resources.[16]

As seen up to now, the massive use of water in proposed mining projects always implies power struggles between winners and losers. But the background to the conflict in Real de Catorce also centers on certain edible plants. One of these, the cactus, grows naturally in this region but is threatened. Additionally, red tomatoes have begun to be cultivated in low-lying valleys due to market interests. Let us consider how the ongoing environmental justice struggle in Wirikuta intersects with the symbolic importance of endemic cactus.

An entire fifth of the world's cactus species grow in the San Luis Potosí desert. The round, spineless *peyote* cactus, which is only found in this desert between southern Texas and northern Mexico, contains hallucinogenic mescaline alkaloids. It is referred to by the Wixaritari as *hikuri*, or simply as medicine. Consuming the *peyote* in strictly regulated occasions (e.g. after a fast and celibacy or during a pilgrimage, and in guided communion) is foundational to the Wixárika culture. This practice closely mirrors ceremonial use of *peyote* among other Native peoples on both sides of the U.S.-Mexico border, even those who traditionally reside hundreds of miles from the desert itself. The right to consume mescaline directly from *peyote* entailed hard-fought confrontations for religious freedom within the U.S. courts. Popularized in the 1950s and 1960s by countercultural figures in Anglo societies, the *peyote* cactus is "subject to special protection" and at risk of extinction due to habitat destruction, illegal poaching and unsustainable harvesting practices. It takes 15 years for the *peyote* to grow from seed to plant. The increased demand for the psychedelic experience among tourists who descend on the desert, combined with the potential for gold mining in Wixárika territory as well as agribusiness, is identified by the Wixárika and their allies as an existential threat to their identity as a people.[17]

In addition to over-exploitation of *peyote*, tomato farming in the Chihuahuan Desert combines issues of unsustainable industrial water use and capitalist globalization with unaccountable political processes that impact Wirikuta's sacred lands. Fragile desert aquifers are being strained by historic levels of drought and reduced rainfall combined with mismanagement of water resources (this is the case not only for Mexican states like San Luis Potosí in the southeastern Chihuahuan Desert, but for the U.S. Southwest as well).[18] State policies in 2012 set into motion the growth of private investments in San Luis Potosí's high plateau region. The State Development Plan of San Luis Potosí (*Plan Estatal de Desarrollo*) for 2015–2021 announced "The promotion of the agricultural sector and agro-industrial activity is a priority to promote jobs, income and social well-being."[19] The State Plan also noted that San Luis Potosí's production of tomatoes is Mexico's second highest in terms of volume and value. This pro-growth policy has resulted in the installation of industrial tomato farms and greenhouses (some located within the National Protected Area of Wirikuta, and others located outside of it) which extract over 4 million cubic meters of non-renewable aquifer water.

Additionally, the pork and poultry industries, both well-known for being highly water-intensive, have established themselves in the desert of San Luis Potosí. Not only have these projects amplified deforestation in the supposedly protected refuge of Wirikuta, but they have increased the use of pesticides, fertilizers and other waste products which in turn degrade soil quality. Since the Wixaritari's identity rests on the ritual use of the already-endangered *peyote* cactus, unrestrained agribusiness can also be seen through the tripartite lens of ecocide, extractivism and ethnocide. A publication by the Wixárika Research Center calls attention to environmental degradation in this underappreciated and over-exploited desert landscape: "These dramatic transformations have begun to overpower the special botanical aroma of the high plateaus. Agrochemicals and animal waste now exude a fetid odor, and flies are unbearable in the proximities of the San Juan egg facilities."[20]

Wixaritari community members affiliated with the FDW movement have widely used the internet and social media to draw continued attention to their aims and demands, to frame issues and to garner national and international support.[21] The FDW comprised Wixaritari members as well as *mestizos* and other non-Indigenous environmentalist allies. But after the protests in 2010, the Wixárika decided to form an all-Indigenous organization, which is still called the Consejo Regional Wixárika (CRW). This group is formed by a structure based in many Indigenous communities known as *cargos*, which are positions appointed by direct election among members of the Wixárika community itself in Durango, Jalisco and Nayarit. Agricultural producers and *campesinos* who are part of cargo systems must report to the CRW. In this way, the CRW has the express intent of protecting sacred sites and controlling the narrative of engagement through press events and legal battles.[22]

There is evidence of a broad Indigenous-led environmental justice movement that ranges beyond anti-mining campaigns. Some small landholders, collectives and residents have published communiques that express opposition to the aforementioned agro-industrial projects. For example, one such group of Indigenous activists took issue with being excluded from government decisions involving water use. This published statement was promoted by other environmental and human rights websites. It reads in part: "Around our municipalities, we see the atrocious advance of agroindustry that disassembles enormous surfaces and extracts underground water that belongs to all and employs agrochemicals over large terrains that harvest tomatoes, peppers, alfalfa and squash which also pollute our soil, air and subterranean water" [original in Spanish].[23]

Large-scale mining by First Majestic Silver has been placed on hold thanks to a mobilization by the Wixárika and their supporters throughout 2010–2015 that won a legal injunction against the project. However, other regional projects such as peyote exploitation, tomato farming and animal agriculture are drastically altering the region's environment and social relations under the pretext of economic development. Wirikuta's Indigenous people and its environmental struggles have awakened and reinvigorated other movements and campaigns throughout Mexico.

Cherán: A Town Defends Its Forests and Expels Narcotraffickers (Western Mexico)

Many additional flashpoints throughout Mexico exemplify dynamics of ecocide, extractivism and ethnic rights mobilizations, but one particularly important and noted case of the past decade is that of Cherán. Cherán is a municipality with a population of 20,000, at an elevation of 7,385 feet in the western state of Michoacán. The community is ethnically homogenous, populated by the P'urhépecha Indigenous people who speak P'urhépecha as well as Spanish. Social ties are close-knit: since residents are overwhelmingly born in the community, people express a high degree of trust and empathy for one another.[24]

Cherán borders ten other P'urhépecha municipalities in this region of northwestern Michoacán. The state of Michoacán continues to have one of the highest rates of violent crime and internal displacement in Mexico owing to organized crime's role in extortion and illegal logging. Well-organized

gangs also control the longstanding drug trade (particularly methampheta-
mine production, opium poppy cultivation, and fentanyl smuggling in the
Tierra Caliente region bordering the state of Jalisco).[25] However, the small
Indigenous community of Cherán now has one of the lowest violent crime
rates in all of Mexico, with zero homicides since 2011. What happened?
Cherán's success story is about self-mobilization, an insurrection by the
poor and campesinos to stop illegal logging, and the achievement of local
autonomy based on direct democratic self-rule.

Gangs have been permitted to flourish in this socio-politically marginalized region of the country. Before 2011, Cherán's troubles resembled those
of its rural peer communities in Michoacán: kidnappings, murders, deforestation, drug smuggling and corruption were fixtures of daily life. Based
on studies from satellite images and soil vegetation maps, 71 percent of the
original vegetation area had been lost between 2006 and 2012. This equals
9,069.35 hectares, or 35 square miles of forest area. Some of this land was
deliberately burned by illegal loggers as a way to intimidate the population as well as to accelerate an official change in land use from forestry to
commercial exploitation.[26]

On April 15, 2011, community defense committees stood up to organized crime and resisted the Mexican government's drug war model, which
began in 2006 under conservative President Felipe Calderon. These civilian
volunteers temporarily blocked lumber trucks and detained ten men who
had gotten "permission" from the compromised local government to chop
down trees in a barrio called El Calvario—a problem that had persisted in
the whole municipality since the 1990s. Importantly, one immediate trigger
to this uprising appears to have been dozens of kidnappings, rapes and ex-
trajudicial executions by alleged associates of the Knights Templario cartel
who had been involved in logging. This motivated a small group of women,
later dubbed *Las Mujeres del Calvario* (Women of the Calvario) to lead one
of the first human blockades against lumber trucks. One of these women,
Doña Chucha, had previously witnessed her husband (a local leader of the
communal lands commission) shot in his home by cartel hitmen.[27]

The loggers were reportedly backed by La Familia Michoacana cartel, and they responded to the civilian blockade with armed violence that
killed three protesters. When the local/state/federal governments failed
or refused to intervene to protect the villagers and their precious oak forests, the villagers eventually formed a *Ronda Comunitaria* (also called
autodefensas, armed self-defense, community watch groups or civilian
militias).[28] The four major neighborhoods of the community were block-
aded in 180 separate locations by men, women and children day and night
so as to restrict outsiders' ability to enter. Up to the present day, the armed
Ronda Comunitaria continues to patrol the three access points with the
objective of keeping organized crime and illegal loggers out.

Like many environmental justice struggles throughout Mexico to varying degrees, it is important to recognize that this particular uprising "involved the valorization of a collective and territorially rooted identity."[29] In other words, Cherán's inhabitants have a well-developed sense of community. Their collective identity has enabled them to protect and recover their territory and democratic forms of decision-making throughout many twists and turns since 2011. They uphold a sense of K'umanchikua, a P'urhépecha word meaning "home."[30] This movement is known for taking extraordinary steps to reclaim its people's home in the broadest sense.

In 2011, a Cherán resident and participant was interviewed about the motivations, demands and goals of this environmental justice movement. Their testimony gives form to the local understanding and content of K'umanchikua:

> We rose up because our children deserve to have a forest, to have something in the future. As I am older, I can say that we have had that opportunity, but they deserve it too. The forest is our sustenance, it is our life, it is what allows us campesinos and Indigenous to progress ... We demand Justice, Security, and Peace for the town of Cherán. No more loggings, no kidnappings. And an end to organized crime.[31]

Self-defense groups in states like Michoacán, particularly in Cherán, can be interpreted as both a cause *and a result* of successful collective action in high-risk circumstances. When "an at least *partially* armed civilian group occupies the public realm" in order to combat criminal organizations in particular, political scientists have called this "insurgent vigilantism."[32] In Cherán, people utilized *fogatas* as a key communicative space for expressing their anger and their desire for justice. In doing so, they were adapting a tradition that had survived for hundreds of years in Michoacán's highlands. *Fogatas* are campfire assemblies where community members meet to plan and discuss and resolve the conflicts of the day. While embracing the *fogatas* as a foundation of the local culture, the villagers eventually evicted the timber mafia and its complicit politicians. This is captured in the testimony from a participant: "It was at the *fogatas* where people started saying they were fed up, that they were tired of being afraid. That's where we decided we wanted to fight, and we started to organize."[33]

The movement has had remarkable success in forming a lasting system of *cargos*, or self-governance structures directly accountable to the people and elected by consensus. Based on Article 33 of the Mexican Constitution, majority-Indigenous municipalities may exercise their right to be governed by *usos y costumbres*, or traditional Indigenous customs outside of the dominant political system. Along with these achievements come some impressive advances in sanitation—most especially, a large rainwater collector on top of Mount Kukundicata with a capacity for 20 million liters of water. This rainwater collector is actually the largest in Latin America and has been operational since 2016.

Scholars will continue to debate the reasons why Cherán has been able to realize its vision of a social and environmental justice to this extent. We can conclude with the insights of one researcher who collected data there for years. Italian anthropologist Giovanna Gasparello pinpointed how the impunity enjoyed by criminal gangs eventually precipitated a massive, planned response from below: "Although the conflict regarding the use of land and communal resources is old, the connection between extractive interests and the direct violence of organized crime determined the organized reaction against dispossession."[34] Its achievements remain palpable today, over a decade after the initial uprising. They have been widely disseminated by committed journalists, national and international environmental/ pro-Indigenous networks and localized activists in other parts of Mexico who take inspiration from Cherán for their own campaigns.

Biodiversity and Farm Laborers vs. Export Agriculture in the Valley of San Quintín (Northwestern Mexico)

The Valley of San Quintín resides in the northern and western (Pacific) side of Baja California Norte, a peninsular Mexican state that borders the U.S. state of California. The Valley lies 185 miles south of the U.S./Mexico border (San Diego/Tijuana). With its six small towns, and roughly 30,000 inhabitants, the San Quintín Valley is the most important agricultural area in the state. It was also the epicenter of a major labor strike by farmworkers, most of whom are migrants from Indigenous communities in Mexico's interior.

Agriculture in the Valley remains dependent on four underground aquifers. Its year-round production of high-quality tomatoes, strawberries and raspberries continues to expand. Other crops include onions, avocados, wheat and cut flowers. Before the arrival of export agriculture, San Quintín was only sparsely populated. Tomato farmers from the states of Sinaloa and Mexicali began to produce in the San Quintín region by the 1970s and 1980s. Growers of tomatoes in Baja, and buyers in the U.S., benefited from the completion of Baja California's Transpeninsular Highway in 1973. This region did not have an extensive local workforce prior to the construction of the highway. It emerged as a new export zone since then, attracting tens of thousands of workers from Southern Mexico. About half of all farmworkers in San Quintín were born in another state. With 41 percent from Oaxaca and 13 percent from Guerrero, the workers who pick these crops that U.S. consumers buy in supermarkets are often speakers of Indigenous languages such as Mixtec and Zapotec. Reports suggest that 64 percent lack adequate housing and access to water, power and sanitation.[35] Today, agriculture utilizes 80 percent of the San Quintín Valley's water supply.[36] Groundwater extraction has been extracted at highly unsustainable levels and soil-water intrusion has damaged soil quality for over two decades.[37]

The wetlands, lagoons and sand dunes of Baja California attract a variety of rare migratory birds. The region is also host to many endemic plants, but its biodiversity is *critically endangered* by industrial agriculture and the associated urbanization. Various mammals are facing extinction in the area due to habitat loss, while over-exploitation of water sources threatens amphibians and plant species.[38] The immediate source of these significant ecological pressures in Baja California lies in the current profit-driven farming practices. However, this same neoliberal economic logic has also led to (1) recruitment of low-wage and undereducated workers from Indigenous regions in the interior of Mexico, and (2) pro-union mobilizations by these workers who (in spite of having some of the highest agricultural wages in Mexico) are squeezed by high costs of living, poor working conditions and substantial individual costs of running water/sewer services.[39] One 27-year-old worker, Celina Sierra, was quoted about conditions in the field where she had worked half her life: "We get up at 4 in the morning to wait for the truck to take us to the fields ... We get to the fields at 6:30 even though we're not supposed to start work till 7 ... They never pay overtime, they tell us we have to finish a harvest and work until four or five o'clock in the afternoon, bent over and sometimes without water."[40]

In March of 2015, the hidden exploitation and hardships of farm laborers in San Quintín began to grab headlines as several thousand farmworkers went on strike. Workers who were earning an average of $10 per day initially demanded doubled wages (300 pesos), state-healthcare and an independent union.[41] They marched on the Baja Peninsular highway for 15 days, shutting down the export of berries and tomatoes. For nine months, political conflict ensued among workers, growers (including larger farms and smaller farms), politicians in Baja California and U.S.-based importers such as Driscoll's. After almost two months, some farm owners even attempted to use the police to force workers into trucks and back onto the fields. Indignant workers reacted by throwing rocks at police, but those who suffered greatly were the workers themselves, as two workers died and five others were arrested. The governor of the state used armored vehicles to break up road blockades. He claimed that the freedom to protest would be respected while illegal acts (i.e. "crimes" committed by the striking farmworkers) would be punished.

The powerful strike in San Quintín just a few hours' drive from the U.S. border with San Diego, and the unresolved socio-ecological problems it concentrates, should remind us that high-profit margins in export agriculture necessitate a "vulnerable, largely invisible workforce and cheap unprotected nature."[42]

After almost three months of negotiations and a boycott of related crops on both sides of the border, the government of Baja California reached a settlement with workers and growers. The deal included a minimum daily wage of $12 (150 pesos per day), which was significantly less than what the workers were initially pushing for. Federal, state and local governments pledged to improve public services that these migrant workers depend on, and many workers voted to retain an independent union.

As other analysts on the case of San Quintín have forcefully pointed out, "the exploitation of natural resources transcends the negative local consequences and impacts nature as a global system. There is a need for new, more equitable, and postcapitalist systems of production directed toward the greater good ... it will take a much larger logistic effort to ensure the overall balance between agricultural lands and native flora and fauna, water tables, and soil biodiversity."[43] While labor-based conflicts in the Valley are ongoing, the 2015 strike illustrates how the expansion of capitalist accumulation, with the support of competing government factions, engenders serious environmental problems.

Southern and Southeastern Mexico, from 1990s to Present Day: *No es Desarrollo, es Despojo!*

In December of 2021, I arrived in Oaxaca City (capital of the Mexican state of Oaxaca) to spend my winter break visiting my partner and reconnecting with some research contacts who reside in different areas of the state. We learned via social media that grassroots organizations working for Indigenous rights and environmental justice were holding a press conference the very next day in the center of the city. We made our way to Plaza Santo Domingo to attend this gathering in front of the baroque cathedral constructed in 1608 under Dominican order. A few dozen people (some members of NGOs and activist collectives) had congregated to deliver and hear statements and speeches outlining their concerns about a range of government-supported projects. They voiced dissent against mining, hydroelectric dams, fracking, agro-industry, refineries and water privatization. These industrial works remain in motion, and have even received a boost, in Oaxaca and Southern Mexico throughout the allegedly center-left presidential administration of Andrés Manuel López Obrador (AMLO) which won federal elections in December of 2018.

Mexico's southern and southeastern states indeed have the highest rates of poverty in the nation, and AMLO is certainly not the first Mexican leader to put forward audacious economic plans for states like Chiapas,

Yucatan and Campeche. But instead of investing in health and public transportation, AMLO assumed office while banking on his most massive undertaking: a 1,000-mile passenger railway known as the Mayan Train. Since then, government spokespeople for the project have repeated AMLO's talking points, claiming it will be a "great detonator" for the region. Fernando Vasquez of FONATUR, Mexico's national tourism agency, has been ebullient: "President Andrés Manuel sees the Mayan Train as an important engine of economic development that seeks to close inequality gaps." After what was announced by AMLO to be a four-year project has proven much more environmentally damaging and costly to taxpayers, many engineers, architects and urban planners who once supported the project have warned of its undemocratic and potentially disastrous planning process.[44] But the grassroots organizations that we met on December 13, 2021 dissented even more stridently. Under a colorful banner that hung on trees in the Plaza Santo Domingo, a young activist with tattoos and face piercings originating from the ethnic Zapotec city of Juchitán de Zaragoza spoke calmly with confidence:

> What is happening to us in the communities, to our forests, jungles and rivers, will not only have a direct effect on those of us who inhabit the territory. It will also entail everything that supplies the big cities: the food they consume from the farmers and fishermen … these projects will endanger all of the forms of life that we know … this project of the Inter-Oceanic Corridor is the handover (*la entrega*) of Mexican territory, of the Tehuantepec Isthmus, to the great economic powers that are engaged in a power dispute, principally the USA. It will not only affect the Isthmus and Oaxaca. A series of projects and privatizations are coming, that we had with prior governments, and that *this* government with its discourse of being a 'Left' government, is doing everything that the Right was not able to do.[45]

The speaker identified himself as Mario and passed out copies of a written statement authored by *Asamblea Oaxaqueña en Defensa de la Tierra y el Territorio* (Oaxacan Assembly in Defense of Land and Territory). The title of the flyer duplicated the main slogan that was printed on the hanging banners that decorated this impromptu press conference: *No Es Desarrollo—Es Despojo!* which translates to *It's not Development, It's Dispossession!*

Here I can only offer a cursory overview of this "series of projects and privatizations" that has been a long time in the making. Some neoliberal policies and initiatives in Southern Mexico have received intense opposition by the more organized sectors of Indigenous communities. As these projects have rolled out in phases by several federal administrations,

Figure 7.1 Press conference attended by the author: December 13, 2021. Oaxaca City (in Plaza Santo Domingo). Speakers from diverse social movements and non-governmental organizations denounce corruption and megaprojects. Banner: "People in Struggle and Resistance. Permanent Campaign: It's Not Development, It's Dispossession!" Photo taken by Author, December 13, 2021.

resistance has become increasingly well-networked with other campaigns and movements across Mexico—culminating in the current civil dissent against the Mayan Train.

Since the presidential administration of Carlos Salinas de Gortari (1988–1994), "southern and southeastern Mexico has been a constant preoccupation for the Mexican government," but the enormous resources dedicated have certainly not "resolved poverty or reduced social discontent" in the region.[46] In early 2001, then-President Vicente Fox announced the Plan Puebla Panama (PPP) aimed at Mexico's nine southeastern states and all seven Central American republics. The PPP aimed to "build, or improve, large infrastructure projects (toll highways, airports, deep-water ports, electrical and telecommunications grids), that, together with on-going projects (hydroelectric dams, 'dry' trans- isthmus canals), would motivate large private companies to locate there."[47] In fact, the $10 billion project was forced to backtrack due to local, regional, national and transnational bodies of grassroots opposition, which included coalitions like

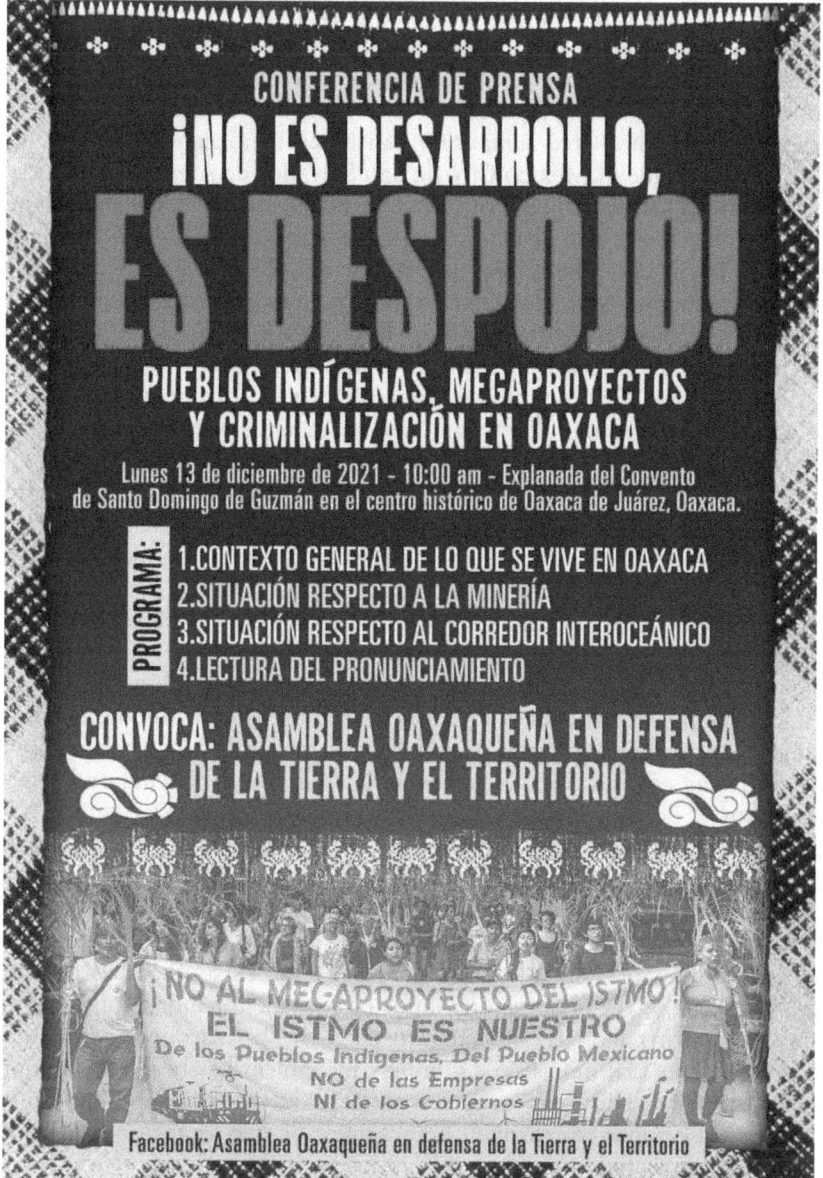

Figure 7.2 Poster by Asamblea Oaxaqueña en defensa de la Tierra y el Territorio ("Oaxacan Assembly in defense of Land and Territory"). Advertising a press conference on "Indigenous Peoples, Megaprojects, and Criminalization in Oaxaca." Photo taken by Author, December 13, 2021.

AMAP (the Mexican Alliance for People's Self Determination), the EZLN (Zapatista Army of National Liberation), plus "a multitude of forums on dams, biodiversity, water, agrotoxins, genetically-modified substances, militarization, autonomy, grassroots economics and others."[48]

Along with the fierce *campesino* resistance to a planned airport on farmlands outside of Mexico City (in San Salvador de Atenco), this flurry of protest rocked back much of President Fox's economic agenda. At the same time, these same civil uprisings and sophisticated networks laid the basis for much of the counter-hegemony of Indigenous groups and their supporters that has since flowered. In spite of having few if any allies in national office, these coalitions continue to challenge more recent iterations of these same extractivist programs in Southern Mexico.

Plan Mesoamerica (or the Mesoamerica Project) can be viewed as the PPP with a changed moniker throughout the remainder of the early 2000s under President Felipe Calderon. Plan Mesoamerica continued neoliberal restructuring by channeling FDI (foreign direct investment) mostly from the U.S. but also from other wealthy nations into Mexico's manufacturing and finance sectors to boost trade and growth. This dynamic expanded into a legitimacy crisis for Mexico's government once a contested and likely fraudulent election in 2006 gave Calderon the justification to launch a war against drug cartels. This has been an immensely violent affair which has turned large parts of the country into conflict zones, all while fueling attacks on land defenders, journalists and activists who oppose corrupt and unconstitutional policies. Criminologist Steven Osuna explains how the U.S. has advised Mexico to invest in militarized policing and securitization of borders precisely in order to maintain globalized capital accumulation: "Rather than communists or subversives as the national and transnational security threat, criminality and drug trafficking is the substitute that justifies the deployment of state forces, paramilitary groups, and U.S. intervention."[49]

The Isthmus of Tehuantepec located in Oaxaca's southeastern region served as one of the next major rounds of struggle for Indigenous autonomy, land rights and environmental justice—this time around the politics of wind energy. An intense and complex movement has been playing out in the Isthmus against massive wind energy parks that privatized the lands of Zapotec *campesinos* through highly questionable legal mechanisms.[50] In 2003, USAID published a report on this region's strategic international potential for generating wind energy. Almost two decades later, the Isthmus is dotted with some 2,000 wind turbines by wind energy companies which have invested $7 billion in Oaxaca. Organizations of ethnic Zapotec and *Ikoots* communities have, to varying degrees, risen up in nonviolent *and* physical resistance to these so-called wind factories and to stop their continued expansion.[51] Throughout my ethnographic interviews and conversations in the Isthmus, I learned that many of these activists have endured the common experience of death threats, kidnapping attempts, jail, defamation and police violence. Groups such as the APPIIDT and APPJ have at times led hundreds of community members to employ civil disobedience (as well as other less dramatic tactics) to challenge the legality of these corporate complexes which generate privatized energy for transnational corporations.[52]

One prolific field researcher who has studied the wind energy conflict in great depth has noted: "In towns like La Ventosa in the Northern Coastal Istmo, many of these promises [by the wind energy companies] remained unfulfilled, limited and benefited a minority of the population."[53] Additionally, in spite of being viewed by many as some "clean" solution to the climate crisis, corporate wind parks in Oaxaca have some harmful environmental impacts: "Wind energy still requires large mining and processing facilities to refine iron, stainless steel, dysprosium, oil lubricants, sealing resins, fibreglass, concrete as well as the construction of transportation and electrical infrastructure networks."[54] The ecological harms of corporatized wind farms have long been identified and protested by activists on the ground.

In the wake of the quiet collapse of the PPP and the rise of the Indigenous-led movement against these coastal Oaxacan wind energy parks, a new panorama of extractivism began coming into focus. This inspired a more intense level of activist networking in Southern Mexico and nationally. Companies who aim to extract gold, silver and other metals have become the personified figureheads of what land defenders and human rights workers call *proyectos de muerte* (projects of death), though mining is of course not the only industry which threatens farmland, forests, water, air quality and social peace.

In just three presidential administrations (those who held office from 1994 to 2012), *nearly half* of Mexico's national territory has been auctioned off to mining companies. That is, over 95 million hectares of land were being explored and prepared for exploitation in the year 2014 by mining companies, three fourths of which are Canada-based. Such intense extractivism represents a continuation of the centuries of social and ecological warfare that we referred to in the introduction of this chapter. However, capital technologies and the globally competitive rush to access the last of cheap minerals have quickened the pace of the conflict. For example, during the 310 years of outright Spanish colonialism, 190 tons of gold were extracted from what is present-day Mexico. Today, in just a single decade (2000–2010), corporations managed to extract 420 tons of gold.[55]

This new wave of mineral extractivism (*el nuevo extractivismo minero*) has divided communities between those who favor the mining industry and those who oppose it. It has been well documented how community members and local officials receive monetary bribes to allow mining companies access to land, while those who continue to protest the mining companies are extremely vulnerable to assaults and even murder. In one early example, the company Fortuna Silver Mines Inc. arrived in Oaxaca's Central Valley region in 2006. When their silver mine in San Jose del Progreso became active in 2010, hundreds of peaceful anti-mine protesters blocking roads were attacked by police and military forces. In 2012, two leading opponents of Fortuna Silver were killed by paramilitary attacks that remain unresolved and unpunished to this day: Bernardo Mendez, followed by Bernardo Vásquez several days later.[56]

Some communities in the state of Oaxaca have succeeded in stopping open-pit mining activities by multinational corporations before full-scale mining operations could even begin. Each of these campaigns was supported along the way by human rights organizations and environmental justice advocates within Mexico and abroad. At the local level, these efforts are motivated by agrarian grievances, environmental concerns, health and Indigenous worldviews, while they interact with the government's ongoing legitimacy crisis and the distrust of corporations. These Oaxacan victories against the mining industry include the communities of Capulalpam de Mendez (located in the Sierra Norte), Magdalena Teitipac (in the Central Valley region) and Ixtepec (in the Isthmus of Tehuantepec, just north of Juchitán).

In order for minerals and other raw material to be exploited, infrastructure must be modernized, and fiscal/legal measures must offer greater financial incentives. Meanwhile, the more that infrastructure such as highways, ports and telecommunications are established, the more that natural resources and cheap labor can be accessed and tapped. This dynamic can be seen in the policies that were attempted, and partly fulfilled, by President Enrique Peña Nieto (2012–2018). Much like his predecessors, he continued policies that closely wedded Mexico to the U.S. in terms of "energy reform," foreign policy, militarization and macroeconomics. Early in his term, he issued decrees outlining his plan for "Special Economic Zones" (SEZ's) in seven sites of Southern Mexico. This was yet another effort to encourage Mexican as well as foreign investors to move their operations to Southern Mexico. The federal laws on SEZ's included such corporate giveaways as exemption of income tax for ten years, and zero value-added tax for investors who acquire goods in an SEZ.[57]

Community members in Oaxaca whom I have known since 2016 and interviewed about their motivations for persisting in environmental justice work, have long associated these state-corporate partnerships as inimical to collective land rights. One Catholic school coordinator and Zapatista sympathizer in the Isthmus of Oaxaca (which was then slated to become one of Mexico's main SEZ's) addressed this point head-on at a town assembly: "We live in difficult times. Defenders of their rivers and lands have been killed throughout our nation. The government is planning a Special Economic Zone on our lands. This will directly impact our lives, mostly in Indigenous areas."[58]

I attended many meetings in which speakers and organizers were not only talking about how Special Economic Zones would accelerate the extraction, export and destruction of natural resources—they were also referencing and screening videos about similar movements and affectations throughout Mexico and Latin America. These communal tactics helped raise consciousness to the greater social injustice of extractivism on an international scale. This offers more evidence that movements in a particular region build on and learn from prior struggles—such as how Oaxaca's campaign against wind energy parks primed participants and community members to scale up the fight against Special Economic Zones. It also suggests that human rights defenders are highly engaged, aware and linked with other circles of activists and news sources. This ongoing connectivity helps provide analysis and coordinated guidance about other movements which are happening hundreds of miles away.

At the same time, most movement participants exhibit clear strategic thinking about why such policies bring environmental risks as well as more militarism. A young educated woman of Zoque ethnic background, at her home in San Miguel Chimalapa (a mountain municipality in Oaxaca near Chiapas which is targeted by mining companies) had this to say: "It seems to me that, it is quite logical to read it here in Mexico, that the military supports investments, but on the one hand, in these conflictive zones of Chimalapas, there is the illicit issue—the issue of migrant smuggling and drug trafficking."[59]

This woman has been a member of a local feminist activist group called Colectivo Matza (which means morning star). The group resolutely opposes proposals by Canadian mining companies to enter this biodiverse mountain municipality that they grew up in. I ask her whether stopping mining is Colectivo Matza's main goal, or whether they have a larger platform they work toward. She answered: "It could be thought of in terms of the Earth. Of our Mother Earth, protection of water, of the love for our territory. And if mining or any project threatens our space, our territory ... then yes, our effort is concentrated against such projects."[60]

But grassroots efforts of which my respondent speaks, such as the campaign to close off the Chimalapa Forest of southeastern Oaxaca and Chiapas to corporate incursions, are very risky. Consider the fact that since 1985, there has been a massacre in some Indigenous community of Oaxaca under every governor since Jesús Martínez Álvarez, according to Francisco López Barcenas. Despite the progressive promises by President López

Obrador to crack down on corruption and protect Indigenous rights, territorial disputes in Mexico's south drag on as powerful actors continue to stake claims on this large and still relatively under-exploited multiethnic region. This is partly because many of the same functionaries from prior neoliberal governments have been recruited to top positions by AMLO, and indeed some Indigenous political actors have been co-opted. This situation is argued in an editorial to a major daily newspaper:

> Government officials who call themselves neoliberals have been recycled, and the indigenous people who operate the policies have ended up being functional to the government that they fought against in previous years. In this sense ... megaprojects promoted by previous governments were not only not suspended in the current one, but other similar projects were added. Although it is denied, the policies of dispossession continue to be practiced.[61]

Above, López Barcenas is referring to how 15 people were tortured and brutally killed in an outlying village of the ethnic *Ikoots* municipality, Santa Maria del Mar, on June 21, 2020. Santa Maria del Mar is a coastal lagoon municipality in the Oaxacan Isthmus that for years has challenged megaprojects like the wind farms which would cut off access to fishing and contaminate the sea. The massacre took place in Huazantlán, a smaller satellite actually closer to the industrial city of Salina Cruz that has long been implicated in trans-Isthmus economic plans. Santa Maria del Mar has undergone some complicated electoral disputes, and the tragedies of violence in 2020 were an escalation of these tensions.[62] Disagreeing with President AMLO's "official" line that this was merely a typical inter-communal dispute, López Barcenas and his activist-researcher colleagues conclude that the loss of life in Santa Maria del Mar conforms to a disturbing pattern of organized crime, or "informal operators," facilitating the implementation of government projects in Oaxaca. This is the analysis that AMLO and official media outlets will not air. The Special Economic Zones of former President Peña Nieto have been renamed or otherwise abandoned, but in their place, AMLO has been promoting the Trans-Isthmus Corridor (*Corredor Transístmico*) which has very similar aims of turning the area between Veracruz and Oaxaca (from the Gulf of Mexico south toward the Pacific Coast) into a hub for trade, production and raw materials. For this reason, critics point to a strained social fabric resulting in greater conflict, disputes and corruption.

Some small but prolific and dedicated organizations have continued documenting all of this, while helping cohere much of the grassroots environmental and pro-Indigenous activism. Throughout the course of my fieldwork, the Oaxaca City-based group EDUCA consistently comes up in conversations, lawsuits and workshops. EDUCA (*Servicios Para Una*

Educacion Alternativa, a civil association founded in 1997) works to make land defenders visible and advise a great many grassroots environmental justice campaigns. In one example among hundreds of others from different parts of Mexico, EDUCA publicized the aforementioned atrocity in San Mateo del Mar in a detailed press release.[63]

The documents revealed by a group of pro-environmentalist hackers ("Guacamaya") in October of 2022 as part of 400,000 leaked military and police emails from Mexico and other Latin American nations provide further evidence of numerous organized crime groups operating in Oaxaca. Local cliques part of larger cartels, such as Los Zetas and their rivals Cártel de Jalisco Nueva Generación, rely on several of Oaxaca's regions for cultivation, processing and transport of drugs by air, land and sea.[64] This large data leak paints a grim picture of Southern Mexico's penetration by organized crime. Documents expose army weapons sold to criminal groups, and the possible continued collusion of officials in top government agencies like SEDENA (National Defense). This of course results in a failure on the part of military personnel to prevent homicides based on intelligence they're receiving.[65]

Another recent social-environmental conflict in Oaxaca's Isthmus of Tehuantepec, connected to the planned "inter-oceanic corridor" (*Corredor Interoceánico del Istmo de Tehuantepec*), has been the criminalization of protest in the municipality of San Blas Atempa since 2021. The small community of Puente Madera (with just over 700 inhabitants, 98 percent of whom are Zapotec) rejects the construction of an industrial park (one of three planned for the Isthmus) on 300 hectares of their communal lands known as El Pitayal. Other activists seem to accuse the municipal president of being a narcotrafficker and call out all branches of government for threatening local inhabitants who dissent against related land use policies.[66]

One of my respondents who lives in the eastern Isthmus community of Santo Domingo Zanatepec (Daniel Cirilo) has consistently been involved in environmentalist efforts to prevent mining and other extractive industries. According to Daniel, who regularly travels to attend workshops and forums by EDUCA, the work of EDUCA has been very helpful: "It's allowed us to dream. These issues are now being recognized."[67] Zapotec men and women have marched with banners and signs to denounce "irregularities" in the consultation process which violate communal land statutes. They argue that the proposed industrial park threatens wild animals (armadillos, deer, rabbits) and the lands that residents use for gathering firewood and cultivating crops.[68] EDUCA has supported these efforts, which have informed and further energized people from nearby towns such as Daniel. Recently, he has been working with a small committee of neighbors to create an educational/agricultural community space in his fraction of

Zanatepec. On separate occasions, Daniel has welcomed me to visit this project which he calls "Lixi Bandaga" (*house of leaves* in Zapotec).

Daniel also offered me his opinion about the Mayan Train project. He fears the destruction of forest lands, underground caves (*zenotes, cavernas*) and animal species. Daniel commented that while the government was digging and clearing lands for the train, they found temples that were part of the ruins of an ancient Mayan city. He feels that since information in Mexico is so centralized, "the government only shows what they want you to see," but that "social movements and grassroots media reveal more of the truth."[69] Daniel is certainly not alone in these sentiments. Environmental and Indigenous activist networks in Mexico have for the past several years been publicizing and protesting the Mayan Train. Besides professional working scientists, a great deal of ecological knowledge exists among volunteer groups like *Las Chelemeras*—a collective of 18 women who have been working to restore the mangroves of the Yucatán Peninsula.[70]

Back in Oaxaca, the sixth annual *Día de Rebeldía Contra la Minería* (Day of Rebellion Against Mining) took place in Oaxaca's capital city in July of 2022. Organized and hosted by numerous communities and organizations, the call for this day of rebellion read in part: "we are united by the cause of the defense of territories and our common goods against extractivism, and the dispossession that threatens our territorial space, as well as against the false discourse of progress and development, with which they have deceived our communities in order to plunder our riches."[71]

Conclusion: No Sustainable Future without Native Peoples and Ecological Revolution

As global climate change, militarism, authoritarianism and extreme weather patterns reinforce one another in the early 21st century, many who have adopted a passive consumer mentality have fallen victim to bleak cynicism about humanity's future. These case studies showcase a counter-example. Even with the difficulties and sacrifices they've endured, environmental justice struggles in Mexico may help point the way toward a collectively just and ecological future, with coalitions among diverse social sectors and regions. The understandings forged about ancestral methods of farming, food supply, resource conservation and reverence for non-human landscapes is compelling. It can certainly be important when famous persons with a large platform act and speak in solidarity with Indigenous peoples and environmental justice struggles. But the movement depends on the willingness of many ordinary individuals and communities, like those mentioned in this chapter, to directly disrupt corrupt political processes and the environmental violence of profit-driven industries.

Whether a more openly extractivist, violent and conservative leader replaces AMLO, or whether some other candidate from AMLO's MORENA party continues to carry the Mexican presidency for another six years, the Mayan Train could result in the destruction of millions of acres of tropical rainforest. The current government seeks to maneuver around its political problems such as Central American migrants, tourism dependence, narcotraffickers and the need to earn foreign currency. Zapatista communities in the state of Chiapas have paid a price for their strident criticisms of the Mayan Train, seen in multiple murders and other armed attacks by pro-government paramilitaries.[72] While co-optation of dissent by the state and elites is certainly not new in Mexico, it seems that much of civil society has been quiet about how federal policies under AMLO are not a rejection of his predecessors' unpopular neoliberalism, but "a nationalist reencoding of it."[73] The groups and movements outlined in this chapter provide some counter-examples of environmental justice and dignity that can be further built upon—though certainly they are not the only grassroots campaigns and networks within Mexico that can be appreciated in this way.

Mexico's grassroots environmental movements share many of the same challenges as others in the Americas. While autonomous localized campaigns in regions of Latin America "may manage to halt the advance of extraction at the local level, they have certain limitations in challenging the legitimacy of the region's extractive projects and the deepening of dependency. Only if these incipient forms of local resistance manage to become more widespread will it be possible to formulate a counterhegemonic project that changes the articulation between the extractivism of transnational capital and national development projects."[74]

Notes

1 Daniel Faber, "The Ecological Crisis of Latin America: A Theoretical Introduction," *Latin American Perspectives* 19, no. 1 (1992): 3–16.
2 Nemer E. Narchi et al., "Environmental Violence and the Socio-environmental (de) Evolution of a Landscape in the San Quintín Valley," *Latin American Perspectives* 47, no. 6 (2020): 103.
3 Nemer E. Narchi, "Environmental Violence in Mexico: A Conceptual Introduction," *Latin American Perspectives* 42, no. 5 (2015): 11.
4 Sailesh Mehta and Prisca Merz, "Ecocide- A New Crime against Peace?" *Environmental Law Review* 17, no. 1 (2015): 3.
5 David Ziegler, *The Invention of Ecocide: Agent Orange, Vietnam, and the Scientists Who Changed the Way We Think about the Environment* (Atlanta: University of Georgia Press, 2011); Avi Brisman and Nigel South, *Green Cultural Criminology: Constructions of Environmental Harm, Consumerism, and Resistance to Ecocide* (New York: Routledge, 2014); Martin Crook, Damien Short, and Nigel South, "Ecocide, Genocide, Capitalism and Colonialism: Consequences for Indigenous Peoples and Global Ecosystems Environments,"

Theoretical Criminology 22, no. 3 (2018): 298–317; Mark Allan Gray, "The International Crime of Ecocide." In *International Crimes* (New York: Routledge, 2017): 456–511.

6 *The Economist* (no author), "It is time for ecocide to become an international Crime" (2021). https://www.economist.com/international/2021/02/28/is-it-time-for-ecocide-to-become-an-international-crime

7 Jorge J. Gómez de Silva Cano, *El derecho agrario mexicano y la Constitución de 1917. Colección INEHRM* (Mexico City: Instituto de Investigaciones Jurídicas, Universidad Nacional Autónoma de México, 2017).

8 Francisco López Barcenas, "Pueblos indígenas y megaproyectos en México: las nuevas rutas deldespojo," Mexico: Patrimonio biocultural, 2011.

9 Barcenas, "Pubelos indígenas y megaproyectos en México," 3, 6.

10 Eduardo Gudynas, "Diez tesis urgentes sobre el nuevo extractivismo," *Extractivismo, política y sociedad* 187 (2009): 187–225.

11 Emiliano López and Francisco Vértiz, "Extractivism, Transnational Capital, and Subaltern Struggles in Latin America," *Latin American Perspectives* 42, no. 5 (2015): 152.

12 Paul M. Liffman, "Mining and the Huichol Ancestral Landscape," *Frente en Defensa de Wirikuta*, March 29, 2012. https://frenteendefensadewirikuta.org/p1176.html

13 Dawn Paley, "A Canadian Mining Company Prepares to Dig Up Mexico's Eden." *This* (2011) https://this.org/2011/09/15/first-majestic-silver-wirikuta/

14 Andrew Boni, Claudio Garibay, and Michael K. McCall, "Sustainable Mining, Indigenous Rights and Conservation: Conflict and Discourse in Wirikuta/Catorce, San Luis Potosí, Mexico," *GeoJournal* 80, no. 5 (2015): 774.

15 Salvador Chava Contreras, personal correspondence with the author, October 2022.

16 Mariana Leon, "Real Bonanza: mina de la Discordia," *El Universal*, 2012 https://archivo.eluniversal.com.mx/notas/836671.html

17 For an excellent critical analysis of the contradictions of the area's expanding "shamanic tourism" which some locals around Real de Catorce have attempted to harness with mixed results, see Vincent Basset, "New Age tourism in Wirikuta: Conflicts and Rituals," in *Peyote: History, Tradition, Politics, and Conservation* (Santa Barbara, CA: ABC-CLIO/Praeger Publishers, 2016): 191–210.

18 Priyadarsi D. Roy, Sekar Selvam, Selvaraj Gopinath, Natarajan Logesh, José L. Sánchez Zavala, and Chokkalingam Lakshumanan, "Geochemical Evolution and Seasonality of Groundwater Recharge at Water-Scarce Southeast Margin of the Chihuahuan Desert in Mexico," *Environmental Research* 203 (2022): 1–13. https://doi.org/10.1016/j.envres.2021.111847

19 Gobierno del Estado de San Luis Potosí, "Plan Estatal de Desarrollo", 2016. https://slp.gob.mx/sitionuevo/DocumentosPLAN/plan2016_eje1.pdf

20 Diana Negrín, "Water and Power in Wirikuta." Wixárika Research Center. https://www.wixarika.org/water-and-power-wirikuta 2021.

21 Alexandra Engelsdorfer, "Producing In-Between Spaces of Resistance: Nexus-Thinking and the Virtual and Physical Spaces of Protest around the 'La Luz' Mining Project in Wirikuta, Mexico," *IReflect: Student Journal of International Relations* 4, no. S1 (2017): 97–116.

22 Consejo Regional Wixárika, "En Defensa de Wirikuta," (2022). http://consejoregionalwixarika.org/

23 Altiplano Wirikuta, "Excluyen a wixárikas de decisiones sobre megaproyectos y problemática del Altiplano Wirikuta en San Luis Potosí." Biodiversidad LA,

(2019). https://www.biodiversidadla.org/Documentos/Excluyen-a-Wixárikas-de-decisiones-sobre-megaproyectos-y-problematica-del-Altiplano-Wirikuta-en-San-Luis-Potosí

24 Michael J, Wolff, "Insurgent Vigilantism and Drug War in Mexico," *Journal of Politics in Latin America* 12, no. 1 (2020): 37.

25 Vanda Felbab-Brown, "Criminal Violence, Politics, and State Capture in Michoacán," *Mexico Today*, 2021. https://www.brookings.edu/opinions/criminal-violence-politics-and-state-capture-in-michoacan/

26 María Luisa España-Boquera, and Omar Champo-Jiménez, "Proceso de deforestación en el municipio de Cherán, Michoacán, México (2006–2012)," *Madera y bosques* 22, no. 1 (2016): 141–53; Giovanna Gasparello, "Communal Responses to Structural Violence and Dispossession in Cherán, Mexico," *Latin American Perspectives* 48, no. 1 (2021): 42.

27 Wolff, "Insurgent vigilantism," 37.

28 Clayton Conn, "Cherán: Community Self Defense in Mexico's Drug War." In *North American Congress on Latin America*, 2011. https://nacla.org/blog/cher%25C3%25A1n-community-self-defense-mexico%25E2%2580%2599s-drug-war-photo-essay

29 Gasparello, "Communal Responses to Structural Violence," 42.

30 Emmanuel Guillén, Cherán, el pueblo mexicano que expulsó a criminales y politicos. *Vice,* 2021. https://www.vice.com/es/article/4av58b/cheran-el-pueblo-mexicano-que-expulso-a-criminales-y-politicos

31 Conn, "Cherán: Community Self Defense in Mexico's Drug War."

32 Wolff, "Insurgent Vigilantism," 37.

33 Wolff, "Insurgent Vigilantism," 39–40.

34 Gasparello, "Communal Responses to Structural Violence," 50.

35 Omar Millan, "Mexico's Baja Farmworkers Strike for Better Conditions," *San Diego Union Tribune,* March 24, 2015. https://www.sandiegouniontribune.com/sdut-mexicos-baja-farmworkers-strike-for-better-2015mar24-story.html

36 Rural Migration News, (no author) "Farm Workers in San Quintín," *RMN Blog* 254 (2021). https://migration.ucdavis.edu/rmn/blog/post/?id=2676

37 Alfonso Aguirre-Muñoz, Robert W. Buddemeier, Víctor Camacho-Ibar, José D. Carriquiry, Silvia E. Ibarra-Obando, Barbara W. Massey, Stephen V. Smith, and Fredrik Wulff, "Sustainability of Coastal Resource Use in San Quintin, Mexico," *AMBIO: A Journal of the Human Environment* 30, no. 3 (2001): 142–9; George Zalidis, Stamatis Stamatiadis, Vasilios Takavakoglou, Kent Eskridge, and Nikolaos Misopolinos, "Impacts of Agricultural Practices on Soil and Water Quality in the Mediterranean Region and Proposed Assessment Methodology," *Agriculture, Ecosystems & Environment* 88, no. 2 (2002): 137–46.

38 Narchi et al., "Environmental Violence and the Socio-environmental (de) Evolution," 109.

39 Rural Migration News, "Farm Workers in San Quintín."

40 Millan, "Mexico's Baja farmworkers strike for better conditions."

41 Tomas Vilagrim, "Thousands of Mexican Farmworkers March 15 Miles as Strike Talks Continue," *Los Angeles Times,* March 26, 2015. https://www.latimes.com/world/mexico-americas/la-fg-baja-farmworkers-strike-20150326-story.html

42 Narchi et al., "Environmental Violence and the Socio-environmental (de) Evolution," 113.

43 Narchi et al., "Environmental Violence and the Socio-environmental (de) Evolution," 113–4.

44 Maria Abi-Habib, "Over Caves and Over Budget, Mexico's Train Project Barrels toward Disaster." *New York Times*, August 28, 2022. https://www.nytimes.com/2022/08/28/world/americas/maya-train-mexico-amlo.html

45 Speech by Mario (last name unknown), personal transcription by the author, December 2021.

46 Jaime Ornelas Delgado, "El Plan Puebla-Panamá y la globalización neoliberal," *Aportes* VII, no. 21 (2002): 143.

47 Miguel Pickard, "The Plan Puebla Panama Revived: Looking Back to See What's Ahead," *La Chronique des Amériques* 12, no. 7 (2004): 1–7.

48 Molly Talcott, *Claiming Dignity, Reconfiguring Rights: Gender, Youth, and Indigenous-Led Politics in Southern México* (Santa Barbara: University of California, 2008); for quotation see Pickard, "The Plan Puebla Panama Revived."

49 Steven Osuna, "Securing Manifest Destiny," *Journal of World-Systems Research* 27, no. 1 (2021): 16.

50 Sofia Avila-Calero, "Contesting Energy Transitions: Wind Power and Conflicts in the Isthmus of Tehuantepec," *Journal of Political Ecology* 24, no. 1 (2017): 992–1012.

51 Alexander Dunlap and Martín Correa Arce, "'Murderous energy' in Oaxaca, Mexico: wind Factories, Territorial Struggle and Social Warfare," *The Journal of Peasant Studies* 49, no. 2 (2022): 455–80.

52 Alessandro Morosin, "Paramilitaries in Oaxaca, Mexico: Enforcing Accumulation in a Geo-Strategic Region," in *Paramilitary Groups and the State under Globalization: Political Violence, Elites, and Security*, ed. J. Hristov, J. Sprague, and A. Tauss (New York: Routledge, 2021), 88–106.

53 Alexander Dunlap, "The 'solution' Is Now the 'problem': Wind Energy, Colonisation and the 'genocide-ecocide nexus' in the Isthmus of Tehuantepec, Oaxaca," *The International Journal of Human Rights* 22, no. 4 (2018): 558.

54 Dunlap, "The 'solution' Is Now the 'problem,'" 560.

55 Ricardo M. Reséndez, *Entregaron la mitad de méxico a mineras.* https://razacero.com/?p=724

56 Pedro Matias, "'Toman' Minera en San José del Progreso, Oaxaca: Exigen Salida de Empresa Canadiense." *Proceso*, 2016. https://www.proceso.com.mx/nacional/estados/2016/5/6/toman-minera-en-san-jose-del-progreso-oaxaca-exigen-salida-de-empresa-canadiense-163917.html

57 Santiago L. Zapatero, and Alberto A. Pineda, "Where Are Mexico's Special Economic Zones?" *International Tax Review,* October 12, 2018. https://www.internationaltaxreview.com/article/2a68vp2bhdes26sg0okxs/where-are-mexicos-special-economic-zones

58 Morosin, "Paramilitaries in Oaxaca."

59 Interview with a young woman in San Miguel Chimalapa (anonymous) conducted by the author.

60 *Ibid.*

61 Francisco López Barcenas, "Pueblos indígenas y megaproyectos en México: las nuevas rutas deldespojo." Mexico: Patrimonio biocultural, 2011.

62 Daniela Rea, San Mateo del Mar, "nos quieren matar y ponerle precio a la tierra," *Pie de Página,* 2020. *https://piedepagina.mx/san-mateo-del-mar-nos-quieren-matar-y-ponerle-precio-a-la-tierra/*

63 Oaxaca Denuncia, "La violencia sistemica criminaliza a pueblo indigena ikoots" (2020). https://denunciaoaxaca.org/la-violencia-sistemica-criminaliza-a-pueblo-indigena-ikoots/

64 Agencia Zona Roja, "*Confirma la Sedena presencia de delincuencia organizada en Oaxaca*" (2022). http://www.zonaroja.com.mx/?p=46742

65 Chris Dalby, "Three Criminal Revelations from Mexico's Defense Minis try Leaks," *Insight Crime,* October 10, 2022. *https://insightcrime.org/news/ three-criminal-revelations-from-mexico-defense-ministry-leaks/*

66 Asambela de Pueblos Indígenas del Istmo Oaxaqueño en Defensa de la Tierra y el Territorio, (APPIDDTT) (2021) https://tierrayterritorio.wordpress.com/ 2021/06/23/alto-a-la-criminalizacion-hostigamientos-y-amenazas-por-parte-del-gobierno-municipal-estatal-y-federal-a-nuestros-companeros-indigenas-defensores-de-la-tierra-y-el-territorio-del-pitayal-puente-mad/

67 Daniel Cirilo, personal correspondence with the author, October 2022.

68 EDUCA, "Protestan zapotecas de Puente Madera contra parque industrial de Corredor Interoceánico," 2022a. https://www.educaoaxaca.org/protestan-zapotecas-de-puente-madera-contra-parque-industrial-de-corredor-interoceanico/

69 Daniel Cirilo, personal correspondence with the author, October 2022.

70 Caitlin Cooper, "Las mujeres que reviven el manglar en la Península de Yucatán: 'Nuestro trabajo es recuperar las áreas muertas'" *Mongbay,* 2022. https:// es.mongabay.com/2022/06/mujeres-reviven-el-manglar-en-la-peninsula-de-yucatan-entrevista/

71 EDUCA, Realizan calenda "contra la minería, el despojo y el falso discurso del progreso" en Oaxaca, 2022b. https://www.educaoaxaca.org/realizan-calenda-contra-la-mineria-el-despojo-y-el-falso-discurso-del-progreso-en-oaxaca/

72 Autonomous University of Social Movement, "The Fourth Transformation and the Future of Mexico," *The Bullet,* 2022. https://socialistproject.ca/2022/02/ the-fourth-transformation-and-the-future-of-mexico/

73 Richard W. Coughlin, "The Fourth Transformation and the Trajectory of Neo-liberalism in Mexico," *Latin American Perspectives* 50, no. 2 (2022).

74 López and Vértiz. "Extractivism, Transnational Capital, and Subaltern Struggles," 165.

Bibliography

Abi-Habib, Maria. "Over Caves and Over Budget, Mexico's Train Project Barrels toward Disaster." *New York Times,* August 28, 2022. https://www.nytimes. com/2022/08/28/world/americas/maya-train-mexico-amlo.html

Aguirre-Muñoz, Alfonso, Robert W. Buddemeier, Víctor Camacho-Ibar, José D. Carriquiry, Silvia E. Ibarra-Obando, Barbara W. Massey, Stephen V. Smith, and Fredrik Wulff. "Sustainability of Coastal Resource Use in San Quintin, Mexico." *AMBIO: A Journal of the Human Environment* 30, no. 3 (2001): 142–49.

Altiplano Wirikuta. "Excluyen a wixárikas de decisiones sobre megaproyectos y problemática del Altiplano Wirikuta en San Luis Potosí." Biodiversidad LA, 2019. https://www.biodiversidadla.org/Documentos/Excluyen-a-Wixárikas-de-decisiones-sobre-megaproyectos-y-problematica-del-Altiplano-Wirikuta-en-San-Luis-Potosí

Asambela de Pueblos Indígenas del Istmo Oaxaqueño en Defensa de la Tierra y el Territorio (APPIDDTT) 2021 https://tierrayterritorio.wordpress.com/2021/06/23/ alto-a-la-criminalizacion-hostigamientos-y-amenazas-por-parte-del-gobierno-municipal-estatal-y-federal-a-nuestros-companeros-indigenas-defensores-de-la-tierra-y-el-territorio-del-pitayal-puente-mad/

Autonomous University of Social Movements. "The Fourth Transformation and the Future of Mexico." *The Bullet*, 2022. https://socialistproject.ca/2022/02/the-fourth-transformation-and-the-future-of-mexico/

Avila-Calero, Sofia. "Contesting Energy Transitions: Wind Power and Conflicts in the Isthmus of Tehuantepec." *Journal of Political Ecology* 24, no. 1 (2017): 992–1012.

Barcenas, Francisco López. *Pueblos indígenas y megaproyectos en México: las nuevas rutas del despojo*. Mexico: Patrimonio biocultural, 2011.

Basset, Vincent. "New Age Tourism in Wirikuta: Conflicts and Rituals." In *Peyote: History, Tradition, Politics, and Conservation*, 191–210. Santa Barbara: ABC-CLIO/Praeger Publishers, 2016.

Brisman, Avi, and Nigel South. *Green Cultural Criminology: Constructions of Environmental Harm, Consumerism, and Resistance to Ecocide*. New York: Routledge, 2014.

Boni, Andrew, Claudio Garibay, and Michael K. McCall. "Sustainable Mining, Indigenous Rights and Conservation: Conflict and Discourse in Wirikuta/Catorce, San Luis Potosí, Mexico." *GeoJournal* 80, no. 5 (2015): 759–80.

Conn, Clayton. "Cherán: Community Self Defense in Mexico's Drug War." In *North American Congress on Latin America*, 2011. https://nacla.org/news/cher%C3%A1n-community-self-defense-mexico%E2%80%99s-drug-war-photo-essay

Consejo Regional Wixárika. "En Defensa de Wirikuta." 2022. http://consejoregional Wixárika.org/

Cooper, Caitlin. Las mujeres que reviven el manglar en la Península de Yucatán: "Nuestro trabajo es recuperar las áreas muertas." *Mongbay*, 2022. https://es.mongabay.com/2022/06/mujeres-reviven-el-manglar-en-la-peninsula-de-yucatan-entrevista/

Coughlin, Richard W. "The Fourth Transformation and the Trajectory of Neoliberalism in Mexico." *Latin American Perspectives* 50, no. 2 (2022). https://doi.org/10.1177/0094582X221103875

Crook, Martin, Damien Short, and Nigel South. "Ecocide, Genocide, Capitalism and Colonialism: Consequences for Indigenous Peoples and Glocal Ecosystems Environments." *Theoretical Criminology* 22, no. 3 (2018): 298–317.

Dalby, Chris. "Three Criminal Revelations from Mexico's Defense Ministry Leaks." *Insight Crime* 2022. https://insightcrime.org/news/three-criminal-revelations-from-mexico-defense-ministry-leaks/

Delgado, Jaime Ornelas. El Plan Puebla-Panamá y la globalización neoliberal. *Red Aportes*, 2002. http://www.eco.buap.mx/aportes/revista/21%20Ano%20VII%20Numero%2021,%20Septiembre-Diciembre%20de%202002/08%20El%20Plan%20Puebla%20Panama%20y%20la%20globalizacion%20neoliberal-Jaime%20Ornelas%20Delgado.PDF

Denuncia, Oaxaca. "La violencia sistemica criminaliza a pueblo indigena ikoots." 2020. https://denunciaoaxaca.org/la-violencia-sistemica-criminaliza-a-pueblo-indigena-ikoots/

Dunlap, Alexander. "The 'solution' Is Now the 'problem': Wind Energy, Colonisation and the 'genocide-ecocide nexus' in the Isthmus of Tehuantepec, Oaxaca." *The International Journal of Human Rights* 22, no. 4 (2018): 550–73.

———, and Martín Correa Arce. "'Murderous energy' in Oaxaca, Mexico: Wind Factories, Territorial Struggle and Social Warfare." *The Journal of Peasant Studies* 49, no. 2 (2022): 455–80.

EDUCA. Protestan zapotecas de Puente Madera contra parque industrial de Corredor Interoceánico, 2022a. https://www.educaoaxaca.org/protestan-zapotecas-de-puente-madera-contra-parque-industrial-de-corredor-interoceanico/
_____. Realizan calenda "contra la minería, el despojo y el falso discurso del progreso" *en Oaxaca*, 2022b. https://www.educaoaxaca.org/realizan-calenda-contra-la-mineria-el-despojo-y-el-falso-discurso-del-progreso-en-oaxaca/

Engelsdorfer, Alexandra. "Producing In-Between Spaces of Resistance: Nexus-Thinking and the Virtual and Physical Spaces of Protest around the 'La Luz' Mining Project in Wirikuta, Mexico." *IReflect* 4, no. S1 (2017): 97–116.

España-Boquera, María Luisa, and Omar Champo-Jiménez. "Proceso de deforestación en el municipio de Cherán, Michoacán, México (2006–2012)." *Madera y bosques* 22, no. 1 (2016): 141–53.

Faber, Daniel. "The Ecological Crisis of Latin America: A Theoretical Introduction." *Latin American Perspectives* 19, no. 1 (1992): 3–16.

Felbab-Brown, Vanda. "Criminal Violence, Politics, and State Capture in Michoacán." *Mexico Today*, 2021. https://www.brookings.edu/opinions/criminal-violence-politics-and-state-capture-in-michoacan/

Gasparello, Giovanna. "Communal Responses to Structural Violence and Dispossession in Cherán, Mexico." *Latin American Perspectives* 48, no. 1 (2021): 42–62.

Gobierno del Estado de San Luis Potosí. "Plan Estatal de Desarrollo." 2016. https://slp.gob.mx/sitionuevo/DocumentosPLAN/plan2016_eje1.pdf

Gómez de Silva Cano, Jorge J. *El derecho agrario mexicano y la Constitución de 1917. Colección INEHRM.* Mexico City: Instituto de Investigaciones Jurídicas, Universidad Nacional Autónoma de México, 2017.

Gray, Mark Allan. "The International Crime of Ecocide." In *International Crimes*, 456–511. New York: Routledge, 2017.

Gudynas, Eduardo. "Diez tesis urgentes sobre el nuevo extractivismo." *Extractivismo, política y sociedad* 187 (2009): 187–225.

Guillén, Emmanuel. "Cherán, el pueblo mexicano que expulsó a criminales y politicos." *Vice*, 2021. https://www.vice.com/es/article/4av58b/cheran-el-pueblo-mexicano-que-expulso-a-criminales-y-politicos

Leon, Mariana. "Real Bonanza: mina de la Discordia." *El Universal*, 2012. https://archivo.eluniversal.com.mx/notas/836671.html

Liffman, Paul M. "Mining and the Huichol Ancestral Landscape." *Frente en Defensa de Wirikuta*, 2012. https://frenteendefensadewirikuta.org/p1176.html

López, Emiliano, and Francisco Vértiz. "Extractivism, Transnational Capital, and Subaltern Struggles in Latin America." *Latin American Perspectives* 42, no. 5 (2015): 152–68.

López Barcenas, Francisco. "Indigenismo, violencia y despojo." *La Jornada*, 2021. https://www.jornada.com.mx/2021/06/22/opinion/016a2pol

Matias, Pedro. "'Toman' Minera en San José del Progreso, Oaxaca: Exigen Salida de Empresa Canadiense." *Proceso*, 2016. https://www.proceso.com.mx/nacional/estados/2016/5/6/toman-minera-en-san-jose-del-progreso-oaxaca-exigen-salida-de-empresa-canadiense-163917.html

Morosin, A. "Paramilitaries in Oaxaca, Mexico: Enforcing Accumulation in a Geo-Strategic Region." In *Paramilitary Groups and the State under Globalization:*

Political Violence, Elites, and Security, edited by Jasmin Hristov, Jeb Sprague, and Aaron Tauss, 88–106. New York: Routledge. 2021.

Millan, Omar. "Mexico's Baja Farmworkers Strike for Better Conditions." *San Diego Union Tribune*, 2015. https://www.sandiegouniontribune.com/sdut-mexicos-baja-farmworkers-strike-for-better-2015mar24-story.html

Narchi, Nemer E. "Environmental Violence in Mexico: A Conceptual Introduction." *Latin American Perspectives* 42, no. 5 (2015): 5–18.

_____, Sula E. Vanderplank, Jesús Medina-Rodríguez, and Enrique Alfaro-Mercado. "Environmental Violence and the Socio-Environmental (de) Evolution of a Landscape in the San Quintín Valley." *Latin American Perspectives* 47, no. 6 (2020): 103–18.

Negrín, Diana. "Water and Power in Wirikuta." Wixárika Research Center, 2021. https://www.wixarika.org/water-and-power-wirikuta

Osuna, Steven. "Securing Manifest Destiny." *Journal of World-Systems Research* 27, no. 1 (2021): 12–34.

Paley, Dawn. "A Canadian Mining Company Prepares to Dig Up Mexico's Eden." *This*, 2011. https://this.org/2011/09/15/first-majestic-silver-wirikuta/

Pickard, Miguel. "The Plan Puebla Panama Revived: Looking Back to See What's Ahead." *La Chronique des Amériques*, 12, no. 7 (2004). https://www.irn.org/files/pdf/mesoamerica/Plan_Revived.pdf

Rea, Daniela. San Mateo del Mar: "nos quieren matar y ponerle precio a la tierra". *Pie de Página*, 2020. https://piedepagina.mx/san-mateo-del-mar-nos-quieren-matar-y-ponerle-precio-a-la-tierra/

Reséndez, Ricardo M. "Entregaron la mitad de méxico a mineras." 2014. https://razacero.com/?p=724

Roja, Agencia Zona. *Confirma la Sedena presencia de delincuencia organizada en Oaxaca*. 2022. http://www.zonaroja.com.mx/?p=46742

Roy, Priyadarsi D., Sekar Selvam, Selvaraj Gopinath, Natarajan Logesh, José L. Sánchez Zavala, and Chokkalingam Lakshumanan. "Geochemical Evolution and Seasonality of Groundwater Recharge at Water-Scarce Southeast Margin of the Chihuahuan Desert in Mexico." *Environmental Research* 203 (2022): 111847.

Rural Migration News, (no author) "Farm Workers in San Quintín." *RMN Blog* 254, (2021). https://migration.ucdavis.edu/rmn/blog/post/?id=2676

Talcott, Molly. *Claiming Dignity, Reconfiguring Rights: Gender, Youth, and Indigenous-Led Politics in Southern México*. Santa Barbara: University of California, 2008.

The Economist, (no author). "It Is Time for Ecocide to Become an International Crime." 2021. https://www.economist.com/international/2021/02/28/is-it-time-for-ecocide-to-become-an-international-crime

Vilagrim, Tomas. "Thousands of Mexican Farmworkers March 15 Miles as Strike Talks Continue." *Los Angeles Times*, 2015. https://www.latimes.com/world/mexico-americas/la-fg-baja-farmworkers-strike-20150326-story.html

Wolff, Michael J. "Insurgent Vigilantism and Drug War in Mexico." *Journal of Politics in Latin America* 12, no. 1 (2020): 32–52.

Zalidis, George, Stamatis Stamatiadis, Vasilios Takavakoglou, Kent Eskridge, and Nikolaos Misopolinos. "Impacts of Agricultural Practices on Soil and Water Quality in the Mediterranean Region and Proposed Assessment Methodology." *Agriculture, Ecosystems & Environment* 88, no. 2 (2002): 137–46.

Zapatero, Santiago L., and Alberto A. Pineda "Where Are Mexico's Special Economic Zones?" *International Tax Review*, 2018. https://www.internationaltaxreview.com/article/2a68vp2bhdes26sg0okxs/where-are-mexicos-special-economic-zones

Ziegler, David. *The Invention of Ecocide: Agent Orange, Vietnam, and the Scientists Who Changed the Way We Think about the Environment.* Atlanta: University of Georgia Press, 2011.

8   Plundered Paradise

The Puerto Rican Struggle against Environmental Colonialism

A. J. Hudson

Colonialism hardly ever exploits the whole of a country. It contents itself with bringing to light the natural resources, which it extracts, and exports to meet the needs of the mother country's industries, thereby allowing certain sectors of the colony to become relatively rich. But the rest of the colony follows its path of under-development and poverty, or at all events sinks into it more deeply.

(Frantz Fanon, *The Wretched of the Earth*[1])

The deepest roots of Latin American environmentalism come from resistance to conquest.

(Barbara Lynch, *The Garden and the Sea*[2])

Ni ayer, ni hoy, ni mañana, ni nunca nuestra patria dejará de ser nuestra.
(Eugenio María de Hostos, Puerto Rican Revolutionary[3])

Not a "Part" of the United States

Understanding colonialism in the Caribbean is essential to understanding the Caribbean struggle for environmental justice. Puerto Rico, which remains an active colony, reveals even more nakedly than many of its neighboring islands how the historical power dynamics of Caribbean colonialism stretch firmly into the modern day. The "island" of Puerto Rico is, in actuality, an archipelago including the Spanish Virgin Islands, consisting of three large inhabited islands: San Juan or Isla Grande (the largest rectangle-shaped island people think of as synonymous with "Puerto Rico") and the two smaller island municipalities of Vieques and Culebra, as well as 140 surrounding islets.[4] Environmentally speaking Puerto Rico is rich in biodiversity, natural resources, strategic military geography, productive soil, and plentiful fresh water. For imperial powers looking to

DOI: 10.4324/9781003214380-11

stake their claims in the Caribbean, it has served as an ideal location for plantation slavery, political domination, and colonial extraction.

During his second voyage to the "New World" on November 19, 1493, Christopher Columbus landed on the island of Borinquen (Spanish corruption of the Indigenous Taino word), home to a people who called themselves the Boricua and who had settled there more than 1,000 years before he arrived.[5] The Spanish murdered, starved, decimated through Smallpox, and enslaved the island's native people to mine resources and build European settlements. Then Spanish governors brought enslaved people stolen from Africa to develop a plantation economy for cash crop production and build military forts.[6] In the 530 years since that landing, Puerto Rico has been under the imperial power of a colonial ruler. Its resources have been extracted, and its peoples displaced and disinherited to build the wealth of colonizers living a world away, and while roughly half of its Caribbean neighbors have struggled to and through independence during the 20th century, Puerto Rico has instead remained a dependent territory of the United States.[7]

Puerto Rico was taken as plunder during the Spanish-American war (April 21–August 13, 1898) along with Guam, the Philippines, and Cuba. While Cuba (1898) and the Philippines (1946) both won their independence, for the last 125 years, Puerto Rico and Guam have remained under the control of the United States.[8] This means that the economic, sociopolitical, and environmental rights of Puerto Ricans have been decided in Washington, DC for over a century. This fraught history with the United State is crucial to understanding the archipelago's struggle for environmental justice. Politically speaking, Puerto Rico exists in a legal limbo of US federal law where some constitutional rights apply, while others do not.[9] The most famous example, while still not common knowledge, is Puerto Rican citizenship granted by Congress in 1917.[10] One hundred years later in 2017, the *New York Times* published the results of a poll taken to understand the public's mixed reactions to Hurricane Maria, revealing that only an estimated 54% of Americans knew that Puerto Ricans were American citizens, casting into considerable doubt whether they even know that the colony is a "part" of the United States.[11] Perhaps their ignorance can be forgiven considering that they will not see results from the archipelago on election day: Puerto Ricans are subject to federal laws and taxation but have no representatives in Congress and cannot vote in presidential elections as long as they live in Puerto Rico.[12]

This second-class limbo citizenship finds its legal justification under the Territorial Clause of the US Constitution.[13] Although the constitution makes no mention of colonialism, the words "Territory or other Property belonging to the United States" were interpreted by the Supreme Court to give Congress full power over Puerto Rico and other colonies, and this seemingly unconstitutional treatment of its citizens was allowed through the decisions of the *Insular Cases*. This series of Supreme Court decisions concerning the US island territories was issued between 1901 and 1922 and established legal precedent on the status of its colonial possessions.[14] These judicial decisions, which relied on the infamous "separate but equal" logic of *Plessy v. Ferguson*, effectively handcrafted an understanding of the constitution which both allowed further imperial expansion and cemented the sociopolitical limbo state of constitutional exception for people living in US territories.[15] That legal limbo still governs the territories' fate today.[16]

The *Insular Cases* decisions justify the non-recognition of fundamental constitutional rights from voting to the right of a trial by jury in Puerto Rico (and other US colonies) because they establish that Puerto Rico "is a territory ... *belonging* to the United States, *but not a part* of the United States," and is "foreign to the United States in a domestic sense."[17] This confusing and incoherent legal distinction crafted a new form of statehood, the "unincorporated territory," and a matching new form of citizenship, the "non-citizen national" that still guide colonial relations in the United States to different degrees in each territory.[18] Although *Plessy v. Ferguson* and race-based segregation were overturned domestically with *Brown v. Board of Education*, varying levels of segregative and disparate treatment are still fully legal under the *Insular Cases*.[19] To appreciate the stakes of environmental degradation and lack of environmental agency in Puerto Rico, one must also confront the second-class citizenship that is legally allowed in the US territories. In 1980, the *Insular Cases* were used to legally justify disparate funding levels of social welfare programs like Medicare and Food-stamps in the territories.[20] This legal inequality has been affirmed as recently as 2022 in a SCOTUS case which determined it was constitutional to deny blind and disabled Puerto Ricans Social Security funds that other citizens are entitled to receive.[21]

In 1952 Congress established Puerto Rico as a "Commonwealth." While granting many self-governing powers, this also enabled a reinforcement of the status-quo as "commonwealth" itself is yet another legal loophole, which removes Puerto Rico from the United Nations list of "active colonies" and consequently from much of the world's attention, while allowing it to remain under the full power of the Territorial Clause, and the Insular Cases which govern colonies.[22] Despite now having a constitution and an elected governor, its imports and exports are determined by the Jones Act, a 100-year-old US shipping law, its federal courts are conducted in English instead of Spanish, and its economy is governed by an unelected board appointed by the US President. Even as a "commonwealth" Puerto Rico remains firmly under the plenary powers of Congress, which govern colonies, Indian Reservations, and immigration.[23] The myriad ways in which Puerto Rico exists in a limbo state of possession, wherein its people have some rights but not others, are quintessential aspects of Caribbean colonial rule and governance by imperialism which long predate the US itself. Territories have always been claimed by nations for their resources, whether that be strategic military positioning, or the nutrients of rich soil, and their governance has centered on that extraction. Because the benefits of that extraction are never distributed evenly, instead going principally to the colonizers and secondly to the colonial appointees, elites, and local insiders facilitating colonization, colonial intervention, and development is rife with environmental justice concerns.

Environmental Colonialism in the Caribbean

Colonialism relies on the exploitation of people and place as "other" to facilitate the extraction of wealth in the form of human bodies, minerals, luxury, sustenance, energy, and other natural resources.[24,25] The destruction and alteration of the environment that goes hand-in-hand with this exploitative process is often taken for granted when we discuss the Caribbean: we know that the Europeans brought horses, pigs, rats, an entire infectious microbiome, fruits, and vegetables from Asia. They leveled forests, tore down traditional homes, and replaced them with haciendas built out of quarried stone. They changed the language and the culture of the people. Countless humans died and were enslaved. What is less often stated is that these colonial processes are inherently environmental processes depending on and determined by geography and control of resources, and that because they were embedded within systems of oppression and hierarchies of agency, they can also be understood as environmental justice issues.

The degradation of the environment in the colonial Caribbean was not felt equally, nor were its benefits. The destruction of the environment made the empires of Western Europe rich, while decimating the Indigenous cultures and stolen people of Africa by exposing them to disease, starvation, violence, and the horrific environmental realities of slave labor. This chapter posits that colonialism itself is an act of profound environmental injustice, as it is the ultimate perversion of an equitable and self-determined relationship with the land and its natural resources. It builds the literal opposite of a just environment: one where resources are extracted by the highly exposed, destroying their own home, ruining their health, and threatening their future livelihoods for generations, all for the benefit of another group thousands of miles away and their local appointed accomplices. Many scholars argue that the concepts of race and caste themselves, central pillars of understanding environmental justice, were forged by the governors of colonialism to facilitate extraction.[26,27] This chapter examines the history of environmental justice in Puerto Rico as a question of colonialism and considers historical acts of imperial intervention as nearly always having environmental justice implications.

This concept, especially apt for understanding environmental power dynamics in the Caribbean, is often called environmental colonialism and is understood both to be a dimension of environmental justice while also being an independent field of inquiry. It is an essential frame to understand the central questions of Puerto Rico's environmental history: of who has access to and control over resources, of who has political power to determine who benefits from extraction, and of who constructs the dominant view or cultural understanding of the Caribbean landscape. The legal scholar Ugo Mattei and anthropologist Laura Nader call the ongoing governance dynamics of environmental colonialism "plunder," a legal structure or "rule of law" devised entirely to separate colonial peoples from their natural resources, created by Europe and perfected by the Global North worldwide.[28] In a striking historical analysis of the legality of colonialism from the discovery of the "New World" to more recent imperialism, they show that the overall political and economic strategies used by colonizers to justify and rationalize the treatment of colonial subjects have evolved and shifted, while the primary goal of the extraction of resources and exploitation of people within colonized territories for profit has not.

Importantly this legal theory of plunder as being at the heart of colonial governance does not mean that independent Caribbean colonies are not still faced with environmental colonialism. *Plunder* argues that colonial-capitalist practices and policies imposed by colonial governments, commonwealths, and post-colonial governments can all constitute the legal model of plunder.[29] In modern times plunder often comes instituted by the economic elites within a nation or colony, facilitating the extraction and commodification of colonial and post-colonial environments alike all for the benefit of the Global North, making many of the nations of the Caribbean independent in name only. In a survey of Caribbean environmentalisms, Puerto Rican anthropologist Manuel Valdés highlights the ways in which modern Caribbean development policies have come with "benefits" that only seem to accrue for the wealthy elites of a territory or post-colonial nation, and that these policies from agricultural production to urban expansion and tourism industries consistently destroy the environment.[30] Therefore, commodification-driven environmental politics which disenfranchise entire nations can tragically continue long after independence.

While this chapter is concerned principally with those who have suffered and resisted against the injustices of environmental colonialism, it is crucial to remember that at every stage of Puerto Rico's history, there has been a set of colonial elites, whether foreigners directly appointed by the federal government or locals elected by the populace, which has facilitated and benefitted from the exploitation of Puerto Rico's resources. These people, even when native-born Puerto Ricans, have historically been racially white (of predominately European lineage) or white presenting and part of the educated and wealthy upper class. While they are also robbed of agency by the colonial relationship, due to their privileged position within colonial power, they suffer less exposure to the injustices of environmental colonialism than the poor and Afro and Indigenous descended peoples of the archipelago. As Puerto Rico has shifted to a tourism economy, a significant number of this insulated class of elites are foreigners from the United States, who have moved to the archipelago seeking paradise and they encounter few if any consequences of the colonial relationship.

A critique of "paradise" itself is a central task of environmental colonialism. Paradise is enacted at the ecosystem level through romantic and two-dimensional depictions of Caribbean nations and their landscapes.[31] This means that classical Caribbean vacations "other" at the landscape scale, exoticizing and simplifying entire locales to justify shallow behavior and cultureless interactions which often involve subordination and strict service hierarchies with the locals. Compare the idea of a Caribbean vacation with a visit to Paris to see the Louvre, and the "othering" of the colonial landscape becomes even more stark. Jamaica Kincaid's *A Small Place* about Antigua and the neocolonialism of tourism underlines this idea: paradise is an inherently transactional and exotic concept for most people, it is a vacation far away, and it is unrelated to and unconcerned with the lives or struggles of the locals who live in the lush tropical environments.[32] Political ecologist Sherrie Baver has argued that the Caribbean paradise landscape which drives tourism is carefully crafted and socially produced for the benefits of overwhelmingly outside/foreign developers, and at the expense of the agency and self-determined development of the islanders.[33] The hawking of paradise to tourists can be understood as a form of plunder, in that it commodifies the colonial and post-colonial environs by selling the experience of them primarily for the benefit and wealth of the global north.

Even worse, paradise-based tourism veils countless other aspects of modern environmental colonialism, shielding tourists from the environmental realities facing Caribbean citizens. The people of Caribbean islands are faced with unsafe drinking water, vacation-rental-driven gentrification, towering coastal development, fragile storm and climate damaged ecosystems, unplanned and chaotic urbanization, toxic pesticides flooding from cash crop agriculture, rising seas, and yet have little to no say in how their environments are ultimately managed.[34] This is because in order for their environment to be paradise (instead of a complex struggling community where real people live), it must be managed for foreigners, for tourists, and ultimately for plunder. Other eras of extraction were driven by different understandings of the Caribbean landscape; plantations were in part a response to the perception of islands' tropical fertility and lush vistas covered in exotic fruits, and cheap exploitative industrial labor regimes were a response to third world visions of the Caribbean as impoverished and undeveloped. For centuries, global powers sculpted the landscapes of the Caribbean to suit their needs, producing sugar, coffee, rum, and other luxury goods, but now they sculpt it to produce vacations.

From revolts on colonial-era slave plantations to protests of modern condo development in fragile mangroves, the struggle against plunder-based legal-regimes has driven much of the reaction to environmental injustice in Puerto Rico and the broader Caribbean. Sociologist José Atiles-Osoria has said that instead of a tension between geopolitical liberation and environmental agency, it is only natural that anti-colonial resistance has married these two domains on the archipelago.[35] In Puerto Rico, anti-colonial movements have been involved in the struggle for environmental justice from the start, with many seeing the colony as birthing an Environmental Justice movement long before the civil rights movements of the 1960s, instead born in struggles for greater political autonomy.[36] In turn, Puerto Rican emancipatory struggles against colonialism have also been environmental in that they are meant to take the agency to extract and destroy away from colonizing powers and give environmental autonomy to the people of Puerto Rico. This also means that unlike in the United States where the disparate evolution of conservation and environmental justice has led to a starkly divided environmental movement, traditional conservation and development concerns in Puerto Rico are often considered justice and access issues. Because of the constant and continuing struggle against US colonialism in Puerto Rico, movements for environmental justice cannot be pulled apart from decolonial movements against the development of Puerto Rico's natural resources but must be understood together within the same historical lens. Environmental colonialism then is not just a helpful framing, but a necessary one. What follows is an examination of the history of environmental colonialism in Puerto Rico as a study of environmental justice and resistance. It is by no means exhaustive but instead paints with broad chronological strokes four overlapping eras of environmental colonialism undertaken by the United States in Puerto Rico: agricultural, industrial, militarism, and tourism.

Agriculture: Sugar's Sweet Syrupy Stranglehold

Environmental colonialism reveals that the project of colonial domination is not merely power over a people, or power over the place itself, but power over the relationship and agency between people and place that would otherwise exist without interference. The process of colonization asserts dominion over the environment and its natural resources, dictating how the conquered people consume those resources, how they work and interact with those resources, and ultimately who receives the benefits and consequences of resource extraction. Accordingly, the history of Puerto Rico's environment has been determined by colonizers for more than 500 years, starting with the Spanish. Spain brought slavery, coffee, tobacco, sugar, and rum to Puerto Rico in order to fund its military campaigns at home and in the Caribbean.[37] The relationship between people and the land was driven by extraction, but Puerto Ricans remained small-scale cash crop and subsistence farmers into the 20th century because the lack of easily accessible silver and gold deposits in the colony meant its agricultural development among Spanish colonies was relatively neglected.[38] The most imposing Spanish extraction of resources in Puerto Rico was martial; the use of slave labor to build the massive fort surrounding the Puerto Rican capital of San Juan, El Castillo San Felipe del Morro, and the fortified governor's palace La Fortaleza.[39] The Spanish called Puerto Rico "the key" to the Caribbean because of its military significance and invested heavily in the archipelago's fortifications.[40] After taking over the territory in 1898, the United States would make the same estimation, further militarizing the colony with new forts and turning the castle of El Morro into an active Army base until 1961.[41] It is now a national park.

The first economic change initiated by the United States was the transformation from subsistence farming, cattle ranching, and small-scale cash crop-based economy to sugarcane monoculture in the early 20th century.[42] The shift was rapid; a report by the Federal Reserve found that in 1900, Puerto Rico produced less than a fifth of the sugar grown in Louisiana, but by 1930 at its height, the industry was growing five times more sugar than Louisiana.[43] By the 1950s, even as the sugar industry began to flag and fail, it was still responsible for around 25% of the total wages paid in the colony, making Puerto Ricans extremely dependent upon it.[44]

The decision to turn Puerto Rico into a giant sugar plantation was not driven merely by climate or economic decisions, but also by imperialist opportunism. The first civilian governor of Puerto Rico, appointed in 1900 by President McKinley, was the Massachusetts born Charles Herbert Allen. In his first report to the US President, he spoke at length about sugar but provided little to no updates on democracy, education, or the well-being of his Puerto Rican subjects. "The yield of sugar per acre is greater than in any other country in the world," he wrote to President Theodore Roosevelt. "A large acreage of lands which are now devoted to pasturage, could be devoted to the culture of sugarcane. Molasses and rum, the incidental products of sugarcane, are themselves sufficient to pay all expenses of the sugar planters and leave the returns from this sugar as pure gain."[45] It was not only the geography of the place that excited him, but the affordability of the labor which he determined would make sugar $47 cheaper

Figure 8.1 Bags of refined sugar in the warehouse of the South Puerto Rico Sugar Company. Puerto Rico Ponce Municipality Ensenada, 1942. Jan. Photograph by Jack Delano. https://www.loc.gov/item/2017798259/.

per ton than sugar produced in Louisiana. He only served for 17 months, but by the time, he left office virtually the entire executive cabinet of the colony and more than half of its government appointments had gone to American Businessmen and investors instead of Puerto Ricans.[46]

Allen's political appointees and future governors facilitated Puerto Rico's sugar transformation by selling land to American corporations from under Puerto Ricans who had lived there for generations, granting major tax subsidies, foreclosure sales, and land easements, effectively consolidating a sugar colonial monopoly run by Allen's wall street affiliated capitalists that would turn 50% of Puerto Rico's arable land into sugar plantations.[47] Allen would go on to found Domino Sugar, the largest sugar syndicate in the world, owning over 80% of Puerto Rico's sugar, and affiliated with banks that held the Puerto Rican railroads, postal service, and seaport of San Juan.[48] In the span of 30 years, and the shift from one imperial power to another, the entire economy of the colony was transformed. Nearly all the profits of this monoculture economy left Puerto Rico to make US businessmen rich. The US population was not entirely ignorant about this situation either, an Indianapolis Newspaper from 1934 about an upcoming presidential visit to the archipelago described the situation succinctly: "Roosevelt will find a crazy economic system built by American capital, which bought up vast acreages for sugarcane cultivation and left Puerto Rico without enough to eat ... Half the heads of families are counted as unemployed."[49]

Such a momentous shift to monoculture in such a short time left several environmental scars across the islands of Puerto Rico. Sugarcane is a thirsty crop, and the sugar syndicate had to build three massive dams across San Juan island, all of which are now at risk of catastrophic failure from age, sedimentation, and the battering of hurricanes.[50] In 2017, after hurricane Maria, Guajataca Dam suffered a partial collapse and an entire community had to be evacuated from the destruction.[51] The communities living in the shadows of these dams are predominantly the impoverished ancestors of workers left behind when sugar companies abandoned Puerto Rico from the 1940s to 1970s. The sugar mills themselves were a source of pollution and deforestation; wood was burned to boil the sugarcane on site because otherwise it would spoil, and charcoal made from local mangroves was used as fuel, causing ecosystem degradation from mountains to the coast.[52] As mills modernized oil and molasses spills became common problems, and since mills were often located in coastal communities for easy access to shipping, much of the damage was aquatic. A local environmental history undertaken in the predominantly Afro-Puerto Rican community of Aguirre, Salinas (home to one of the largest sugar mills in Puerto Rico), reveals that molasses spills killed countless fish, leading to food insecurity and lasting coastal damage.[53]

Sugar monoculture discouraged education, which was unneeded in most sugarcane labor, as well as public investment because company towns provided the infrastructure surrounding sugar mills, and encouraged staggering wealth and racial inequality between the agrarian and mill workers and the white US-based landowners who profited off the exported sugar.[54] Because the sugar economy of Puerto Rico began collapsing in the

1950s with the last major mills closing around 2,000, many Puerto Ricans whose families had cut sugarcane for generations now live impoverished in environments diminished by intensive sugarcane farming and processing.

While depending on the sugar syndicate jobs for a living, Puerto Ricans did not passively accept their exploitation. From the start of the syndicate's hold on the archipelago until the next economic shift and the eventual collapse of agriculture, Puerto Ricans organized labor unions and leveraged developing US labor laws in pursuit of greater workers' rights and autonomy, as well as investment in the colony's neglected infrastructure. Resistance to the political and economic order on the archipelago became a shared motive for agricultural communities who also fought for a higher living wage and a more significant share of sugar's profits, eventually forming the Confederación General de Trabajadores (General Union of Workers).[55] Part of the reason why wages were critical to sugarcane workers was because of their lowered environmental autonomy; unlike coffee and tobacco growers whose crops had retained an artisanal quality, industrial sugar workers were not allowed to grow subsistence crops on sugarcane plantations to supplement their income.[56] This denial of access to the land they worked for sustenance, a tradition of Spanish agriculture, deepened their poverty and raised the stakes for their meager wages.

Puerto Rico's food insecurity would increase over time due to colonial intervention. For a colony that started out producing virtually all food domestically through subsistence farming, by the collapse of sugarcane, it was only growing 40% of the food consumed locally. Currently the number hovers around 10–15%, a devastating reminder of the agricultural agency, lost under US rule, and the forced erasure of a traditional land-relationship between the archipelago and its people.[57] Because forced dependence on expensive imported food was a byproduct of the growing sugar industry, the already low wages of sugar workers became even less tenable over time. The response to this rising tension was the 1941 sugar strike, the largest in Puerto Rico's history, that began January 19 and lasted until February 6. Puerto Ricans demanded that US labor laws and standards be applied to Puerto Rico, and that they could renegotiate their wages. The strike was successful and led to the formation of Puerto Rico's first minimum wage committee, and the sugar industry's recognition of their union.[58] Because it was a response to the poverty created through the new colonial land-relationship, this worker strike was inherently an environmental issue. By requesting that the imperial powers of the United States follow their own domestic rules, Puerto Ricans were demanding a more just relationship with a colonial power and resisting the colonial industry that prioritized profits over their lives. Many historians see the 1941 sugar strike as a disruption which in part led to the immense land reform of the next era, a shift away from the economic chokehold of monoculture sugar, and the eventual industrialization of Puerto Rico.[59]

Figure 8.2 Delano, Jack, photographer. *Yabucoa, Puerto Rico. Strikers on the picket line at a sugar mill.* Yabucoa, Puerto Rico, 1941. Dec. Photograph. Library of Congress https://www.loc.gov/item/2017798974/.

Industry: Displacement and Diaspora

Even before sugarcane's eventual collapse, the Great Depression devastated Puerto Rico's monoculture economy and revealed the fragility of the first colonial development experiment.[60] Chicago journalist John Gunther brought this to light for US audiences in patrician terms through his *Inside Latin America* series. "I saw, in short, misery, disease, squalor, filth. It would be lamentable enough to see this anywhere ... to see it on American territory ... is a paralyzing jolt to anyone who believes in American standards of progress and civilization."[61] The next major shift was the federal government's attempt at living up to these standards of progress by "modernizing" Puerto Rico through industrialization and development programs aimed at fixing the abject poverty and public neglect caused by monoculture.[62] "Operation Bootstrap," in reference to pulling oneself up by the bootstraps to achieve the American Dream, brought textile factories, oil refineries, and other polluting industries starting after World War II.[63]

The federal government, in partnership with local leaders hoping to weaken the hold of the sugar syndicate in Puerto Rico, enacted the Industrial Incentives Act of 1947, effectively eliminating all corporate taxation in a bid to bring US private capital to Puerto Rico.[64] The Act gave ten-year waivers from all the taxation involved in running most corporations, from licensing fees and property tax to income tax. This industrialization by invitation was meant to make Puerto Rico modern, to shift away from the agrarian monoculture society that US involvement had established for the previous 40 years, but also to create products for export to the US market. Operation Bootstrap's allure depended on colonial aspects of the relationship between the United States and Puerto Rico beyond tax breaks: labor was much cheaper on the archipelago, and Puerto Ricans were forced to learn English through a school curriculum imposed by Congress.[65] The manufacturing industries of Operation Bootstrap also led to further dependence upon the United States, much like sugarcane did, as nearly all the products manufactured by Puerto Ricans were sold to US markets, and nearly all the wealth was reaped by US corporations.[66]

This development experiment, like monoculture before it, left further lasting marks of environmental colonialism on the archipelago. By 1952, coinciding with the new political status of commonwealth, 152 new industrial factories had opened, overwhelmingly producing highly polluting consumer goods: footwear, electronics, artificial flowers, plastics, and metal.[67] Consequently, Puerto Rico is now home to 18 Superfund sites.[68] Massive demographic shifts were to follow the creation of these new factories. During the height of the sugarcane syndicate, the 1930 census showed that only 27% of Puerto Ricans lived in urban areas; by 1980, 69% of Puerto Ricans lived in cities, while the 2010 census recorded 94% of Puerto Ricans as urban dwellers, with 76% living in the capital San Juan.[69] The movement was not only from the rural to the urban, as Operation Bootstrap effectively killed the sugar industry and there were nowhere near enough of the new better paying factory jobs to go around, as a result, around 25% of the population left Puerto Rico entirely. During "La Gran Migración," lasting from 1950 to 1970, more than 600,000 Puerto Ricans, predominantly from post-sugarcane communities, emigrated to the United States seeking jobs and prosperity after being left behind by the archipelago's latest colonially mandated economic experiment.[70] Their descendants are spread in cities from Chicago to Hartford but are predominantly centered around New York City and Orlando.[71] With a diaspora of nearly 6 million Puerto Ricans, there are now almost twice as many Puerto Ricans in the United States as there are in Puerto Rico. The astonishing displacement effects of Operation Bootstrap on Puerto Rico have in turn been continued by the impact of hurricanes on Puerto Rico, themselves likely exacerbated by climate change. In 2017, Hurricane Maria in a single act killed 3,000–4,600 people and forced the migration of an estimated 200,000 Puerto Ricans (around 5% of the archipelago) to the communities already established in the United States by the displacement of Operation Bootstrap.[72]

From an environmental justice perspective, these movements of people are incredibly important, and not only from the environmental colonialism aspect which highlights the lack of agency and self-determination of Puerto Ricans to guide their own development. Movement to cities meant exposure to the pollution of the new Bootstrap industries, of the growing emissions from traffic and explosive urban sprawl, and a further collapse of the traditional subsistence food systems which fed the islands for generations. Emigration exposed Puerto Ricans and their descendants to the industrial pollution of the United States where they became enmeshed in the well-known environmental justice communities of the Eastern and Midwestern states. Migrating Puerto Ricans were also exposed to the horrors of Jim Crow segregation and ghettoization faced by domestic Black communities, hemmed in by highways and industrial zones, and confronted with the environmental racism that plagues cities like New York and Philadelphia. This was due both to the direct racism against people of color and the indirect racism of where immigrating Puerto Ricans could afford to settle. Sunset Park, a typical Puerto Rican diaspora community in Brooklyn, NY is home to three heavily polluting peaker power plants, two solid waste transfer stations, the Gowanus Expressway, several brownfields, and a large industrial park, making it heavily overburdened by pollution and undesirable land use.[73] Because US intervention effectively forced Puerto Ricans to migrate to environments that were damaging to their health and then developed and isolated those areas in the same manner as it did segregated domestic Black communities, the entire human geography history of Puerto Rico raises continuing issues of environmental justice.

Puerto Ricans in the diaspora also resisted environmental injustices through movements throughout the 1960s and 1970s which echoed the civil rights movements across the country at the time and presaged the environmental justice movement of the 1980s. Often called the Puerto Rican Black Panthers, the Young Lords were primarily Black and Brown Puerto Ricans inspired both by the Black Panther Party and the Vietnamese liberation struggle to build their own youth-organized anti-colonial civil rights movement. Founded in Chicago, the Young Lords Party quickly spread to nearly every Puerto Rican diaspora community.[74] In the summer of 1969 after the first Earth Day, a chapter of the Young Lords launched a "Garbage Offensive" movement to clean up El Barrio, a major site of the Puerto Rican Diaspora, in East Harlem. This radical group of young people of color confronted the environmental inequities of Harlem, a historically Black neighborhood long segregated and denied public services, through direct action, political organizing, community engagement, and militant marches.[75] The Garbage Offensive was in response to disastrous levels of urban decay in the segregated portions of Harlem, where garbage was seldom picked up by sanitation services and empty lots and dilapidated buildings created constant environmental hazards. In response to government inaction, the movement culminated when the Young Lords created street barricades out of the garbage, abandoned cars, and detritus from collapsed buildings which plagued their lives.[76] By blocking the streets and creating traffic jams, these barricades turned the garbage into the city's problem and forced passersby to confront the daily consequences of segregation that were normal for Puerto Ricans living in New York. The Young Lords Party remains a steadfast inspiration for movements fighting racism across the United States, and in New York organizations like UPROSE continue this radical Puerto Rican-led environmental justice tradition.[77] While the Puerto Rican diaspora has had a profound impact on the US struggle for civil rights and environmental justice, the opposite is also true.

The diaspora resulting from this displacement has become a strength of Puerto Rico, giving the colony a displaced voice in the first Puerto Rican congresspeople, like Herman Badillo who was elected by Bronx voters in 1970. This tradition has continued to the present day with congressional leaders like Alexandria Ocasio Cortez who in 2022 was essential to the House of Representatives passing a Puerto Rico Self-Determination bill.[78] The legislation acknowledges the colonial wrongs of the federal government and transfers the power to vote for independence or statehood from the federal government to Puerto Rico. The bill does not have Senate approval, and until its very unlikely passage, the power of determining Puerto Rico's future rests squarely within the halls of Congress. While this proxy of representation has done little to stymie environmental colonialism in Puerto Rico, it has created heightened levels of government accountability,

Figure 8.3 The New York chapter of the Young Lords protesting with a Puerto Rican flag during their Garbage Offensive. Michael Abramson, Young Lords Party, 1969.

brought attention to injustices in Puerto Rico, and created support for lasting political movements. Consequently, the diaspora has been influential in environmental justice struggles in Puerto Rico from protesting a polluting trash incinerator in Arecibo, to mounting domestic support to shut down US military base in Vieques in 2003, to hurricane recovery and government accountability in the aftermath of disastrous Hurricane Maria in 2017, and the rising impacts of climate change.[79]

Militarism: Enduring Sickness

Concurrent with Operation Bootstrap's push for industrialization, Puerto Rico's strategic position at the entrance to the Caribbean made it ripe for military expropriation of land. Spanish forts were almost immediately converted into US military bases upon the colony's transfer in 1898. Eager to protect the Panama Canal and guard the entrance to the Caribbean, while also wary of German U-boats and European powers encroaching on the Americas, the United States began a process of extended militarization starting in 1940 that stretched across the Caribbean with Puerto Rico at its center.[80] The militarization of the Caribbean remains one of the greatest acts of environmental colonialism of the past century as it displaced thousands from their homes to turn their stolen land into military bases, ammunition storage, and explosive target ranges. Broad swaths of the Puerto Rican archipelago were expropriated by the US military and Puerto Ricans living on these lands were deemed "squatters" and forced to move to tent colonies until they were resettled.[81] Militarization in Puerto Rico led to organized movements demanding the return of lands to displaced people and the end of dangerous military exercises, but despite success stories of these environmental justice activists eventually closing bases on both Culebra (in 1975) and Vieques (2003), both islands remain Superfund sites requiring intense cleanup.[82] While the impact of militarism stretched across Puerto Rico and the Caribbean, Vieques and its people (Viequenses) have particularly suffered from its consequences: from 1941 to 2003, Vieques served as a live firing range for the US Navy.[83]

Vieques is an island municipality of Puerto Rico just 7 miles off the eastern shore of the larger Isla Grande. The small isolated island has an area of about 51 square miles (about twice the size of Manhattan), and it can only be reached by plane or a historically unreliable ferry in Fajardo.[84] In 1941, the US Navy expropriated around 70% of Vieques and built military training facilities to be used for live firing practice and ammunition storage.[85] Thousands of Vieques 10,000 residents were removed from their homes during the military takeover, about 3,000 Viequenses were moved out of Puerto Rico to the nearby Virgin Islands territory St. Croix, and an untold number left for resettlement on Isla Grande.[86] The entire island

Figure 8.4 Delano, Jack, photographer. *Bayamon, Puerto Rico vicinity. Tent colony of 25 families who had been squatters on land taken over by the Navy. They are to be resettled by the land authority.* Bayamon, Puerto Rico, 1941. Dec. Photograph. Library of Congress https://www.loc.gov/item/2017797111/.

would have been removed to St. Croix in 1947 and again in 1961 by the Department of Defense if not for the intervention of local government officials and activists.[87] Those who remained (an estimated 6,000–7,000) were eventually settled into a small strip in the middle of Vieques reserved for civilians and sandwiched on both sides by the new military bases. They were forced to sign agreements stating that the military could evict them from their land at any time, and that they were not landowners. A copy of the letter received by Viequenses Ramon Rodriguez is reproduced below:

> The house and land which you occupy in the municipality of Vieques was acquired by the United States under judgment of the Federal Court which granted the right of immediate possession. You will be required to vacate this property within ten days from the date of this notice. Should you wish to move to another site on federal property you will be assigned a suitable area by the Officer-in-Charge of the project upon execution by you of an agreement setting forth the terms upon which your occupancy of the site is permitted.
> (Yours very truly, J.C. Gebhard, Captain USN[88])

At the time of this land seizure and even before the creation of thousands of landless and homeless people, Vieques was already the poorest municipality in Puerto Rico (it still is), but after this massive displacement, the remaining Viequenses were even more dispossessed. The military takeover shut down Vieques's sugarcane economy, and the Navy's security commandeered trade, water usage, coastal development, and civilian zoning.[89] Their constant bombing of the sea killed off vast populations of fish that community members depended on for food and wages and contaminated the local environment, leaving a displaced people poisoned.

From 1970 to 1975 sustained physical resistance by the residents of Culebra, also home to a Navy bomb range expropriated in the 1940s, forced the closure of that base but those explosive activities moved almost immediately to Vieques.[90] During the Navy's time on Vieques, they conducted daily bombings, deployed massive numbers of troops to the island to practice amphibious landings, hit the island with missiles from battleships at sea and torpedoes from submarines, and set its forests ablaze with napalm.[91] Both Agent Orange and depleted uranium weapons were tested on Vieques. Over a span of 60 years, Vieques was showered by more than 80 million pounds of ordnance and some 2 million pounds of toxic military waste and industrial byproducts were dumped into its sensitive wetland ecosystems.[92] The extent of environmental damage done, although massive, is largely unknown due to the Navy's past lies and secrecy. To this current day, the Navy's limited remediation activities do not include plans to restore or decontaminate the habitats destroyed and endangered by their explosions and pollution.[93] Vieques has been called "one of the most highly contaminated sites in the world" by environmental health professor John Wargo, who served as an expert witness during litigation.[94]

The military occupation of Vieques is one of the quintessential environmental justice struggles of Puerto Rico and the broader Caribbean, as it perfectly illustrates the displacement, disenfranchisement, and exposure to hazards that define the consequences of colonialism. The people of Vieques were displaced, alienated from their traditional relationship with their environment, forced to contend with severe economic hardship, and exposed to the toxic byproducts of militarization. All for the sake of strategic resources which did not benefit them, and decisions made in Washington a world away. Nevertheless, Viequenses did not passively accept their island's destruction, instead resisting through swells of protests from 1960s when they occupied the bombing range with their fishing boats, to the 1990s when Congress demanded that the Navy close the base.[95] However, real progress toward the Navy's removal was not made until the tragic death of David Sanes Rodríguez, a civilian Navy employee who was hit by a stray missile. Rodriguez's death sparked a monumental grassroots movement, uniting Puerto Ricans across the political divisions and outraging the Puerto Rican diaspora living in the United States.[96] Tens of thousands of Puerto Ricans flooded the streets of San Juan to protest the military occupation, and the face of Milivi Adams, a two-year-old Viequense, diagnosed with cancer made the front page of Puerto Rican Newspapers. This was the birth of the Peace for Vieques movement, perhaps the first Puerto Rican environmental justice movement to gain broad US and international support; several prominent American activists including Reverend Al Sharpton were imprisoned for joining the protests in solidarity.[97] In 2001, President George W. Bush responded to the mounting activist pressure with an executive order halting all military activity on the island. This resulted in the base closing and the Navy's permanent departure from Vieques in 2003.[98]

After 60 years of occupation, the Navy's departure was not the end of Vieques' struggle for environmental justice. The island remains deeply scarred by bombardment on both land and underwater, covered in unexploded ammunition, home to a poisoned community brimming with health problems, and is still the poorest in Puerto Rico.[99] Now even after two decades of achingly slow remediation efforts, the former military zones remain off limits and dangerously contaminated, and the community remains frustrated by the Navy's approach.[100] The Navy estimates it will take them until 2032 to make the areas safe for human access again. Meanwhile, their decontamination techniques of setting swaths of the forest on fire, and exploding ordnance in the open air, leave much to be desired for locals.[101] The post-Navy environmental justice movement of Vieques calls for cleaning and decontamination of the ruined military zones and ultimately the return of the land to the people of Vieques.[102] Instead of fully demilitarizing and returning the land upon the Navy's departure, the federal government transferred it to the Department of the Interior under the management of the US Fish and Wildlife Service and established it as Vieques National Wildlife Refuge, with the most dangerous portions of Vieques marked as "Wilderness Areas," prohibiting public access entirely.[103]

Many Viequenses believe that the wilderness area is merely an excuse to absolve the Navy of responsibility for cleanup and allow the United States to hold onto the stolen land with lowered liability.[104] The fact that the Navy's "cleanup" involves open air detonation of bombs raises questions about the new toxins being released as a byproduct of remediation. This cheap clean-up technique itself, while lessened in magnitude, is still a direct continuation of the dangerous and careless military practices and unjust exposures to toxins that Viequenses fought to end. Despite promises to begin contained explosions, as of this writing open detonation remains the norm.[105] The wildlife refuge system is itself a significant injustice to Viequenses because it continues to restrict land use, limit agriculture and precludes community participation in determining the future of Vieques.[106] The fortress model of conservation is handy for preventing tourists from getting hurt by chemical weapons and unexploded bombs, but it steeply limits autonomy for Viequenses. The result is that the economic situation on the island is little changed despite the Navy's departure: more than 75% of Vieques residents are below the poverty line, twice that of the rest of Puerto Rico. The population of Vieques,

Figure 8.5 In this April 17, 2008, file photo, an unexploded ordinance is blown up in a controlled demolition at the former Vieques Naval Training Range, on Vieques Island, Puerto Rico. The cleanup of Vieques currently requires unexploded ordinance to be detonated the same way it always was. These controlled explosions remain a major pollution concern (AP Photo/Brennan Linsley, File).

still hovering around 10,000, remains sandwiched between federal lands, with limited opportunities for development and expansion, and is forced to accept growing concessions to embracing tourism with all its colonial drawbacks.[107]

The legacy of Vieques bombing is also embodied in its residents. According to Nazario, an epidemiologist at the University of Puerto Rico, Viequenses are eight times more likely to die of cardiovascular disease and seven times more likely to die of diabetes than the rest of Puerto Ricans. A study released upon the Navy's departure revealed that cancer rates in Vieques are 30% higher than the rest of Puerto Rico, and that asthma and hypertension rates are also at their highest in Vieques.[108] In 2010, thousands of Viequenses accused the Navy's activities of causing illness in continuing lawsuits and their activism was successful in getting the Navy to finally admit that they had used heavy metals, radioactive materials, and chemical weapons on Vieques, which they had previously vehemently denied.[109] Despite this success, the Navy and the government refuse to recognize any link between their bombing of the island for 60 years and illness rates on Vieques. Many of those suffering from these illnesses must leave Vieques to receive care, as the only hospital on Vieques was destroyed by Hurricane Maria in 2017 and a new one isn't scheduled to open until 2024.[110] Vieques remains the poorest and sickest municipality in Puerto Rico, a direct consequence of colonialism, but Viequenses persistently resist colonialism and demand greater autonomy through continued organizing. Ultimately the resistance to military presence on Vieques built a lasting environmental justice movement, a point of pride for many Puerto Ricans, and a symbol of the union among anti-colonial struggle, self-determination, and environmental justice.

Tourism: The Price of Paradise

The displacement and dispossession of Puerto Rico's native population, the consequences of industrialization and militarization, is directly connected to the most recent economic experiment of environmental colonialism: tourism and repopulation. Fallow agricultural land, abandoned beach front properties, and derelict homes with absurdly affordable price tags have left Puerto Rico primed for tourism development. Despite the endurance of the tax breaks and economic philosophy of Operation Bootstrap, manufacturing as an economic mode declined steadily on the archipelago and since Puerto Rican citizens are now legally required to earn the federal minimum wage, globalization opened far cheaper labor markets in Latin America and Asia for US corporations to exploit. By 1990, the colony's economic model had become predominantly post-industrial, other than the pharmaceutical and electronics industries, based chiefly on service and tourism.[111] Much as Operation Bootstrap had dealt the deathblow to agriculture, the push for tourism led to underdevelopment and underinvestment in industry, leaving infrastructure crumbling and many factories as empty as the hollowed-out sugar mills. The modern era of tourism, retirement, and relocation, intensifying in the 1990s, was an attempt to make Puerto Rico more like Hawai'i : a place wealthy white Americans would want to visit and even retire in, a profitable paradise.

Puerto Rico's stark turn toward service and tourism as a primary economic model raised many familiar environmental justice concerns around agency. Once again Puerto Ricans, due to federal intervention on the archipelago, were unable to guide the development and extraction of their own environment. Now instead of arable land turned into sugar plantations or vast rural regions emptied out to use peoplepower toward producing factory goods, the archipelago's beaches, mangroves, and expansive coastal zones were transformed into paradise for foreigners. While Puerto Ricans themselves have a storied and vital culture of beach-based leisure and sport, people also depend on coastal lands for traditional subsistence fishing, and for their ecosystem services from water purification to hurricane amelioration.[112] The transformation of sweeping biologically diverse, aesthetically beautiful, and ecologically valuable coastal corridors into casinos, towering hotels, sprawling golf courses, and exclusive condominium enclaves for the primary benefit of United States and foreign interests means the direct removal of environmental resources from Puerto Ricans without their meaningful participation.

Even under true democratic representation, we could consider this style of development an environmental justice issue because it has wildly uneven impacts and poorly distributed benefits. Puerto Rico remains poorer than all 50 states despite yet another economic experiment, so the past 30 odd years of tourism, economy has done little to alleviate the archipelago's poverty or improve the lives of the native population.[113] Coastal populations abandoned first by the sugarcane syndicate and then by industry had long depended on the bounties of the sea as a relatively constant source of nourishment and income. As tourism megastructures have been built, Puerto Ricans have continued to be displaced from their environment and disinherited from the environmental access to coasts, mangroves, fisheries, and forests that has defined and sustained their culture for generations. Further relocations of Puerto Ricans either through physical removals or economic ones, like restrictions on fishing or limits on beach access, facilitated this development at every step, alienating Puerto Ricans from their environment for the double benefit of elites: the vacationers and those who profited from and owned the hotels.[114]

In Puerto Rico, where the law of the land since Spanish rule has always been that no one can own the sea or the marine zone, a historically stratified culture was allowed an equalizer through the beach.[115] Puerto Rico is technically the world's oldest remaining colony; the public beach with its equal opportunity for enjoying the benefits of Puerto Rico's environment holds great meaning in a place where access and control of the environment has never genuinely been in the hands of the people. Despite this equalizing law of the land, tourism developers and tourists have for decades been flouting its authority. Beach access has become a prominent Puerto Rican environmental justice issue stretching back to the 1970s–1980s when massive tourism development began in Rio Grande, PR, famously restricting fisherman from mooring their boats off a small strip of coastal land named Las Picuas.[116] This development follows a clear line connecting the issues this chapter has discussed: the collapse of the sugar syndicate due to the industrial policies of Operation Bootstrap left vast portions of Rio Grande undeveloped and inaccessible to local Puerto Ricans, then these areas were sold to wealthy investors for tourism development. Despite a court order and landmark legal decision preserving Puerto Rico's public beach access laws and declaring the restriction of the fishermen and building of vacation homes unlawful, the land was still developed into luxury housing for tourists and access by local Puerto Ricans remains restricted due to the entire community being gated off. At the time of the Las Picuas decision in 1992 at the height of Puerto Rico's transition toward tourism, this public beach was one of the only accessible areas left undeveloped by tourism for fishermen to rest their boats, and its development for tourism was devastating to local fishing communities.[117] The movement became known as "Las Playas Para El Pueblo" (Beaches for the People) and still inspires the environmental justice and anti-colonial movement of Puerto Rico today, as a story of resistance to colonial development, and the success of a community in restating their rights to access their native environment.

During the shift toward tourism, the tax breaks which had been used by the Puerto Rican and Federal government to invite the tycoons of the sugar syndicate, and then to invite polluting industrial corporations, was now turned toward inviting tourists. In 2005 the federal government terminated Section 936 of the tax code, the spiritual corporate tax break successor to the original legislation of Operation Bootstrap.[118] This led to a near total collapse of what was left of the archipelago's industrial economy, as corporations fled Puerto Rico.[119] This policy commitment to the new tourism economy was cemented by the Individual Investors Tax Act of 2012, or Act 22, which effectively gave the same tax-free status from Operation Bootstrap to extremely wealthy people, this time as individuals, by making all their passive income (capital gains, dividends) tax free.[120] Act 22 was further empowered by Act 60 in 2019 creating a series of other tax benefits for the businesses that wealthy investors would bring and real estate that they would occupy.[121] From 2012 to 2019 an estimated 4,500 individual investors and their families have moved to Puerto Rico to avoid paying taxes, relocating their homes and businesses. Since the benefits require individual investors to truly live in Puerto Rico, and to permanently move their lives and businesses there, Act 22 has been derided by local activists as neo-settler-colonialism and predatory gentrification.[122] Puerto Ricans themselves are effectively barred from obtaining these tax benefits. Investors looking to move to Puerto Rico cannot have lived there in the past ten years, and their businesses cannot have had any prior dealings and connections with the archipelago. This policy in effect encourages people with zero relationship to Puerto Rico to settle it and develop it tax-free.

Considering the archipelago has been hollowed out due to economic collapse from intervention and continuing extreme hurricanes, there could be some justice in encouraging those people who recently left to move back to their home, but instead they are actively excluded from the legislation. Over 200,000 people left Puerto Rico after Hurricane Maria due to the botched federal relief and recovery process, the dilapidated infrastructure, and failing economy. A total of 130,000 never returned.[123] An estimated

Figure 8.6 This beach in one of the richest communities of Puerto Rich was advertised to residents as "private." In response hundreds of Puerto Ricans from across the archipelago assembled for a party as protest to beach privatization. Latino Rebels/Carlos EdillBerríos Polanco, 2022.

3,000–4,600 people died due to the impacts of Hurricane Maria. The phrase a "Puerto Rico without Puerto Ricans" has been used by activists for years as rallying cry against the continuing development of Puerto Rico by outside forces, but under this context and combined with yet another tax intervention that encourages displacement and foreign repopulation, it seems the fear is well-founded.[124] Puerto Rico continues to be emptied out due to the policies of the colonial government, the PROMESA board of unelected appointees which manages Puerto Rico's economy, and the territorial goals of the United States. All while wealthy elites (you must be very wealthy to benefit from Act 22) from the 50 states are invited to Puerto Rico to reap the benefits and pay next to nothing back to the community.[125] This leaves the archipelago further tax-starved and indebted, as nearly half of the population remains in poverty. Meanwhile, the home prices in Puerto Rico have increased 63% in the decade since Act 22 passed and jumped 24% in just the past two years. This has raised the price of long-term rentals as well, making Puerto Rico less affordable than ever for the Puerto Ricans who have lived there for generations, and blocking them out of the communities with the most public resources which are most likely to be gentrified.[126]

The Individual investors of Act 22 have been attracted predominantly to coastal communities, especially those surrounding San Juan; in their creation of large enclaves of foreigners and gated communities, this population has led to ongoing environmental tension surrounding coastal resources. In a rebirth of the coastal environmental justice politics of Las Playas Pal' Pueblo, large movements stretched across the spring and summer of 2022 in response to attempts by Act 22 investors to privatize beaches and restrict coastal access for the elite. In January 2022 in Ocean Park, a historically Afro-Puerto Rican neighborhood where Arturo Schomburg lived and now home to multi-million-dollar ocean-front dwellings, a couple (beneficiaries of Act 22) roped off what they perceived as their portion of the beach to prevent locals from crossing in front of their property and playing sports in front of their house. Activists responded with a massive party called "Beach Olympics" where they protested by taking part in recreational tennis and volleyball tournaments and simply soaking up the sun.[127] Hundreds of local Puerto Rican activists turned a leisure activity, the enjoyment of natural coastal resources, into an environmental justice protest by concentrating their leisure in an area where they felt those activities were threatened.

Ocean Park is the site of severe levels of gentrification, as is much of San Juan, and by reclaiming the public space, Puerto Rican activists used the environment as a means of proclaiming their resistance to displacement and erasure, revealing once again how in a colony decolonial resistance is tied to environmental justice. The Movimiento en Defensa de Nuestras

Playas (Movement to Defend Our Beaches), a new branch in the long history of Puerto Rican beach activism, was born. Two weeks later this resistance-leisure continued in the ultra-elite enclave of Dorado, an area marketed explicitly to Act 22 investors through real estate programs.[128] For decades Dorado, built on the remains of a collapsing fisherman village, has precluded local access to the beach through massive, gated Ritz Carlton condos and other elite brand-name resort communities. They have relied on the legal loophole that locals can technically get to the beach at their own peril, over sharp volcanic rocks, and swells of turbulent water, all while advertising their beaches to foreigners and tourists as "private."[129]

Despite its relative isolation, Puerto Rican environmental activists once again held a massive beach party as protest, scaling the jagged rocks and rough water to the large luxurious beach, and making claim to public space through leisure and joy as resistance. In an act of defiance, they also renamed the beach, including its Geotag on Google maps, to "Ghetto Beach" in reference to the leading developer of Dorado, Federico Stubbe, stating that abolishing Act 22 would create a grand "Ghetto" in Puerto Rico.[130] In the words of Victor García, a local fisherman who attended the protest, "We're here reclaiming what's ours both as fishermen, as citizens, and as contributors to our country." Beach activist Eliezer Molina made a clear connection to colonialism: "We're here so that people have the opportunity to see how our maritime territory assets have been stolen from us." While doing little to directly stymie the flow of Act 22 investors to Puerto Rico, this resistance to the latest act of environmental colonialism brought a great deal of political attention to the tax policy and the protests started a new environmental justice movement to prevent the segregation of beaches by income, by race, and chiefly by colonial status.[131]

Puerto Rico and the US government's tourism-encouraging policies, much like the acts of environmental colonialism before them encouraging agriculture and industry, led to further environmental inequities and unequal distribution of environmental benefits. Land speculation and redevelopment led to increased cost of living, Airbnb proliferation led to increased rents, the purchase of massive homes by wealthy elites led to skyrocketing home prices, and beaches and coastal resources continued to be reserved for the foreigners, tourists, and local elites.[132] Hurricanes have also played into the vulnerabilities of the tourist economy in a vicious cycle. Hurricane Maria in 2017 devastated the tourism industry, leaving the Puerto Rico economically starved for almost two years, yet also left many homes empty for investors to purchase or turn into short-term rentals. The tourism industry itself also continues to destroy the important coastal ecosystems that traditionally protect Caribbean islands from hurricanes. Puerto Rico has always been in the hurricane belt of the Caribbean, but as storms worsen due to climate change coming more often and with more power, the continued development of "paradise" along the coast and the persistent neglect of inland infrastructure will further compound the impacts of storms.[133] When Puerto Ricans push back against beach development, beach privatization and structures built in the marine zone which devastate ecosystems and prevent beach access in the same stroke, they also contest environmental colonialism and the future injustice and chaos that will be wrought by storms and climate change.

Limbo and Exploitation, Resistance and Hope

Nearly every economic shift taken by the federal government in partnership with the colonial government has left Puerto Rico more hollowed out and left its people with less access and agency over their environmental resources. These policies have also left Puerto Rico with fewer of its people period, opening its islands to resettlement and further exploitation. The story of environmental colonialism in Puerto Rico, the details of a series of legislative and policy acts determining the archipelago's economy and future, is inherently a history of environmental justice. It is a history of unequal access to the benefits of a rich environment, and unequal exposure to the burdens of that environment's extraction. From the start of US occupation of Puerto Rico to its development as an industrialized commonwealth, and its modern day incarnation as a tourist destination, Puerto Rico has been the site of consistent colonial interventions by the United States, in partnership with the Puerto Rican elites, that exploit the resources of the archipelago for the benefits of outsiders and create serious environmental justice concerns for native Puerto Ricans, especially those most vulnerable to the colony's historic yet intact race and class-based hierarchies.

The massive theft of land by sugar syndicate and the military, the poisoning of its people through pollution and war games, the destruction of its natural resources for the sake of vacations by the Global North, and the continuing displacement and replacement of its ancestral people: each of these eras of colonial intervention has been a reassertion of US imperial power and the colonial status of the archipelago. Yet each of these interventions has also been followed by profound restatements of Puerto Rican sovereignty though movements resisting colonialism through the ongoing struggle for self-determination and environmental justice. At its core, the struggle for environmental justice in Puerto Rico, the fight against ongoing environmental colonialism, is a struggle for sovereignty. Looking forward into the face of continued environmental degradation, worsening storms, a warming climate, rising seas, and mounting tourism development pressures, one must ask who will control the land of Puerto Rico, and for what purpose?

Will it continue to be the US model of legal limbo, exploitation, and economic experimentation, or will recent political movements for decolonization in the United States result in meaningful change and self-determination for Puerto Ricans? Even as the struggle for statehood and independence heightens in recent years, with either option seeming like a reality by some measures and a distant dream by others, environmental colonialism will continue. In both independent nations like Jamaica and fully incorporated domestic states like Hawai'i, environmental colonialism is still possible due to the frameworks and power structures left over by histories of land theft, displacement, disenfranchisement, and the ongoing hierarchies of white supremacy. Many Puerto Ricans fear statehood, a whitewashing of the archipelago, an erasure of the Spanish language from streets and homes. Others fear independence, an archipelago left to stand on its own amid the chaos of climate change, after it was pillaged for all it was worth. Even if the prospect for continued environmental colonialism persists in all near-future visions of Puerto Rico, so too will the resistance of the Puerto Rican people. The long struggle against environmental colonialism in Puerto Rico provides hope that every future injustice will be answered, every act of colonialism resisted, and that the very sovereignty of Puerto Rico's people will remain firmly rooted in environmental justice.

Notes

1 Frantz Fanon, *Voices of Liberation* (Chicago, IL: Haymarket Books, 2016), 163.
2 Barbara Deutsch Lynch, "The Garden and the Sea: U.S. Latino Environmental Discourses and Mainstream Environmentalism," *Social Problems* 40, no. 1 (February 1993): 115, https://doi.org/10.2307/3097029
3 In English: "*Not yesterday, not today, not tomorrow, not ever will our homeland stop being ours.*"
4 Historically the largest island was named San Juan, and its capital city was named Puerto Rico. Over time, their usage swapped, and now "Puerto Rico" is synonymous with the largest island, while San Juan is the capital. Out of respect for the environmental justice struggles of the two smaller islands, and in resistance to calling Puerto Rico an "island" which erases people living on the smaller islands of the archipelago, the largest island of San Juan will be called "Isla Grande" in this chapter.
5 Irving Rouse, *The Tainos: Rise and Decline of the People Who Greeted Columbus* (New Haven, CT: Yale University Press, 1993).
6 Ricardo E. Alegría, *Historia y Cultura de Puerto Rico: desde la época Pre-Colombina hasta Nuestros Días* (San Juan: Fundación Francisco Carvajal, 1999).
7 For a frequently updated list of Caribbean sovereignty, see: Caribbean Elections, "Independence in the Caribbean," 2023.
8 On the Spanish-American war in this context, see, Thomas Schoonover, *Uncle Sam's War of 1898, and the Origins of Globalization* (Lexington: University Press of Kentucky, 2013); and Virginia Marie Bouvier, *Whose America?* (Westport, CT: Praeger Pub Text, 2001).

9 Jose Atiles-Osoria, "The Criminalization of Anti-Colonial Struggle in Puerto Rico," in *Counter-Terrorism and State Political Violence*, eds. Scott Poynting and David Whyte (London: Routledge, 2012), 156–77.

10 Jones-Shafroth Act, Pub.L. 64-368, 39 Stat. 951. Passed March 2, 1917.

11 Whether or not the archipelago is "part" of the United States, or merely "owned" by it is an ongoing legal question discussed below. See: Kyle Dropp and Brendan Nyhan, "Nearly Half of Americans Don't Know Puerto Ricans Are Fellow Citizens," *The New York Times*, September 26, 2017. https://www.nytimes.com/2017/09/26/upshot/nearly-half-of-americans-dont-know-people-in-puerto-ricoans-are-fellow-citizens.html

12 Decided in *Igartua De La Rosa v. United States*, 229 F.3d 80 (1st Cir. 2000). *see:* Amber Cottle, "Silent Citizens: United States Territorial Residents and the Right to Vote in Presidential Elections," *University of Chicago Legal Forum* 1995, no. 1 (December 7, 2015): 315–38.

13 US Constitution, art. IV, § 3, clause 2.

14 Tom Lin, "Americans, Almost and Forgotten," *California Law Review* 107, no. 4 (September 2019): 1249; *see also:* Edgardo Meléndez "Citizenship and the Alien Exclusion in the Insular Cases: Puerto Ricans in the Periphery of American Empire," *Centro Journal* 25, no. 1 (March 2013): 106–45.

15 Doug Mack, "The Racist Supreme Court Cases That Cemented Puerto Rico's Second-Class Status," *Slate Magazine* (October 9, 2017).

16 Juan R. Torruella, Commentary: "Why Puerto Rico Does Not Need Further Experimentation with Its Future: A Reply to the Notion of 'Territorial Federalism,'" *Harvard Law Review Forum* 131, no. 3 (2018): 65–104.

17 *Downes v. Bidwell*, 182 U.S. 244 (1901).

18 Puerto Ricans are now citizens by congressional statute, but while living in Puerto Rico lose several constitutional citizenship rights. This is also true of any US citizen who moves to Puerto Rico. The remaining territories have statutory citizenship, except people living in American Samoa who remain non-citizen nationals.

19 *Plessy v. Ferguson*, 163 U.S. 537 (1895); and *Brown v. Board of Education of Topeka* 347 U.S. 483 (1954).

20 *Harris v. Rosario*, 446 U.S. 651 (1980).

21 Ariane de Vogue, "Supreme Court Rules Puerto Ricans Don't Have Constitutional Right to Some Federal Benefits," *CNN*, April 21, 2022.

22 United Nations, "Non-Self-Governing Territories," www.un.org, September 22, 2020, https://www.un.org/dppa/decolonization/en/nsgt

23 Natsu Taylor Saito, "Asserting Plenary Power over the 'Other': Indians, Immigrants, Colonial Subjects, and Why U.S. Jurisprudence Needs to Incorporate International Law," *Yale Law & Policy Review* 20, no. 2 (2002): 427–80.

24 Edward Cavanagh and Lorenzo Veracini, *The Routledge Handbook of the History of Settler Colonialism* (New York: Routledge, 2017).

25 Jacqueline Jones, *A Dreadful Deceit: The Myth of Race from the Colonial Era to Obama's America* (New York: Basic Books, 2013).

26 Oliver Cromwell Cox, *Caste, Class, and Race* (New York: Monthly Review Press, 1948).

27 Dorothy E Roberts, *Fatal Invention: How Science, Politics, and Big Business Re-Create Race in the Twenty-First Century* (New York and London: New Press, 2012).

28 Ugo Mattei and Laura Nader, *Plunder: When the Rule of Law Is Illegal* (Malden, MA: Blackwell, 2008).

29 Mattei and Nader, *Plunder*.

30 Manuel Valdés Pizzini, "Historical Contentions and Future Trends in the Coastal Zones," in *Beyond Sun and Sand: Caribbean Environmentalisms*, ed. Sherri Baver and Barbara Deutsch Lynch (New Brunswick, NJ: Rutgers University Press, 2006), 44–64.

31 Neil Smith, "The Production of Nature" in *Futurenatural: Nature/Science/Culture*, ed. George Robertson (New York: Routledge, 1996), 35–54.

32 Jamaica Kincaid, *A Small Place* (New York: Farrar, Straus, and Giroux, 2000).

33 Sherrie Baver, "Environmental Struggles in Paradise: Puerto Rican Cases, Caribbean Lessons," *Caribbean Studies* 40, no. 1 (2012): 15–35.

34 Sherrie Baver and Barbara Deutsch Lynch, "The Political Ecology of Paradise," in *Beyond Sun and Sand: Caribbean Environmentalisms*, 3–15.

35 José M. Atiles-Osoria, "The Criminalisation of Anti-Colonial Struggle in Puerto Rico," in *Counter-Terrorism and State Political Violence*, 156–77.

36 Carmen M. Concepción, "The Origins of Modern Environmental Activism in Puerto Rico in the 1960s," *International Journal of Urban and Regional Research* 19, no. 1 (1995): 112–28.

37 Olga Jiménez, *Puerto Rico: An Interpretive History from Pre-Columbian Times to 1900* (Princeton, NJ: Markus Wiener Publishers, 2006).

38 Jiménez, *Puerto Rico*.

39 Albert Manucy and Ricardo Torres-Reyes, *Puerto Rico, and the Forts of Old San Juan* (New York: Devin-Adair Pub, 1973).

40 Van Middeldyk and Martin Grove Brumbaugh, *The History of Puerto Rico: From the Spanish Discovery to the American Occupation* (London: Appleton, 2018/1903).

41 Jovanni Reyes, "Puerto Rico: A Launchpad for Empire in the Caribbean," *Responsible Statecraft*, November 25, 2022.

42 César J Ayala, *American Sugar Kingdom* (Chapel Hill: University of North Carolina Press, 2009).

43 Benjamin Bridgman et al., "What Ever Happened to the Puerto Rican Sugar Manufacturing Industry?" *Federal Reserve Bank Staff Report*, December 27, 2012.

44 "What Ever Happened to the Puerto Rican Sugar Manufacturing Industry?"

45 Charles Herbert Allen, *First Annual Report of Charles H. Allen, Governor of Porto Rico: Covering the Period from May 1, 1900, to May 1, 1901* (San Juan: Fundación Puertorriqueña de las Humanidades, 2005), 39–41.

46 Manuel Maldonado-Denis, *Puerto Rico: A Socio-Historic Interpretation* (New York: Random House, 1972), 70–76.

47 Nelson A Denis, *War against All Puerto Ricans: Revolution and Terror in America's Colony* (New York: Bold Type Books, 2016).

48 Denis, *War against All Puerto Ricans*.

49 Library of Congress, *The Indianapolis Times*. June 29, 1934, Home Edition, *Chronicling America*.

50 Matthew P. Johnson, "'Thirsty Sugar Lands': Environmental Impacts of Dams and Empire in Puerto Rico since 1898," *Environment and History* 27, no. 3 (2019): 337–65.

51 Samantha Schmidt, "'Thousands of People Could Die': 70,000 in Puerto Rico Urged to Evacuate with Dam in 'Imminent' Danger," *Washington Post*, September 22, 2017.

52 Hilda Lloréns and Carlos García-Quijano "From Extractive Agriculture to Industrial Waste Periphery: Life in a Black-Puerto Rican Ecology," *Black Perspectives AAIHS*, June 22, 2020.

53 Carlos Garcia-Quijano, *Resisting Extinction: The Value of Local Ecological Knowledge for Small-Scale Fishers in Southeastern Puerto Rica* (Athens: University of Georgia Press, 2010), 58–60.
54 Galvin, Miles. "The Early Development of the Organized Labor Movement in Puerto Rico," *Latin American Perspectives* 3, no. 3 (1976): 17–35.
55 Juan A. Giusti-Cordero, "Labour, Ecology and History in a Puerto Rican Plantation Region: 'Classic' Rural Proletarians Revisited," *International Review of Social History* 41, no. S4 (December 1996): 53–82.
56 Kenneth Lugo del Toro et al., *Nacimiento Y Auge de La Confederación General de Trabajadores, 1940–1945, Catalog.loc.gov Library Catalog*, Primera edición (San Juan: Universidad Interamericana de Puerto Rico, Recinto Metropolitano, 2013).
57 Shir Lerman Ginzburg, "Colonial Comida: The Colonization of Food Insecurity in Puerto Rico," *Food, Culture & Society* 25, no. 1 (March 2021): 1–14.
58 Jorge J. Ruscalleda Reyes, "Cambios en el Sindicalismo Puertorriqueño ante la Segunda Guerra Mundial: El Caso de La Confederación General de Trabajadores," *Revista Brasileira Do Caribe* 17, no. 32 (2016): 159–84.
59 Matthew Edel, "Land Reform in Puerto Rico, 1940–1959: Part One," *Caribbean Studies* 2, no. 3 (1962): 26–60.
60 Geoff Burrows, "The New Deal in Puerto Rico: Public Works, Public Health, and the Puerto Rico Reconstruction Administration, 1935–1955," *CUNY Academic Works*, 2014.
61 John Gunther, *Inside Latin America* (New York: Harper & Bros. Press, 1941), 423.
62 Esteban Bird, et al., "Report on the sugar industry in relation to the social and economic system of Puerto Rico" (San Juan: Puerto Rico Reconstruction Administration, 1937), 43.
63 Sherrie L Baver, *The Political Economy of Colonialism: The State and Industrialization in Puerto Rico* (Westport, CT: Praeger, 1993). And also: Deborah Berman Santana, *Kicking Off the Bootstraps: Environment, Development, and Community Power in Puerto Rico* (Tucson: University of Arizona Press, 1996).
64 Eliezer Curet Cuevas, *Economía Política de Puerto Rico: 1950 a 2000* (San Juan: Ediciones M.A.C., 2004), 22–23.
65 Jorge R Schmidt, *The Politics of English in Puerto Rico's Public Schools* (Boulder, CO: First Forum Press, 2014).
66 James Dietz, *Economic History of Puerto Rico: Institutional Change and Capitalist Development* (Princeton, NJ: Princeton University Press, 1986), 226.
67 Henry Wells, *Modernization of Puerto Rico* (Cambridge, MA: Harvard University Press, 2014), 152.
68 Umair Irfan, "Puerto Rico Is Slipping into an Environmental Crisis," *Vox*, October 26, 2017.
69 US Census Bureau, "Urban Areas Facts," *The United States Census Bureau*, October 8, 2021.
70 José L. Vázquez Calzada, *La población de Puerto Rico y su trayectoria histórica* (Río Piedras: Escuela Graduada de Salud Pública, Recinto de Ciencias Médicas, Universidad de Puerto Rico, 1988), 286.
71 César Ayala, "Puerto Rico; Un Pueblo Diaspórico," www.sscnet.ucla.edu, February 3, 2021.
72 Jorge Morales-Burnett, "Five Lessons from the Aftermath of Hurricane Maria for Communities Preparing for Climate Migration," *Urban Institute*, September 19, 2022.

73 Audrea Lim, "A Just Transition Doesn't Have to Be Top-Down. This Brooklyn Neighborhood Is Proof," *Fix*, May 4, 2021. See also Troy Simpson, "Remediating Sunset Park: Environmental Injustice, Danger, and Gentrification," *The Journal of Public Space* 4, no. 4 (December 31, 2019): 187–210.

74 Justin D. García, "Young Lords," *Multicultural America: A Multimedia Encyclopedia*, ed. Carlos E. Cortés and Jane E. Sloan, vol. 4 (Los Angeles, CA: SAGE Reference, 2014), 2216–17.

75 Johanna Fernández, *The Young Lords: A Radical History* (Chapel Hill: University of North Carolina Press, 2020).

76 Fernández, *The Young Lords*.

77 David Gonzalez, "In Sunset Park, a Call for 'Innovation' Leads to Fears of Gentrification," *The New York Times*, March 6, 2016. https://www.nytimes.com/2016/03/07/nyregion/in-sunset-park-a-call-for-innovation-leads-to-fears-of-gentrification.html

78 Emily Cochrane and Patricia Mazzei, "House Passes Bill That Could Pave the Way for Puerto Rican Statehood," *The New York Times*, December 15, 2022. https://www.nytimes.com/2022/12/15/us/politics/house-puerto-rican-statehood.html

79 Diasporic environmental justice organizations like UPROSE, El Puente Latino Climate Action Network, and the Climate Justice Alliance have helped organize responses to Puerto Rican environmental struggles.

80 Humberto García Muñiz and Gloria Vega Rodríguez, *La ayuda militar como negocio: Estados Unidos y el Caribe* (San Juan: Ediciones Callejón, 2002).

81 Amílcar A. Barreto, *Vieques, the Navy, and Puerto Rican Politics* (Gainesville: University of Florida Press, 2002).

82 Sherrie L. Baver, "'Peace Is More than the End of Bombing,'" *Latin American Perspectives* 33, no. 1 (January 2006): 102–15.

83 Katherine T McCaffrey, *Military Power and Popular Protest: The U.S. Navy in Vieques, Puerto Rico* (New Brunswick, NJ: Rutgers University Press, 2002).

84 Katherine McCaffrey, "Environmental Remediation and Its Discontents: The Contested Cleanup of Vieques, Puerto Rico," *Journal of Political Ecology* 25, no. 1 (2018): 80–103.

85 Katherine T McCaffrey, "Social Struggle against the U.S. Navy in Vieques, Puerto Rico," *Latin American Perspectives* 33, no. 1 (January 2006): 83–101.

86 Barreto, "Vieques," 20–24.

87 Mario Murillo, *Islands of Resistance: Puerto Rico, Vieques, and U.S. Policy* (New York: Seven Stories Press, 2001), 25.

88 Murillo, *Islands of Resistance*, 53.

89 McCaffrey, "Social Struggle against the U.S. Navy in Vieques, Puerto Rico," 83–101.

90 Charles C Walker, "Culebra: Nonviolent action and the US Navy," *Liberation without Violence: A Third-Party Approach*, ed. A. Paul Hare and Herbert H. Blumberg (London: Rex Collings, 1976), 178–95.

91 McCaffrey, "Social Struggle against the U.S. Navy in Vieques, Puerto Rico."

92 McCaffrey, "Environmental Remediation and Its Discontents."

93 United States Government Accountability Office (GAO). (2021). *Defense Cleanup: Efforts at Former Military Sites on Vieques and Culebra, Puerto Rico, Are Expected to Continue through 2032* (Report No. GAO-21-268).

94 Gay Nagle Myers, "P.R. Tourism in Damage-Control Mode after CNN's Vieques Report," www.travelweekly.com, February 5, 2010.

95 Deborah Berman Santana, "Resisting Toxic Militarism: Vieques versus the U.S. Navy," *Social Justice* 29, no. 2 (2002): 37–47.

96 Baver, "Environmental Struggles in Paradise," 15–35.

97 José Paralitici, *La represión contra el independentismo puertorriqueño: 1960–2010* (Cayey: Publicaciones Gaviota, 2006). And also: David E. Sanger and Christopher Marquis, "U.S. Said to Plan Halt to Exercises on Vieques Island," *The New York Times*, June 14, 2001. https://www.nytimes.com/2001/06/14/us/us-said-to-plan-halt-to-exercises-on-vieques-island.html

98 Baver, "Peace Is More than the End of Bombing," 102–15.

99 McCaffrey, "Environmental Remediation and Its Discontents."

100 McCaffrey, "Environmental Remediation and Its Discontents."

101 Carlos Giusti, "Puerto Rico Cleanup by U.S. Military Will Take More than a Decade," *NBC News*, March 26, 2021.

102 McCaffrey, "Environmental Remediation and Its Discontents."

103 Baver, "Environmental Struggles in Paradise."

104 Ana Guzman, et al., "Evaluating the Conservation Attitudes, Awareness and Knowledge of Residents towards Vieques National Wildlife Refuge, Puerto Rico," *Conservation and Society* 18, no. 1 (2020): 13.

105 Guzman, et al., "Evaluating the Conservation Attitudes, Awareness and Knowledge of Residents towards Vieques National Wildlife Refuge, Puerto Rico," 13.

106 McCaffrey, "Environmental Remediation and Its Discontents."

107 Mimi Sheller, "Retouching the 'Untouched Island,'" *Téoros: Revue de Recherche En Tourisme* 26, no. 1 (2007): 21.

108 Valeria Pelet, "There's a Health Crisis on This Puerto Rican Island, but It's Impossible to Prove Why It's Happening," *The Atlantic*, September 3, 2016.

109 Abbie Boudreau, "Island Residents Sue U.S., Saying Military Made Them Sick – *CNN.com*, www.cnn.com, February 1, 2010.

110 Maricarmen Rivera Sánchez, "Gov't: Vieques to Have New Hospital by Mid-2024," *The Weekly Journal*, August 20, 2021. https://www.theweeklyjournal.com/online_features/govt-vieques-to-have-new-hospital-by-mid-2024/article_2acfda32-013c-11ec-ac6f-e79d10397f83.html

111 José M. Atiles-Osoria, "Environmental Colonialism, Criminalization and Resistance: Puerto Rican Mobilizations for Environmental Justice in the 21st Century," *RCCS Annual Review* 6, no. 6 (October 1, 2014): 1–21.

112 Manuel Valdés Pizzini, "Historical Contentions and Future Trends in the Coastal Zones: The Environmental Movement in Puerto Rico," in *Beyond Sun and Sand*, December 31, 2020, 44–64.

113 Simone Foxman, "Puerto Rico Is Living an Impoverished Debt Nightmare Reminiscent of Southern Europe or Detroit," *Quartz*, September 19, 2013.

114 Pizzini, "Historical Contentions and Future Trends in the Coastal Zones."

115 Pamela Pogue and Virginia Lee, "Providing Public Access to the Shore: The Role of Coastal Zone Management Programs," *Coastal Management* 27, no. 2–3 (April 1999): 219–37.

116 *US Industries, Inc. v. Laborde*, 794 F. Supp. 454 (D.P.R. 1992).

117 Pizzini, "Historical Contentions and Future Trends in the Coastal Zones."

118 John W. Schoen, "Here's How an Obscure Tax Change Sank Puerto Rico's Economy," www.cnbc.com, September 26, 2017.

119 MacEwan, Arthur, "Quantifying the Impact of 936," Center for Global Development and Sustainability, Working Paper Series 2016-4, Brandeis University, May 2016.

120 Julia La Roche, "This New Puerto Rican Law Makes Wealthy People Want to Move There to Avoid Taxes," *Business Insider*, March 11, 2013.

121 Coral Murphy Marcos, Patricia Mazzei, and Erika P. Rodriguez, "The Rush for a Slice of Paradise in Puerto Rico," *The New York Times*, January 31, 2022.

122 Mariah Espada, "Influencers, Developers, Crypto Currency Tycoons: How Puerto Ricans Are Fighting Back against the Outsiders Using the Island as a Tax Haven," *Time*, April 16, 2021.

123 John D. Sutter, "130,000 Left Puerto Rico after Maria," *CNN*, December 19, 2018.

124 Fidel Martinez, "Latinx Files: Will Puerto Rico Stop Being for Puerto Ricans Soon?" *Los Angeles Times*, January 13, 2022.

125 Carlos Edill Berríos Polanco, "Act 60 Brings People into Puerto Rico and Pushes Others Out," *Latino Rebels*, January 25, 2022.

126 Damaris Suárezy, "A Nightmare for Puerto Ricans to Find a Home, While Others Accumulate Properties," *Centro de Periodismo Investigativo*, December 19, 2022.

127 Carlos Edill Berríos Polanco, "Puerto Ricans Hold Beach Protest against Privatization," *Latino Rebels*, January 31, 2022.

128 Carlos Edill Berríos Polanco, "Hundreds of Puerto Ricans Take over 'Private Beach' in Dorado to Protest Access," *Latino Rebels*, February 14, 2022.

129 Melissa del Carmen Gomez, "Puerto and Their Beaches: A Modern-Day Colonial Issue," *Voices of Gen-Z*, April 16, 2022.

130 Berríos Polanco, "Hundreds of Puerto Ricans Take over 'Private Beach' in Dorado to Protest Access."

131 Coral Murphy Marcos, "'The Beaches Belong to the People': Inside Puerto Rico's Anti-Gentrification Protests," *The Guardian*, July 23, 2022.

132 Raúl Santiago, "The Impact of Short-Term Rentals in Puerto Rico: 2014–2020," *CNE – Centro Para Una Nueva Economía – Center for a New Economy*, December 12, 2022.

133 Yasmin Velez-Sanchez, "Puerto Rico Moves to Limit Coastal Damage from Hurricanes and Other Threats," *pew.org*, December 7, 2022.

Bibliography

Atiles-Osoria, José M. "Environmental Colonialism, Criminalization and Resistance: Puerto Rican Mobilizations for Environmental Justice in the 21st Century*." *RCCS Annual Review*, no. 6 (October 1, 2014). https://doi.org/10.4000/rccsar.524

Ávila-García, Patricia, and Eduardo Luna Sánchez. "The Environmentalism of the Rich and the Privatization of Nature." *Latin American Perspectives* 39, no. 6 (September 12, 2012): 51–67.

Ayala, César J. *American Sugar Kingdom*. Chapel Hill: University of North Carolina Press, 2009.

Ayala, Israel Meléndez, Alicia Kennedy, and Damon Winter. "How the U.S. Dictates What Puerto Rico Eats." *The New York Times*, October 1, 2021, sec. Opinion. https://www.nytimes.com/2021/10/01/opinion/puerto-rico-jones-act.html

Baver, Sherrie. "Puerto Rico: Colonialism Revisited." *Latin American Research Review* 22, no. 2 (1987): 227–34. https://doi.org/10.1017/s0023879100022135

_____. *The Political Economy of Colonialism: The State and Industrialization in Puerto Rico.* Westport, CT: Praeger, 1993.

_____. "Peace Is More than the End of Bombing." *Latin American Perspectives* 33, no. 1 (January 2006): 102–15.

_____. "Environmental Struggles in Paradise: Puerto Rican Cases, Caribbean Lessons." *Caribbean Studies* 40, no. 1 (2012): 15–35. https://doi.org/10.1353/crb.2012.0011

_____, and Barbara D. Lynch. *Beyond Sun and Sand: Caribbean Environmentalisms.* New Brunswick, NJ: Rutgers University Press, 2005.

Berríos Polanco, Carlos Edill. "Act 60 Brings People into Puerto Rico and Pushes Others Out." *Latino Rebels*, January 25, 2022. https://www.latinorebels.com/2022/01/25/act60displacement/

_____. "Puerto Ricans Hold Beach Protest against Privatization." *Latino Rebels*, January 31, 2022. https://www.latinorebels.com/2022/01/31/beachprotest/

_____. "Hundreds of Puerto Ricans Take over 'Private Beach' in Dorado to Protest Access." *Latino Rebels*, February 14, 2022. https://www.latinorebels.com/2022/02/14/doradobeachprotest/

Boudreau, Abbie. "Island Residents Sue U.S., Saying Military Made Them Sick." CNN.com., February 1, 2010. http://www.cnn.com/2010/US/02/01/vieques.illness/

Bouvier, Virginia Marie. *Whose America?* Westport, CT: Praeger, 2001.

Bridgman, Benjamin, Michael Maio, James Andrew Schmitz, and Arilton Teixeira. "What Ever Happened to the Puerto Rican Sugar Manufacturing Industry?" *Federal Reserve Bank*, December 27, 2012. https://doi.org/10.21034/sr.477

Cavanagh, Edward, and Lorenzo Veracini. *The Routledge Handbook of the History of Settler Colonialism.* New York: Routledge, 2017.

Cochrane, Emily, and Patricia Mazzei. "House Passes Bill That Could Pave the Way for Puerto Rican Statehood." *The New York Times*, December 15, 2022, sec. U.S. https://www.nytimes.com/2022/12/15/us/politics/house-puerto-rican-statehood.html

Cox, Oliver Cromwell. *Caste, Class, and Race.* London: Forgotten Books, 1970.

Denis, Nelson A. *War against All Puerto Ricans: Revolution and Terror in America's Colony.* New York: Bold Type Books, 2016.

Dietz, James L. *Economic History of Puerto Rico Institutional Change and Capitalist Development.* Princeton, NJ: Princeton University Press, 2018.

Dropp, Kyle, and Brendan Nyhan. "Nearly Half of Americans Don't Know Puerto Ricans Are Fellow Citizens." *The New York Times*, September 26, 2017, sec. The Upshot. https://www.nytimes.com/2017/09/26/upshot/nearly-half-of-americans-dont-know-people-in-puerto-ricoans-are-fellow-citizens.html

Edel, Matthew O. "Land Reform in Puerto Rico, 1940–1959: Part One." *Caribbean Studies* 2, no. 3 (1962): 26–60.

Espada, Mariah. "Influencers, Developers, Crypto Currency Tycoons: How Puerto Ricans Are Fighting Back against the Outsiders Using the Island as a Tax Haven." *Time*, April 16, 2021.

Fanon, Frantz. *The Wretched of the Earth.* Cape Town: Kwela Books, 1961.

Feffer, John. "Puerto Rico: The Gibraltar of the Caribbean and Launchpad for Empire – FPIF." *Foreign Policy in Focus*, November 15, 2022.

Foxman, Simone. "Puerto Rico Is Living an Impoverished Debt Nightmare Reminiscent of Southern Europe or Detroit." *Quartz*, September 19, 2013.

Galvin, Miles. "The Early Development of the Organized Labor Movement in Puerto Rico." *Latin American Perspectives* 3, no. 3 (July 1976): 17–35.

Ginzburg, Shir Lerman. "Colonial Comida: The Colonization of Food Insecurity in Puerto Rico." *Food, Culture & Society* 25, no. 1 (March 12, 2021): 1–14. https://doi.org/10.1080/15528014.2021.1884440

Giusti, Carlos. "Puerto Rico Cleanup by U.S. Military Will Take More than a Decade." *NBC News*, March 26, 2021.

Giusti-Cordero, Juan A. "Labour, Ecology and History in a Puerto Rican Plantation Region: 'Classic' Rural Proletarians Revisited." *International Review of Social History* 41, no. S4 (December 1996): 53–82. https://doi.org/10.1017/s0020859000114270

Gonzalez, David. "In Sunset Park, a Call for 'Innovation' Leads to Fears of Gentrification." *The New York Times*, March 6, 2016, sec. New York. https://www.nytimes.com/2016/03/07/nyregion/in-sunset-park-a-call-for-innovation-leads-to-fears-of-gentrification.html

Hudson, Nicholas. "'Hottentots' and the Evolution of European Racism." *Journal of European Studies* 34, no. 4 (December 2004): 308–32.

Irfan, Umair. "Puerto Rico Is Slipping into an Environmental Crisis." *Vox*, October 26, 2017.

Jiménez, Olga. *Puerto Rico: An Interpretive History from Pre-Columbian Times to 1900/Monograph*. Princeton, NJ: Markus Wiener Publishers, 2006.

Johnson, Matthew P. "'Thirsty Sugar Lands': Environmental Impacts of Dams and Empire in Puerto Rico since 1898." *Environment and History* 27, no. 23 (2019). https://doi.org/10.3197/096734019x15631846928701

Jones, Jacqueline. *A Dreadful Deceit: The Myth of Race from the Colonial Era to Obama's America*. New York: Basic Books, a Member of the Perseus Books Group, 2013.

Kincaid, Jamaica. *A Small Place*. New York: Farrar, Straus and Giroux, 2000.

Lim, Audrea. "A Just Transition Doesn't Have to Be Top-Down. This Brooklyn Neighborhood Is Proof." *Fix*, May 4, 2021. https://grist.org/fix/justice/uprose-brooklyn-just-transition-world-we-need-book/

Lloréns, Hilda, and Carlos G Garcia-Quijano. "From Extractive Agriculture to Industrial Waste Periphery: Life in a Black- Puerto Rican Ecology." *AAIHS*, June 22, 2020. https://www.aaihs.org/from-extractive-agriculture-to-industrial-waste-periphery-life-in-a-black-puerto-rican-ecology/

Lynch, Barbara Deutsch. "The Garden and the Sea: U.S. Latino Environmental Discourses and Mainstream Environmentalism," *Social Problems* 40, no. 1 (February 1993). https://doi.org/10.2307/3097029

Mack, Doug. "The Racist Supreme Court Cases That Cemented Puerto Rico's Second-Class Status." *Slate Magazine*, October 9, 2017.

Maldonado-Denis, Manuel. *Puerto Rico: A Socio-Historic Interpretation*. New York: Random House, 1972.

Marcos, Coral Murphy. "'The Beaches Belong to the People': Inside Puerto Rico's Anti-Gentrification Protests." *The Guardian*, July 23, 2022.

_____, Patricia Mazzei, and Erika P. Rodriguez. "The Rush for a Slice of Paradise in Puerto Rico." *The New York Times*, January 31, 2022, sec. U.S.

Mattei, Ugo, and Laura Nader. *Plunder: When the Rule of Law Is Illegal*. Malden, MA: Blackwell Publishing, 2008.

McCaffrey, Katherine T. *Military Power and Popular Protest: The U.S. Navy in Vieques, Puerto Rico*. New Brunswick, NJ, and London: Rutgers University Press, 2002.

_____. "Social Struggle against the U.S. Navy in Vieques, Puerto Rico." *Latin American Perspectives* 33, no. 1 (January 2006): 83–101.

_____. "Colonial Citizenship: Power and Struggle in Vieques, Puerto Rico." *Transforming Anthropology* 19, no. 1 (March 11, 2011): 50–52. https://doi.org/10.1111/j.1548-7466.2011.01116.x

_____. "Environmental Remediation and Its Discontents: The Contested Cleanup of Vieques, Puerto Rico." *Journal of Political Ecology* 25, no. 1 (2018). https://doi.org/10.2458/v25i1.22631

Middeldyk, Van, and Martin Grove Brumbaugh. *The History of Puerto Rico: From the Spanish Discovery to the American Occupation*. New York: Appleton, 2018.

Murillo, Mario. *Islands of Resistance: Puerto Rico, Vieques, and U.S. Policy*. New York: Seven Stories Press, 2001.

Myers, Gay Nagle. "P.R. Tourism in Damage-Control Mode after CNN's Vieques Report." *Travel Weekly*, February 5, 2010. www.travelweekly.com

National Endowment for the Humanities. "The Indianapolis Times. [Volume] (Indianapolis [Ind.]) 1922–1965, June 29, 1934, Home Edition, Second Section, Image 21." *Chroniclingamerica.loc.gov*, June 29, 1934.

Osterhammel, Jürgen. *Colonialism: A Theoretical Overview*. Princeton, NJ: Wiener, 2010.

Pelet, Valeria. "There's a Health Crisis on This Puerto Rican Island, but It's Impossible to Prove Why It's Happening." *The Atlantic*, September 3, 2016.

Peon, Harold. "It Is 2020, and Puerto Rico Is Still a Colony." *Harvard Political Review*, November 22, 2020.

Pizzini, Manuel Valdés. "4. Historical Contentions and Future Trends in the Coastal Zones: The Environmental Movement in Puerto Rico." *Beyond Sun and Sand* (December 31, 2020): 44–64. https://doi.org/10.36019/9780813537528-005

Pogue, Pamela, and Virginia Lee. "Providing Public Access to the Shore: The Role of Coastal Zone Management Programs." *Coastal Management* 27, no. 2–3 (April 1999): 219–37. https://doi.org/10.1080/089207599263848

Reyes, Jovanni. "Puerto Rico: A Launchpad for Empire in the Caribbean." *Responsible Statecraft*, November 25, 2022. https://responsiblestatecraft.org/2022/11/25/puerto-rico-a-launchpad-for-empire-in-the-caribbean/

Roberts, Dorothy E. *Fatal Invention: How Science, Politics, and Big Business Re-Create Race in the Twenty-First Century*. New York and London: New Press, 2012.

Roche, Julia La. "This New Puerto Rican Law Makes Wealthy People Want to Move There to Avoid Taxes." *Business Insider*, March 11, 2013.

Rosa-Aquino, Paola. "130,000: The Number of Puerto Ricans Who Never Returned after Maria." *Grist*, December 20, 2018.

Rouse, Irving. *The Tainos: Rise and Decline of the People Who Greeted Columbus*. New Haven: Yale University Press, 1993.

Saito, Natsu Taylor. "Asserting Plenary Power over the 'Other': Indians, Immigrants, Colonial Subjects, and Why U.S. Jurisprudence Needs to Incorporate International Law." *Yale Law & Policy Review* 20, no. 2 (2002): 427–80.

_____. *Settler Colonialism, Race, and the Law: Why Structural Racism Persists*. New York: New York University Press, 2020.

Sánchez, Maricarmen Rivera. "Gov't: Vieques to Have New Hospital by Mid-2024." *The Weekly Journal*, August 20, 2021. https://www.theweeklyjournal. com/online_features/govt-vieques-to-have-new-hospital-by-mid-2024/article_2acfda32-013c-11ec-ac6f-e79d10397f83.html

Sanger, David E., and Marquis Christopher. "U.S. Said to Plan Halt to Exercises On Vieques Island." *The New York Times*, June 14, 2001, sec. U.S. https://www. nytimes.com/2001/06/14/us/us-said-to-plan-halt-to-exercises-on-vieques-island.html

Santiago, Raúl. "The Impact of Short-Term Rentals in Puerto Rico: 2014–2020." *CNE – Centro Para Una Nueva Economía – Center for a New Economy*, December 12, 2022.

Schmidt, Jorge R. *The Politics of English in Puerto Rico's Public Schools*. Boulder, CO: First Forum Press, Inc, 2014.

Schmidt, Samantha. "'Thousands of People Could Die': 70,000 in Puerto Rico Urged to Evacuate with Dam in 'Imminent' Danger." *Washington Post*, September 22, 2017.

Schoen, John W. "Here's How an Obscure Tax Change Sank Puerto Rico's Economy." www.cnbc.com, September 26, 2017.

Schoonover, Thomas D. *Uncle Sam's War of 1898 and the Origins of Globalization*. Lexington: University Press of Kentucky, 2013.

Sheller, Mimi. "Retouching the 'Untouched Island.'" *Téoros: Revue de Recherche En Tourisme* 26, no. 1 (2007). https://doi.org/10.7202/1070991ar

Simpson, Troy. "Remediating Sunset Park: Environmental Injustice, Danger, and Gentrification." *The Journal of Public Space* 4, no. 4 (December 31, 2019): 187–210.

Solá, José O. "Colonialism, Planters, Sugarcane, and the Agrarian Economy of Caguas, Puerto Rico, between the 1890s and 1930." *Agricultural History* 85, no. 3 (July 1, 2011): 349–72. https://doi.org/10.3098/ah.2011.85.3.349

Suárezy, Damaris, Víctor Rodríguez Velázquez, and Omaya Sosa Pascualy. "A Nightmare for Puerto Ricans to Find a Home, While Others Accumulate Properties." *Centro de Periodismo Investigativo*, December 19, 2022.

Sutter, John D. "130,000 Left Puerto Rico after Maria." *CNN*, December 19, 2018. https://www.cnn.com/2018/12/19/health/sutter-puerto-rico-census-update/index.html

Wells, Henry. *Modernization of Puerto Rico*. Cambridge, MA: Harvard University Press, 2014.

Part III

Environmental Justice, Climate Justice, and Sustainability

9 Indigenous Environmental Justice, Renewable Energy Transition, and the Infrastructure of Sovereignty

Kyle Whyte

Introduction

The U.S. National Oceanic and Atmospheric Administration (NOAA) announced that in 2022, climate change cost the US$165 billion to mitigate damages, especially from extreme weather events.[1] A major factor influencing the destabilization of the climate system is the burning of non-renewable sources of energy: coal, oil, and gas. Transitioning to renewable sources, including wind, solar, and electric vehicles, requires the mining of raw materials that are critical inputs into renewable energy technologies, such as solar panels, wind turbines, and batteries. Mining for natural resources has health, environmental, and economic impacts in the areas where mining takes place.[2]

I have heard a number of scientists, policy-makers, and environmentalists express that mining in the lower 48 U.S. states (the "continental U.S.") must be ramped up substantially to meet the burden of raw materials needed for the transition to renewable energy. Some recent studies have shown the vast majority of potential areas to be mined are within or nearby Indigenous peoples' territories,[3] which threatens to expose these places to the negative impacts of mining. American settlers have already imposed environmental, health, and economic hazards on Indigenous peoples for over two centuries. Uranium, coal, and other metals mining, as well as drilling for oil, have been one of the causes of these hazards.[4]

Renewable energy advocates certainly seek to avoid the repetition of harms against Indigenous peoples. They have nonetheless argued that mining must occur to combat the negative impacts of climate change. In this argument, the best that can be done is an "imperfect" solution that will generate hazards on Indigenous peoples who have already endured generations of similar hazards. This argument is a certain version of a "tough luck" argument. The people who typically say "tough luck" are the ones with the least to lose and who have experienced the least "tough" experiences in their own lives and the lives of their ancestors. Once again,

DOI: 10.4324/9781003214380-13

non-Indigenous persons are telling Indigenous peoples the bad news that Indigenous communities will again be on the frontline of the efforts to address the latest crisis: climate change.

The "tough luck" argument obscures knowledge of the historic context from which Indigenous peoples, especially those living in lands now called the "continental U.S.", approach the potential of infrastructure developments in their territories. Indigenous peoples are not just seeking to mitigate the risks of mining and other large infrastructure in the areas where their communities live. They are working to grow their leadership nationally and internationally so that they can influence the decisions about energy and infrastructure that are made by non-Indigenous governments, industry, and scientific institutions.

Indigenous leadership means that Indigenous peoples can exercise self-determination, consent, and self-governance when it comes to the energy systems that they rely on and are affected by. After generations of oppression by American settlers, it is hauntingly familiar that Indigenous peoples are being asked by some people – "again" – to sacrifice their environmental quality, health, and economic vitality for the sake of mining. The need for this mining arises from a climate change crisis that has been cradled by the industrial energy processes that the U.S. and other forms of colonialism globally made way for by dispossessing Indigenous peoples of their territories.

It is important to understand how U.S. infrastructure development has affected Indigenous peoples negatively and how American settlers deliberately disempowered Indigenous self-governance in energy and infrastructure development. When it comes to renewable energy and other ideas about how to stop or manage climate change, the questions should not be "how can *solution x* be done well and avoid re-igniting historic injustices" or "how can *solution y* be done without harming Indigenous peoples." Rather, the question is whether Indigenous peoples exercise self-determination and consent at all in the major decisions about energy, infrastructure, and development.

Infrastructure Development and Indigenous Peoples

"Infrastructure" refers to the biological, environmental, and organizational systems through which the members of a society seek to sustain and improve their individual and collective sustenance and be prepared to be responsive to emergencies. A self-governance system is an example of infrastructure, as it includes (a) how authority is organized, (b) rules regarding how lands and waters are occupied and used, and (c) institutions that support methods that secure peoples' health, education, and other goods – among other aspects of self-governance.

The word infrastructure has an industrial ring to it, yet infrastructure is wide ranging in form, involving diverse ways in which societies organize their affairs through accessing energy and engaging in subsistence, labor, education, culture, science, and technology. Each Indigenous people has a long history of setting up and relying on their own forms of infrastructure, and changing and adapting their infrastructure during different time periods to meet economic needs and environmental conditions.[5]

Part of Indigenous peoples' heritage today is the infrastructure traditions that sustained them for generations and the lessons learned from forms of infrastructure that are no longer part of everyday life.[6] What's called the Anishinaabe seasonal round is one example of this heritage of infrastructure. Anishinaabe people are one of the largest Indigenous peoples in North America and have numerous Tribes and First Nations on the U.S. and Canadian sides of the Great Lakes region and beyond. Anishinaabe seasonal round in the Great Lakes region is a societal organization that centers adaptability to annual changes in environmental conditions and privileges the knowledge of plants, animals, elements, and flows in the environment. Decisions are made through consensus, clan leadership, and kinship relationships. And there is certainly more to Anishinaabe systems of infrastructure. Anishinaabe people were able to maintain trading and treaty relationships with other people, steward crops like corn and wild rice, build transportation, storage, and shelter technologies, and harness the ecosystem services of the aquatic and terrestrial landscapes they inhabited.[7]

By the time Anishinaabe people engaged with treaties with the U.S., the initial intent was to create a consensus arrangement among Anishinaabe, American settlers, and others to continue their different forms of infrastructure. Anishinaabe people, in treaties such as the treaties of 1836 and 1855, ensured that they secured rights to hunt, fish, and gather in areas that they ceded to the U.S., i.e., areas they originally inhabited but were not part of new reservation lands created pursuant to the treaties. They fully expected to have rights recognized by the U.S. to continue their own infrastructure at the same time that they would maintain relationships with the U.S. and American settlers.[8]

Anishinaabe infrastructure involved active stewardship of ecosystems in entire regions, with highly organized agriculture, subsistence, dwellings, and conservation practices. Anishinaabe infrastructure involves science, technology, education, and practices of intergenerational transfer of knowledge and skills. Today, Anishinaabe peoples continue to draw from their infrastructure heritage. In Indigenous climate change planning processes, for example, they are often based on the Anishinaabe seasonal round as an inspiration for how to track climatic trends and plan for solutions.[9]

Diverse Indigenous peoples everywhere in what is currently called the continental U.S. have histories of their own infrastructure. That is, they have biological, governmental, and environmental systems that were designed and operated to support their communities' sustenance and well-being. Self-determination, consent, and self-governance are concepts that refer to the degree to which societies make decisions about and manage the infrastructure they have and the infrastructure that affects them. Self-determination is the right to make decisions about what infrastructure is needed; consent is whether people are able to choose the ways in which infrastructure of another society can affect them (e.g. consenting or vetoing an oil pipeline proposed by a U.S.-based company); self-governance means the management of infrastructure, the capacity to foster innovation, and the capacity to respond to emergencies.

American settlers sought to undermine Indigenous peoples' self-determination, consent, and self-governance in general and more specifically in relation to infrastructure. Here the U.S. refers to federal and state governments, but also the businesses, non-profit organizations, and individual persons who operate through the U.S. sphere. They sought to install their own infrastructure within Indigenous territories. For example, anyone living in Anishinaabe homelands, such as the state of Michigan, will notice very little apparent Anishinaabe infrastructure in terms of the markers of seasonal round activities.

Instead in the Great Lakes, settler infrastructure is everywhere, from the vast neighborhoods and commercial areas built on top of former wetlands, paved roads and highways, petrochemical and nuclear facilities, dams, and intensive commercial agriculture. Anishinaabe people did not determine or consent to this infrastructure, and they do not govern it either – as settlers govern it by and large. In this way, infrastructure is one of the media through which the U.S. has committed environmental injustice against Indigenous peoples. Settler infrastructure has generated excessive levels of pollution that Indigenous peoples suffered acutely, ecological damage that has destroyed the foundations of Indigenous cultures and economies, and created assimilative schools and education programs that channeled Indigenous persons into low-wage employment and business opportunities (such as farming) without the financial and technological support needed to succeed.

Indigenous and allied scholars have discussed settler infrastructure in North America. Using an Indigenous concept that expresses a desire to cannibalize (Wiindigo), Winona LaDuke and Deborah Cowen discuss how American settlers set up their own infrastructure at the expense of Indigenous people's infrastructure. They write that "at the center of the Wiindigo's violence and destruction is infrastructure's seemingly banal and technical world. Wiindigo infrastructure has worked to carve up Turtle

Island, or North America, into preserves of settler jurisdiction, while entrenching and hardening the very means of settler economy and sociality into tangible material structures. We see this in sharp relief today, with pipelines and dams and roads and prisons" and "energy infrastructures constitute the contemporary spine of the settler colonial nation."[10]

For LaDuke and Cowen on my read, environmental injustice, can be understood as a problem of infrastructure. Infrastructure injustice has at least two harmful dimensions in the U.S. First, the U.S. endorses infrastructure that produces land dispossession, harmful health outcomes, physical abuse, and economic deprivation. Second, the U.S. re-engineers Indigeneity as a different form of dependent infrastructure designed for failure. For example, Gail Small, writing about the Northern Cheyenne Tribe in the early 1990s, explains how "the Cheyenne now find our reservation being surrounded by the largest coal strip mines in this country, and the threat that the mines will encroach on our own land is ever-present. I have been involved in the fight to protect our reservation and southeastern Montana from coal mining since the 1970s, when I was in high school. It was then that the Cheyenne learned the horrifying news that the Bureau of Indian Affairs had leased over half our reservation for strip-mining, at the paltry rate of 17 cents per ton, with no environmental safeguards included in the leases."[11] Small describes a situation where excessive levels of air and water pollution and spoilage of land are being foisted onto the Northern Cheyenne community, at the same time new laborers move into the region and create public safety risks, including human trafficking and sexual violence. The coal industry does not offer an economic opportunity to the Tribe, as the economic payoff is nearly nothing, especially given the risks to health and well-being.

Indigenous peoples have expressed widely the infrastructural aspects of environmental injustice. Anne Spice cites Freda Huston, a leader in the Wet'suwet'en people's resistance to the construction of the TransCanada pipeline in their territory:

When I [Spice] asked Friday to describe the difference between industry conceptions of critical infrastructure, and the infrastructure that sustain Indigenous life of Unist'ot'en *yintah* (territory), she told me this: 'So industry and government always talk about critical infrastructure, and *their critical infrastructure is making money, and using destructive projects to make money,* and they go by any means necessary to make that happen.... So for us, our critical infrastructure is the clean drinking water, and the very water that the salmon spawn in, and they go back downstream and four years, come back. That salmon is our food source; it's our main stable food. That's one of our critical infrastructures.'[12]

Colonizing societies seek to install their own infrastructure within Indigenous territories. Colonial infrastructure may be short-term or permanent. Settler colonial societies, such as the U.S., intended to fully establish themselves in the territories they colonized. The U.S. sought to establish its own infrastructure on Indigenous homelands.[13] This essay will discuss, in broad strokes, some of the ways in which U.S. infrastructure has violated Indigenous self-determination, consent, and self-governance, and how this history frames the discussion about mining and renewable energy.

Mining for Renewable Energy

If the U.S. energy system is going to transition to renewable energy through wind, solar, and electric vehicles, there is going to have to be more mining of the raw materials to make and deploy the technologies. Many scientists, policy-makers, and environmentalists are advancing this claim as a strategy to combat climate change through ending the burning of fossil fuels for energy, which produces greenhouse gases in the atmosphere. The technologies for solar and wind energy, batteries for storing renewable energy, and electric vehicles are industrial. They require raw materials, including copper, rare earth elements, cobalt, and lithium, and the mining of these inputs is associated with negative impacts on people and the environment.[14]

In the case of cobalt mining, a recent study cites that "the [Lifecyle Assessment] results found from the present research validates the fact that cobalt mining is harmful not only for people living near the mining area but also for the cobalt miners as they inhale large amounts of particles which are mixed with the air. From the analysis results of this paper, cobalt mining consumes large amounts of electrical power which is responsible for significant environmental effects."[15] Or copper and lithium mining: "Moreover, metal production itself is energy intensive and difficult to decarbonize. Mining for copper, needed for electric wires and circuits and thin-film solar cells, and mining for lithium, used in batteries, has been criticized in Chile for depleting local groundwater resources across the Atacama Desert, destroying fragile ecosystems, and converting meadows and lagoons into salt flats."[16]

Should wind, solar, and other renewable energy become implemented, there is the potential for human risks and ecological degradation at the locations where mining will occur. If the U.S. is to play a large role in mining the raw materials needed for the renewable energy revolution, analysis shows, however, that the vast majority of reserves of cobalt, lithium, and other needed inputs exist on or nearby Indigenous peoples' lands in the lower 48 U.S. states.[17] Some scientists have demonstrated that if the U.S. were to ramp up its mining to meet the levels of renewable energy needed to curb dependence on fossil fuel emissions, Indigenous locations will have to be mined aggressively. Any mining occurring on U.S. soil will likely pose risks to Indigenous peoples.

Yet it is commonly known that Indigenous peoples in the continental U.S. have been harmed since the 19th century by mining, drilling, and other large infrastructure projects tied to energy, food production, and transportation, including hydroelectric and irrigation dams, coal and nuclear power plants, and oil and gas facilities and pipelines. The legacies continue to this day. According to the U.S. Department of the Interior, "Several thousand orphaned oil and gas wells remain on Tribal lands, jeopardizing public health and safety by contaminating groundwater, seeping toxic chemicals, emitting harmful pollutants including methane and harming wildlife. Some of these wells are underwater, which creates an especially high risk of adverse impacts."[18] Diverse communities across the Navajo Nation still suffer from contamination from uranium mining and waste in their land from the mid-20th century, these impacts coming on the heels of the high rates of cancer found in Navajo miners.[19] There are numerous examples of coal, copper, and other mining affecting Indigenous peoples.

Just as historic mining and infrastructure development were associated by settler Americans with lofty economic and political goals akin – in some but not always – to combating climate change, could history be repeated?

Some scientists who have advocated for mining have said that the main problem is that increased mining harms Indigenous peoples' sacred lands and culture, but that there are no alternatives for those of us who care about stopping dangerous climate change.[20] While such harms to sacred places and cultural practices are certainly risks, such a focus on these risks is too narrow. The focus misses Indigenous peoples' history, perspectives, and knowledge of mining over the course of U.S. history and obscures the leadership roles that Indigenous peoples should be in the position to take in politics, society, and industry in efforts to make progress toward the needed transition to renewable energy. Of all people in the world, Indigenous peoples and any community which has experienced sustained oppression are familiar with having to negotiate dilemmas, such as the trade-offs of building and siting renewable energy.

There is a misunderstanding that the topic of Indigenous environmental justice is confined to the idea of Indigenous peoples as populations who commonly stand to be harmed by industrial projects promulgated by public and private sector organizations and that serve the settler population. There is certainly more to the topic. Indigenous environmental justice is importantly about the continued exclusion of Indigenous peoples from having any leadership roles within the development of infrastructure in the U.S. – including energy infrastructure. Leadership means at least three powers: Indigenous peoples' self-determination, consent, and self-governance in determining how their communities and all communities in North America would be powered, fueled, and energized.

Indigenous peoples have not been at the table since the beginning of the U.S. in making decisions about what forms of energy would be established for the U.S. economy (self-determination). Indigenous communities have not had the opportunity to refuse to be saddled by the harms and risks of mining, drilling, and other energy and industrial infrastructure (consent). And Indigenous peoples\ own or manage very little of the energy infrastructure that operates on their lands or that they depend on (self-governance). Given that Indigenous peoples have a track record of environmental advocacy going back to the earliest phases of the U.S., it is perhaps the case that Indigenous self-determination, consent, and self-governance would have curbed U.S. industrial growth from being carried out as irresponsibly as it has been.

Indigenous self-determination, consent, and self-governance are enshrined in recent White House policy on Indigenous peoples and growing international norms tied to the United Nations Declaration on the Rights of Indigenous Peoples (UNDRIP). The Biden-Harris plan for Tribal Nations includes plans to "strengthen the Nation-to-Nation relationship," "restore tribal lands, address climate change, and safeguard natural and cultural resources," and "expand economic opportunity and community development in Native communities."[21] The administration's 2021 "Memorandum on Tribal Consultation and Strengthening Nation-to-Nation Relationships" states:

> It is a priority of my Administration to make respect for Tribal sovereignty and self-governance, commitment to fulfilling Federal trust and treaty responsibilities to Tribal Nations, and regular, meaningful, and robust consultation with Tribal Nations cornerstones of Federal Indian policy. The United States has made solemn promises to Tribal Nations for more than two centuries. Honoring those commitments is particularly vital now, as our Nation faces crises related to health, the economy, racial justice, and climate change—all of which disproportionately harm Native Americans. History demonstrates that we best serve Native American people when Tribal governments are empowered to lead their communities, and when Federal officials speak with and listen to Tribal leaders in formulating Federal policy that affects Tribal Nations.[22]

Regarding climate change, Indigenous self-determination, in terms of political sovereignty, and self-governance are articulated as solutions.

UNDRIP enshrines that Indigenous peoples have the right to self-determination to "freely determine their political status and freely pursue their economic, social and cultural development" (article 3). UNDRIP describes that "States shall consult and cooperate in good faith with the indigenous peoples concerned through their own representative institutions in order to obtain their free, prior and informed consent before adopting and implementing legislative or administrative measures that may affect them" (article 19). Indigenous self-governance is described in terms of Indigenous institutions and capacities: "Indigenous peoples have the right to maintain and develop their political, economic and social systems or institutions, to be secure in the enjoyment of their own means of subsistence and development, and to engage freely in all their traditional and other economic activities (Article 20.1); Indigenous peoples deprived of their means of subsistence and development are entitled to just and fair redress (Article 20.2)."

Numerous Indigenous peoples approach mining in terms of their rights to self-determination, consent, and self-governance. Historically, colonial mining industries and other large infrastructure projects squashed the possibility that Indigenous peoples would emerge as leaders in infrastructure development. Second, over time, Indigenous peoples' institutions for making decisions about large infrastructure have been largely curtailed, making it impossible for Indigenous peoples to take on their own energy projects as quickly as they need to. The history of American settlement, industrial infrastructure, and Indigenous peoples is critical to being able to understand this moment when decisions are being made about the energy future of the U.S., and where many people are pressing for a rapid transition to renewable energy.

The Infrastructure of Colonialism

American settlement very deliberately divested Indigenous peoples of their self-determination, consent, and self-governance of their own infrastructure with respect to mining and all other infrastructure. To understand how some Indigenous peoples may relate to the prospect of more mining in their territories, it is important to respect this history. Indeed, many Indigenous peoples in the U.S. have a history of being involved in the earliest establishment of mining, drilling, dams, and other industrial infrastructure. Consider some randomly selected historical moments in which Indigenous peoples were confronted with colonial infrastructure. In this discussion of these historical moments, the term "infrastructure" can refer to either Indigenous or settler lifeways, similar to how the term has been defined earlier.

The Meskwaki Nation is an Indigenous people of the Great Lakes and Midwest regions in North America, with communities and governments in what is currently referred to as Iowa, Kansas, Nebraska, and Oklahoma. Historically, Meskwaki people have a long history of small-scale lead mining going back prior to 1650.[23] Lead mining was a livelihood for Tribes in that region, and they maintained social and cultural systems controlling access to mines, including important roles played by Native women.[24] Meskwaki people sought to keep lead mining small scale, until American settlers came in and sought control and profit from the mines. Meskwaki people mined lead until 1822, when armed settler miners entered the mining areas. American settlers overwhelmed the lead mines through violence and squatting, violating Meskwaki control over infrastructure, and eventually causing a war. Ironically, the Meskwaki would have to pay for the Americans' costs of the war by ceding their mining lands. The mines were never returned to Meskwaki people.[25] Importantly, in this example, U.S. settlers literally wrested control over mining away from Indigenous peoples.

For generations, Haudenosaunee people in what is currently known as the eastern and midwestern regions of the U.S. and Canada cultivated corn as one major part of their infrastructure. Corn has been cultivated by Indigenous peoples for thousands of years. Haudenosaunee relationships to corn are integral to their origin story, self-governance, families, and society. Corn planting traditions had dramatic effects on North American landscapes, including a polyculture cropping system of corn, beans and squash. Jane Mt. Pleasant describes her analysis of European accounts of Haudenosaunee corn economies that accounted for "vast and productive cropland that was planted, cared for, and controlled by Native women."[26] U.S. colonialism, including religious institutions, state and federal governments, and businesses, dispossessed Haudenosaunee people of much of their land and stripped them of their traditions, directly impeding the continuance of the polyculture cropping infrastructure. Of course, U.S. settlers appropriated and commodified corn for their own benefit, industrializing the cultivation of corn. Corn commodity crops have been associated with threats to biodiversity and climate resilience, and health harms from the pesticides used to maintain the crops and from the overuse of corn in inexpensive foods and meat production.[27]

In the early 1900s, U.S. settler prospectors traveled into Navajo Nation land in search of radioactive materials, especially carnotite (an ore of uranium). Just before and during World War II, demand for uranium increased greatly, pushing mines in the Navajo Nation to produce more. Post World War II, the U.S. Atomic Energy Commission increased demand for uranium for energy. Settler companies and the U.S. government installed massive transportation and industrial infrastructure to create, access, and ship from mines.[28] The profits from uranium mining did not genuinely benefit the Navajo Nation, and Navajo workers in the mines suffered severe health problems. There are now 900 abandoned uranium mines in the Navajo Nation. The Navajo Nation has had to work through settlement agreements, such as the 2014 Tronox agreement and the 2017 agreement with Cyprus Amax Minerals Company/Western Nuclear, all of which attempt to clean up abandoned and dangerous mines.[29]

The U.S. military has been one of the largest perpetrators of infrastructure injustice. Yupik people of St. Lawrence Island in Alaska have an ancient subsistence lifestyle. The U.S. military established defense sites, primarily during the Cold War era, operated by the Department of Defense. There are roughly 600 formerly used defense sites in Alaska, and it is estimated that many are nearby Native communities and places the communities depend on for subsistence. One study documented that "chlorinated and non-chlorinated solvents, herbicides/pesticides, trace metals, containers of human wastes, chemical warfare agents, unexploded ordnance (UXO) and other toxic materials were not only used during active site operations, but were left behind when the posts were abandoned."[30] Native communities in Alaska have suffered health consequences and there are a large number of studies demonstrating the negative environmental and health outcomes.

The U.S. set up many dam projects that affected Tribal lands across North America. The dams' purposes were diverse, from increasing navigation, to managing floods, to supporting recreation. The rationales were argued for under the auspices that rapid U.S. economic growth required the dams. Moreover, Cold War anxieties demanded the U.S. do whatever it could to remain competitive and secure against the threat of global Communist political and economic power.[31] But the large dam projects flooded Indigenous peoples' territories, upending entire Indigenous communities, and shrinking their land base and economic capacities. In the mid-20th century, the Dalles Dam on the Columbia River destroyed Celilo Falls, one of the largest fisheries in North America for salmon and other species and that had been frequented by diverse Indigenous peoples of the broader region for generations. The Pick-Sloan Missouri Basin Program built many dams in the plains region. One dam, the Oahe Dam, created a reservoir that shrunk the land base of the Standing Rock Sioux Tribe, displacing many Tribal members and destroying quality timberlands and soils for cultivation and wildlife habitats. The Seneca Nation lost nine communities to the Kinzua Dam, a dam project that supports flood control and water quality for the area around Pittsburgh Pennsylvania.[32]

American settlers established oil and gas pipelines across Indigenous peoples' territories without the consent of Indigenous peoples. Oil and gas companies, attempting to traverse states and international boundaries (the U.S. and Canada), installed tens of thousands of miles of oil and gas pipelines. The Enbridge Line 5 oil pipeline extends 645 miles from Wisconsin to Ontario, crossing many Indigenous territories, including reservations and areas where Indigenous subsistence practices are protected by treaty. Indigenous peoples have repeatedly been subject to leakage of oil and gas pipelines. Enbridge's Line 6b pipeline going through what is currently the midwestern state of Michigan spilled 1 million gallons of crude oil into the Kalamazoo River, which is in the traditional land base of the Nottawaseppi Huron Band of Potawatomi, impacting important wildlife, plants, and medicines, ultimately having negative effects on the communities' culture and economy.[33]

There are, of course, many thousands more examples of how American settlers disrupted Indigenous peoples' infrastructure through the forced installation of colonial infrastructure. Importantly, Indigenous peoples did not choose to have this infrastructure unleashed on them, nor did they consent to the specific projects. Indigenous peoples did not have the opportunity to govern or manage the new infrastructure, being excluded from having decision-making authority and from sharing in fair and genuine economic benefits.

Indigenous Self-Governance

As discussed before, infrastructure refers to institutions that make up self-governance. Self-governance refers to the actual capacities needed to exercise self-determination and consent. The capacities include institutional, decision-making, economic, and diplomatic capacities. More practically, self-governance means having leaders who can protect and represent the interests and needs of communities. It means having decision-making protocols at different scales in a society that ensure that collective actions are taken that are in the interest of the communities in the society.

Self-governance involves the capacity to plan and implement plans, including plans that prepare for climate change and the supply for energy. Self-governance is about a society's ability to process and utilize means of exchange, including resources and money, among other means of exchange. Self-governance is a society's ability to be diplomatic with other societies and with nations, remaining capable of making agreements, engaging in consultation, implementing agreements, and stopping exploitation from occurring.

Part of the history of U.S. colonialism is forcing the disintegration of Indigenous peoples' own self-governance so that they would be dependent on the U.S., assimilate, and be unable to pose economic competition to American settlers. In a very short period of time within the history of the U.S., Indigenous peoples went from being in the position to negotiate treaties with the U.S. as sovereign parties to being reduced to being business entities unable to compete with the U.S. and international corporations and the business interests of private citizens.

After the War of 1812, the U.S. grew in power against Indigenous peoples, recognizing it had military and economic advantages. At least on my view, treaties after the War of 1812 begin to have much less favorable terms for Indigenous peoples than those prior. The subsequent treaties called for massive cessions of land. In the early and mid-19th century, the U.S. engaged treaties that forced Indigenous peoples onto areas of land that were a fraction of the size of their original homelands or forced them to move away entirely, often thousands of miles away. The U.S. government attacked the conditions that made it possible for different Indigenous peoples to exercise self-governance.

The 1842 Treaty of La Pointe between the U.S. and the Lake Superior Ojibwe in Michigan and Wisconsin is one example. Treaties often had annuities, or payments from the U.S. to Indigenous peoples that would occur each year. Annuities were critical benefits tied to treaties that supported Indigenous peoples' capacity to survive after having ceded large areas of land. In 1850, the U.S. conveyed to a group of Lake Superior Ojibwe that they would need to travel to an area called Sandy Lake to receive their annuities. When the people arrived, there was no money, and the U.S. only provided spoiled rations. About 400 Ojibwe persons died from these conditions, including starvation.[34]

During the 1850s Gold Rush in California, the U.S. Senate met to discuss 18 treaties with Indigenous peoples in different parts of California to seek agreement on ratification. Larissa Miller observed that "when the treaties came up in executive session of the U.S. Senate, the senators found them problematic. It was unclear if Mexico—from which California was acquired recognized native land titles. If Mexico did not, then Indians in California came under U.S. sovereignty without legal claims to the land. Furthermore, the commissioners' appointments were irregular, and in the wake of the gold rush, white Californians strongly objected to the treaties."[35] The Senate did not ratify the treaties, delivering a major negative impact on those Indigenous peoples' capacities to self-govern in territories with increasing incursions of settlers.

The U.S. continued to undermine Indigenous self-governance. For Potawatomi people who were relocated by the U.S. to a Kansas reservation in the 19th century, the emerging statehood of Kansas and the running of the railroad through the Potawatomi reservation threatened Potawatomi lands. The threat of the railroad created internal divisions among the Potawatomi about what to do: keep the parts of the reservation that would not be overrun by the railroad or accept private property, including moving to Oklahoma. Some Potawatomi signed an 1861 treaty with the U.S. through which they would become private property holding citizens of the U.S. Upon receiving private property, taxes were immediately owed, which led many to sell their land to settlers at a reduced price given they had no time to implement their farming businesses.[36]

The U.S. was supposed to provide support for Potawatomi private property holders to start farming businesses, support which was so delayed that they could not raise the income needed to cover their taxes. They had no capacity to take legal action to counteract taxation because the emerging state of Kansas did not recognize Potawatomi as residents of the state, even though they were supposed to be U.S. citizens at the time. When the opportunity came later in the 19th century for the same Potawatomi people to take up land in the Indian Territory (now Oklahoma), they could not afford to make the trip. The U.S. entangled Potawatomi people in a situation in which Potawatomi institutions could not take advantage of the options available to them.[37]

Over the course of the 19th century and early 20th century, Tribes attempted to create their own laws and policies that would allow them to operate their own economies in relation to the growing U.S. settler economy. The Osage had been forced to move their land base multiple times, which itself creates strain on the organizational structure of any society. When they arrived in Oklahoma, the land only appeared to be usable for basic subsistence homesteading. The Osage worked hard to create new organizational structure to continue their self-governance, including establishing their own legislation, laws, and social order.[38]

The U.S. refused to respect this Osage sovereignty, which created problems when substantive reserves of oil were found below the reservation. The U.S. government and American settlers began finding ways to explore and tap these reserves, pressuring the Osage to privatize their lands. One of the first oil production deals of American settlers on Osage lands involved a deal to which the Tribal council did not consent. While for a short time the Osage grew in wealth, they were subject to violence and economic exploitation, and American settlers ultimately took control of the oil economy in the region.[39]

In the 1930s, a form of respect for Indigenous self-determination resurfaces through the Indian Reorganization Act (IRA). The condition, of course, was that Indigenous peoples could create governments recognized by the U.S. as sovereign if they structured the governments according to U.S. standards, such as democratic elections and boilerplate constitutions. These constitutions had the effect of dividing communities, creating problematic political authorities, and weakening the presence and representation of Indigenous cultures in relations with the U.S. Some Indigenous peoples opted out of IRA governments, which carried negative repercussions for them.[40]

The IRA governments were structured to facilitate extraction industries onto Indigenous lands, but with the intent of lessening previous land dispossession and exploitation. However, the policies and practices governing IRA vested regulatory sovereignty in the hands of U.S. government agencies that did not have the capacity to stop abuses by corporations hungry for extraction on Indigenous lands. Despite some Indigenous peoples being able to take some degree of control over the ownership of extractive industries, the IRA period helped to create political relations in which Indigenous peoples could play a role in shaping the development and fairness of the U.S. energy system.[41]

No different from the other transitions, the IRA period gave way to the termination period from roughly the 1950s to 1970s. Through several notable laws, such as Public Law 280, the U.S. sought to terminate Indigenous peoples' sovereignty if particular communities were deemed to be self-sufficient by U.S. standards. Termination represented an abrogation of whatever political means Indigenous peoples had left to enforce regulations that protected their ways of life. The state of Wisconsin, for example, radically altered parts of the landscape of the Menominee Tribe's reservation during the period in which the Tribe was terminated. When the Menominee was again recognized as sovereign in the 1970s, it had to rebuild its government and forestry business, and some of their lands were permanently lost.[42]

Due to allotment, the Oneida Nation lost control over the majority of its land in what is currently known as Wisconsin. This created a patchwork of land ownership in the Oneida territory of non-Oneida property owners. Two towns were created by the Oneida, Hobart and Oneida, which were to support infrastructure development in the wake of land loss. Settlers divested Tribal members quickly of control of the town of Hobart.[43] After the IRA, the Oneida Tribe sought to actively restore control over its land, creating a Land Committee charged with this purpose.[44] The jurisdictions of the Oneida Nation and Hobart overlapped. This situation has led the village of Hobart attempting to thwart the Oneida Tribe's plans to govern their territory, including to engage in economic development, taxation, the management of services (such as garbage and recycling), the restoration of land, and the practice of festivals and traditions. The Oneida Nation has had to invest substantive resources in legal, economic, and political capital to be able to resist the village's attempt to stop the Nation's economic and cultural aspirations.[45]

In the early 20th century, the U.S. implemented the Rural Electrification Act, which has set up some of the basic energy grids for rural communities. Indigenous peoples were often excluded from being connected to energy through this act. Over time, the lack of energy infrastructure has led to a situation where Tribes' energy costs are high, they rely on fossil fuel burning, and their own future economic development in other areas is thwarted by their not having the energy infrastructure to expand. Many Tribes face gaps in energy infrastructure. Catherine Sandoval writes that "addressing electricity access gaps on Native American reservations requires more than money; it requires making electricity access on reservations a priority. Many tribes such as the Yurok do not have substantial financial resources from gaming and other businesses to fund electric grid expansion projects costing tens of millions of dollars. The federal government, states, the private sector, philanthropy, and non-governmental organizations should support tribal efforts to procure safe, reliable, and sustainable energy access at just and reasonable rates."[46]

In the histories just described, Indigenous peoples' own institutions are blocked from developing in ways that can make it possible for them to have leading roles within major infrastructure decisions. Indigenous peoples not only lost land but were deprived of opportunities to develop their capacities to influence decisions about major infrastructure in the U.S. and from having the institutional (or bureaucratic) and technological capacities to be able to own and operate their own infrastructure.

The impact of history is significant. The National Congress of American Indians' infrastructure report states a "$50 billion unmet need for infrastructure on Indian reservations. The number of 'shovel ready' infrastructure projects in Indian Country remains too many to count, and many of those have been that way for years if not decades."[47] The U.S. Civil Rights Commission has produced two reports on Indigenous infrastructure and U.S. federal spending. In the 2003 report, "A Quiet Crisis," it states "that due to the failure of the federal government's efforts to carry out its promises, 'Native Americans continue to rank at or near the bottom of nearly every social, health, and economic indicator'" and that "... despite significantly increased federal spending between 1994 and 2003, the sums failed to 'compensate for a decline in spending power' or 'overcome a long and sad history of neglect and discrimination', and concluded that 'Native Americans living on tribal lands do not have access to the same services and programs available to other Americans'."[48]

Yet the 2018 report, "Broken Promises," demonstrates that the situation has worsened. "Unfortunately, the Commission's current study reflects that the efforts undertaken by the federal government in the past 15 years have resulted in only minor improvements, at best, for the Native population as a whole. And, in some respects, the U.S. Government has backslid in its treatment of Native Americans, and there is more that must be done compared to when the Commission issued 'A Quiet Crisis'." Moreover, "Federal funding for Native American programs across the government remains grossly inadequate to meet the most basic needs the federal government is obligated to provide. Native American program budgets generally remain a barely perceptible and decreasing percentage of agency budgets."[49]

The history of colonialism in the U.S. involves over time an erosion of Indigenous peoples' self-governance, including the very institutions of self-governance. This makes it so that Indigenous peoples are at a disadvantage in discussions about infrastructure development, whether mining, energy, or hydropower. In the previous section, the focus was on the harms and violence of American settlement and its deployment of infrastructure. The harms are compounded by the erosion of Indigenous self-governance institutions, including those that are particularly important for investing in and managing infrastructure, described in this section.

Indigenous Renewable Energy Development

Should mining of cobalt or copper occur on Indigenous peoples' lands in order to hasten the industrial transition to renewable energy? The question is not the right question, and perhaps it is not best to understand this issue through any question. A better pathway may be as follows.

Indigenous peoples have their own infrastructure traditions, which have been subject to rampant environmental injustice. Indigenous infrastructure has been overthrown by American settler infrastructure. The U.S. government and American settlers have actively undermined Indigenous peoples' capacity to make decisions about infrastructure today and to implement infrastructure plans through self-governance. The various government, corporate, and other architects of U.S. infrastructure today did not design or organize the infrastructure systems for Indigenous peoples to exercise self-determination, consent, and self-governance.

The transition to wind and solar energy and electric vehicles is a matter of decisions that the public and private sectors will make about how to change the design and organization of American infrastructure and its relationships to infrastructure throughout the world. Through organizations such as the National Congress of American Indians or the Affiliated Tribes of Northwest Indians, Indigenous peoples in the continental U.S. have made clear that they want to be at the forefront of efforts to transition to renewable energy and to move away from the burning of fossil fuels. Indigenous leaders, both elected officials and Tribal staff, are knowledgeable about infrastructure and supply chains. They have to govern, create investments for, and foster economic security for the Tribes they work for and represent.

The vision for renewable energy today that Indigenous peoples have combined their traditions, knowledge, culture, respect for future generations, and understanding of their current infrastructure situation today. It is visionary and practical knowledge, knowledge of exactly what needs to be put in motion to lead to a renewable energy transition. But Indigenous visions and knowledge, understood in this way, are not being portrayed when some people settle on simply asking the question of whether the needed mining will affect sacred or culturally significant places on Indigenous peoples' land.

The better pathway forward is one in which Indigenous communities and Indigenous leaders are thoroughly capable of being real members of the private and public sector groups that make and influence decisions about American infrastructure – that is, Indigenous self-determination, consent, and self-governance are fully part of the decisions that will be made.

In 2021–22, various news headlines shared some striking announcements about the relationship between Tribal nations and the U.S. The White House and Congress promised unprecedented investments into Tribal infrastructure. The areas of infrastructure investment range from water sanitation to orphan well clean up to climate change resilience to renewable energy. The new investments are coming after decades of reports that cited the ways in which the U.S. deprived Tribal nations of infrastructure.

Given the new resources available at this possibly temporary moment in time, it is more important to begin to appreciate the challenges that Indigenous peoples are facing when it comes to repairing the infrastructure situation. The 2021 Bipartisan Infrastructure Bill includes 13 billion for infrastructure projects in Tribal nations, including in sanitation facilities, transportation, water rights, broadband, clean water and drinking water, bridges, dams, climate adaptation and relocation, and orphaned wells, sparking some to call it a "once-in-a-generation" investment and "gold rush for Tribal energy projects." These were the largest investments ever. But the investments will not be taken advantage of if Indigenous peoples are not shaping how the resources and technologies in the investments are designed to be fully harnessed by communities, Tribal governments, and Indigenous organizations and businesses. The issue of mining for the raw materials needed for renewable energy is part of a much larger set of infrastructure topics. It is not an isolated or discreet issue that can be treated separately.

Indigenous environmental justice cannot be reckoned with on an issue-by-issue basis. Raw materials' mining for renewables, should it be proposed on Indigenous lands, is connected to much longer histories of Indigenous exclusion from infrastructure, infrastructure injustice, and the deliberate undermining of Indigenous institutions of self-governance by American settlers. Advocates of renewable energy should ask the question of whether Indigenous peoples have had the capacity to exercise self-determination, and are in the position to consent to, the decisions that went into this whole topic of how renewable energy and mining are related in the first place.

If Indigenous peoples continue to not be at key decision-making tables, then they will continue to have to be passive actors resisting the plans devised by American settlers and other non-Indigenous parties to mitigate and manage crises that Indigenous peoples by and large did not cause, climate change being one example. There is no "tough luck" argument that is acceptable to make when discussing the industrial realities of developing renewable energy that will make a difference. Scientists, policy-makers, and others, if they are serious about mitigating climate change, will recognize that strengthening Indigenous peoples' self-determination and consent is the fastest route. For Indigenous peoples will have the empowerment to plan their economies, including the ending of all problematic forms of economic and energy dependence on fossil fuels in their territories, and be able to devise collaborative strategies with others about how to develop renewable energy responsibly.

Notes

1 National Oceanic and Atmospheric Administration, "U.S. 2022 Billion-Dollar Weather and Climate Disasters." *Twitter Post*, 2023. https://www.noaa.gov/sites/default/files/2023-01/Slides-NOAA-NASA-Global-Analysis-2022-011223.pdf. Accessed April 2, 2023.

2 Laura J. Sonter et al., "Renewable Energy Production Will Exacerbate Mining Threats to Biodiversity," *Nature Communications* 11, no. 1 (2020): 1–6.

3 Indigenous peoples are societies that exercised sovereign self-governance prior to the establishment of some nation state that currently lays claim to being the sovereign, such as the U.S. People have asserted Indigenous identity as a rights concept to end slavery, human trafficking, forced labor, economic deprivation, gender-based violence, political persecution, and cultural assimilation. A key aspect of Indigenous rights is that rights violations stem in part from the undermining of self-governance.

4 Paul C. Rosier, "'They Are Ancestral Homelands': Race, Place, and Politics in Cold War Native America, 1945–1961," *The Journal of American History* 92, no. 4 (2006); Donald L. Fixico, *The Invasion of Indian Country in the Twentieth Century: American Capitalism and Tribal Natural Resources* (Niwot: University Press of Colorado, 1998); David Rich Lewis, "Native Americans and the Environment: A Survey of Twentieth-Century Issues," *American Indian Quarterly* 19, no. 3 (1995): 423–450; Winona LaDuke, "Indigenous Environmental Perspectives: A North American Primer," *Akwe: kon Journal* 9, no. 2 (1992): 376–388.

5 Anne Spice, "Fighting Invasive Infrastructures: Indigenous Relations against Pipelines," *Environment and Society* 9, no. 1 (2018): 40–56.

6 Robert M. Figueroa and Gordon Waitt, "Climb: Restorative Justice, Environmental Heritage, and the Moral Terrains of Uluru-Kata Tjuṯa National Park," *Environmental Philosophy* 7, no. 2 (2011): 135–164.

7 Brenda J. Child, *Holding Our World Together: Ojibwe Women and the Survival of Community* (New York: Penguin, 2012); Heidi Bohaker, "'Nindoodemag': The Significance of Algonquian Kinship Networks in the Eastern Great Lakes Region, 1600–1701," *The William and Mary Quarterly* 63, no. 1 (2006): 23–52; Michael Witgen, *An Infinity of Nations: How the Native New World Shaped Early North America* (Philadelphia, PA: University of Pennsylvania Press, 2011); James A. Clifton, et al., *People of the Three Fires: The*

Ottawa, Potawatomi, and Ojibway of Michigan (Grand Rapids, MI: Michigan Indian Press, 1986).

8 Heidi Kiiwetinepinesiik Stark, "Respect, Responsibility, and Renewal: The Foundations of Anishinaabe Treaty Making with the United States and Canada," *American Indian Culture and Research Journal* 34, no. 2 (2010): 145–164; Heidi Kiiwetinepinesiik Stark, "Marked by Fire: Anishinaabe Articulations of Nationhood in Treaty Making with the United States and Canada," *The American Indian Quarterly* 36, no. 2 (2012): 119–149.

9 Tribal Adaptation Menu Team. 2019. Dibaginjigaadeg Anishinaabe Ezhitwaad: A Tribal Climate Adaptation Menu. Great Lakes Indian Fish and Wildlife Commission, Odanah, Wisconsin. 54 p.

10 Winona LaDuke and Deborah Cowen, "Beyond Wiindigo Infrastructure," *South Atlantic Quarterly* 119, no. 2 (2020): 244–49.

11 Gail Small, "War Stories: Environmental Justice in Indian Country," *Daybreak* 4, no. 2 (1994): 2.

12 Spice, "Fighting Invasive Infrastructures," 40–41.

13 Spice, "Fighting Invasive Infrastructures," 45.

14 Benjamin K. Sovacool et al., "Sustainable Minerals and Metals for a Low-Carbon Future," *Science* 367, no. 6473 (2020): 30–33; Saeed Rahimpour Golroudbary et al., "Global Environmental Cost of Using Rare Earth Elements in Green Energy Technologies," *Science of The Total Environment* 832 (2022): 1–12.

15 Shahjadi Hisan Farjana, Nazmul Huda, and MA Parvez Mahmud, "Life Cycle Assessment of Cobalt Extraction process," *Journal of Sustainable Mining* 18, no. 3 (2019): 155

16 Sovacool et al., "Sustainable Minerals and Metals for a Low-Carbon Future," 30.

17 Samuel Block, "Mining Energy-Transition MEtals: National Aims, Local Conflicts," *MSCI Inc.*, June 3, 2021.

18 https://www.doi.gov/pressreleases/biden-harris-administration-releases-draft-guidance-new-tribal-orphaned-well-program

19 Tommy Rock, "Exposing Years of Uranium Water Contamination in a Navajo Community," *Grantee Highlights, National Institute of Environmental Health Sciences*, February 24, 2017.

20 Block Mining Energy-Transition MEtals: National Aims, Local Conflicts.

21 https://joebiden.com/tribalnations/#

22 https://www.whitehouse.gov/briefing-room/presidential-actions/2021/01/26/memorandum-on-tribal-consultation-and-strengthening-nation-to-nation-relationships/

23 Greg A. Ludvigson and James A. Dockal, "Lead and Zinc Mining in the Dubuque Area," *Iowa Department of Natural Resources* (1984); https://www.iowadnr.gov/portals/idnr/uploads/geology/LeadZincMiningDubuqueArea.pdf

24 M.C. Gill, "The Diffusion of Ore-Hearth Smelting Techniques from Yorkshire to the Upper Mississippi Valley Lead Region," *British Mining* 43 (1991); https://www.iowapbs.org/iowapathways/mypath/2659/lead-mining

25 https://www.iowadnr.gov/Portals/idnr/uploads/geology/LeadZincMiningDubuqueArea.pdf

26 Jane Mt. Pleasant, *Traditional Iroquois Corn: Its History, Cultivation and Use* (Ithaca: Plant and Life Science Publishing, 2011), 8.

27 Jane Mt. Pleasant, "A New Paradigm for Pre-Columbian Agriculture in North America," *Early American Studies* (2015): 374–412; Jane Mt. Pleasant, "Food Yields and Nutrient Analyses of the Three Sisters: A Haudenosaunee Cropping

System," *Ethnobiology Letters* 7, no. 1 (2016): 87–98; Jane Mt. Pleasant and Robert F Burt, "Estimating Productivity of Traditional Iroquoian Cropping Systems from Field Experiments and Historical Literature," *Journal of Ethnobiology* 30, no. 1 (2010): 52–79.

28 Shane Wero and Rena Martin, "Łeetso – Working and Living within the Monster: A Cultural Resources Study of Navajo Habitations within Former Uranium Mines in the Cove and Monument Valley Regions," *Kiva* 87, no. 3 (2021): 354–376; Rock Exposing Years of Uranium Water Contamination in a Navajo Community; Doug Brugge, Timothy Benally, and Esther Yazzie-Lewis, eds., *The Navajo People and Uranium Mining* (Albuquerque: University of New Mexico Press, 2006).

29 Wero and Martin, "Łeetso – Working and Living within the Monster"; Rock Exposing Years of Uranium Water Contamination in a Navajo Community; Brugge, Benally, and Yazzie-Lewis, *The Navajo People and Uranium Mining*.

30 Ronald J. Scrudato et al., "Contaminants at Arctic Formerly Used Defense Sites," *Journal of Local and Global Health Science* 2012, no. 1 (2015): 1–12.

31 Katrine Barber, *Death of Celilo Falls* (Seattle: University of Washington Press, 2005); Paul C Rosier, "Dam Building and Treaty Breaking: The Kinzua Dam Controversy, 1936–1958," *The Pennsylvania Magazine of History and Biography* 119, no. 4 (1995): 345–368.

32 Rosier, "Dam Building and Treaty Breaking"; Barber, *Death of Celilo Falls*; Peter Capossela, "Impacts of the Army Corps of Engineers' Pick-Sloan Program on the Indian Tribes of the Missouri River Basin," *Journal of Environmental Law and Litigation* 30 (2015): 143–218; Michael L. Lawson, *Dammed Indians: The Pick-Sloan Plan and the Missouri River Sioux, 1944–1980* (Norman: University of Oklahoma Press, 1994).

33 https://nhbp-nsn.gov/media/the-great-stain-10-years-after-the-kalamazoo-river-oil-spill/

34 Colin Mustful, "The Sandy Lake Tragedy," *Mnopedia*, 2021. https://www.mnopedia.org/event/sandy-lake-tragedy

35 Larisa K. Miller, "The Secret Treaties with California's Indians," *Prologue-Quarterly of the National Archives and Records Administration* 45, no. 3–4 (2013): 39.

36 Kelli Mosteller, "The Cultural Politics of Land: Citizen Potawatomi Allotment and Citizenship in Kansas and Indian Territory, 1861–1891."

37 Mosteller, "The Cultural Politics of Land."

38 Fixico, *The Invasion of Indian Country in the Twentieth Century*.

39 Fixico, *The Invasion of Indian Country in the Twentieth Century*.

40 Paul C. Rosier, *Rebirth of the Blackfeet Nation, 1912–1954* (Lincoln, NE: U of Nebraska Press, 2004).

41 Fixico, *The Invasion of Indian Country in the Twentieth Century*; James Allison, "From Survival to Sovereignty: 1970s Energy Development and Indian Self-Determination in Montana's Powder River Basin," *Environmental Justice* 5, no. 5 (2012): 252–263.

42 Judith Royster, "Oil and Water in Indian Country," *Natural Resources Journal* 37 (1997): 457–490; Michael Dockry and Kyle Whyte. "Improving on Nature: The Legend Lake Development, Menominee Resistance, and the Ecological Dynamics of Settler Colonialism." *The American Indian Quarterly* 45, no. 2 (2021): 95–120.

43 Doug Kiel, "Nation v. Municipality: Indigenous Land Recovery, Settler Resentment, and Taxation on the Oneida Reservation," *Native American and Indigenous Studies* 6, no. 2 (2019): 51–73.

44 Rebecca M. Webster, "This Land Can Sustain Us: Cooperative Land Use Planning on the Oneida Reservation," *Planning Theory & Practice* 17, no. 1 (2016): 9–34.
45 Webster, "This Land Can Sustain Us"; Kiel, "Nation v. Municipality."
46 Catherine JK Sandoval, "Principles to Advance Energy Justice for Native Americans," *EBA BRIEF*, October, 2020; Catherine J.K. Sandoval, "Energy Access is Energy Justice: The Yurok Tribe's Trailblazing Work to Close the Native American Reservation Electricity Gap," in *Energy Justice*, ed. Raya Salter et al. (Cheltenham: Edward Elgar Publishing, 2018).
47 National Congress of American Indians, "Infrastructure in Indian Country," 2017.
48 U.S. Civil Rights Commission, "A Quiet Crisis: Fedearl Funding and Unmet Needs in Indian Country," 2003.
49 U.S. Civil Rights Commission, "Broken Promises: Continuing Federal Funding Shortfall for Native Americans," 2018.

Bibliography

Allison, James. "From Survival to Sovereignty: 1970s Energy Development and Indian Self-Determination in Montana's Powder River Basin." *Environmental Justice* 5, no. 5 (2012): 252–63. http://online.liebertpub.com/doi/abs/10.1089/env.2011.0032

Barber, Katrine. *Death of Celilo Falls*. Seattle: University of Washington Press, 2005.

Block, Samuel. "Mining Energy-Transition Metals: National Aims, Local Conflicts." *MSCI Inc.*, June 3, 2021.

Bohaker, Heidi. "'Nindoodemag': The Significance of Algonquian Kinship Networks in the Eastern Great Lakes Region, 1600-1701." *The William and Mary Quarterly* 63, no. 1 (2006): 23–52.

Brugge, Doug T., Timothy Benally, and Esther Yazzie-Lewis, eds. *The Navajo People and Uranium Mining*. Albuquerque: University of New Mexico Press, 2006.

Capossela, Peter. "Impacts of the Army Corps of Engineers' Pick-Sloan Program on the Indian Tribes of the Missouri River Basin." *Journal of Environmental Law and Litigation* 30 (2015): 143–217.

Child, Brenda J. *Holding Our World Together: Ojibwe Women and the Survival of Community*. New York: Penguin, 2012.

Clifton, James A., et al. *People of the Three Fires: The Ottawa, Potawatomi, and Ojibway of Michigan*. Grand Rapids, MI: Michigan Indian Press, 1986.

Cowen, Deborah. "Following the Infrastructures of Empire: Notes on Cities, Settler Colonialism, and Method." *Urban Geography* 41, no. 4 (2020): 469–86.

Farjana, Shahjadi Hisan, Nazmul Huda, and MA Parvez Mahmud. "Life Cycle Assessment of Cobalt Extraction Process." *Journal of Sustainable Mining* 18, no. 3 (2019): 150–61.

Figueroa, Robert M., and Gordon Waitt. "Climb: Restorative Justice, Environmental Heritage, and the Moral Terrains of Uluṟu-Kata Tjuṯa National Park." *Environmental Philosophy* 7, no. 2 (2011): 135–63.

Fixico, Donald L. *The Invasion of Indian Country in the Twentieth Century: American Capitalism and Tribal Natural Resources*. Niwot: University Press of Colorado, 1998.

Gill, M.C. "The Diffusion of Ore-Hearth Smelting Techniques from Yorkshire to the Upper Mississippi Valley Lead Region." *British Mining* 43 (1991): 118–28.

Golroudbary, Saeed Rahimpour, Iryna Makarava, Andrzej Kraslawski, and Eveliina Repo. "Global Environmental Cost of Using Rare Earth Elements in Green Energy Technologies." *Science of The Total Environment* 832 (2022): 155022.

Green, Michael D. ""We Dance in Opposite Directions": Mesquakie (Fox) Separatism from the Sac and Fox Tribe." *Ethnohistory* 30, no. 3 (1983): 129–40.

Kiel, Doug. "Nation V. Municipality: Indigenous Land Recovery, Settler Resentment, and Taxation on the Oneida Reservation." *Native American and Indigenous Studies* 6, no. 2 (2019): 51–73.

LaDuke, Winona. "Indigenous Environmental Perspectives: A North American Primer." *Akwe: kon Journal* 9, no. 2 (1992): 52–71.

LaDuke, Winona, and Deborah Cowen. "Beyond Wiindigo Infrastructure." *South Atlantic Quarterly* 119, no. 2 (2020): 243–68.

Lawson, Michael L. *Dammed Indians: The Pick-Sloan Plan and the Missouri River Sioux, 1944-1980.* Norman: University of Oklahoma Press, 1994.

Lewis, David Rich. "Native Americans and the Environment: A Survey of Twentieth-Century Issues." *American Indian Quarterly* 19, no. 3 (1995): 423–50.

Ludvigson, Greg A, and James A Dockal. "Lead and Zinc Mining in the Dubuque Area." *Iowa Department of Natural Resources* (1984): 1–7.

Miller, Larisa K. "The Secret Treaties with California's Indians." *Prologue – Quarterly of the National Archives and Records Administration* 45, no. 3–4 (2013): 36–43.

Mosteller, Kelli. "The Cultural Politics of Land: Citizen Potawatomi Allotment and Citizenship in Kansas and Indian Territory, 1861–1891."

Mt. Pleasant, Jane. *Traditional Iroquois Corn: Its History, Cultivation and Use.* Ithaca, NY: Plant and Life Science Publishing, 2011.

_____. "A New Paradigm for Pre-Columbian Agriculture in North America." *Early American Studies* 12, no. (2015): 374–412.

_____. "Food Yields and Nutrient Analyses of the Three Sisters: A Haudenosaunee Cropping System." *Ethnobiology Letters* 7, no. 1 (2016): 87–98.

Mt. Pleasant, Jane, and Robert F Burt. "Estimating Productivity of Traditional Iroquoian Cropping Systems from Field Experiments and Historical Literature." *Journal of Ethnobiology* 30, no. 1 (2010): 52–79.

Mustful, Colin. "The Sandy Lake Tragedy." *Mnopedia*, 2021. https://www.mnopedia.org/event/sandy-lake-tragedy

National Congress of American Indians. "Infrastructure in Indian Country," 2017.

Rock, Tommy, "Exposing Years of Uranium Water Contamination in a Navajo Community." *Grantee Highlights. National Institute of Environmental Health Sciences*, February 24, 2017.

Rosier, Paul C. "Dam Building and Treaty Breaking: The Kinzua Dam Controversy, 1936-1958." *The Pennsylvania Magazine of History and Biography* 119, no. 4 (1995): 345–68.

_____. *Rebirth of the Blackfeet Nation, 1912–1954.* Lincoln NE: University of Nebraska Press, 2004.

_____. ""They Are Ancestral Homelands": Race, Place, and Politics in Cold War Native America, 1945–1961." *The Journal of American History* 92, no. 4 (2006): 1300–26.

Royster, Judith. "Oil and Water in Indian Country." *Natural Resources Journal* 37 (1997): 457–90.

Sandoval, Catherine JK. "Energy Access Is Energy Justice: The Yurok Tribe's Trailblazing Work to Close the Native American Reservation Electricity Gap." In *Energy Justice*, edited by Raya Salter, Carmen G. Gonzalez, Michael H. Dworkin, Roxanna A. Mastor, and Elizabeth Kronk Warner, 166–207. Cheltenham: Edward Elgar Publishing, 2018.

_____. "Principles to Advance Energy Justice for Native Americans." *EBA BRIEF*, October 2020.

Scrudato, Ron L., Jeff R. Chiarenzelli, Pamela K. Miller, Clark R. Alexander, Jr., John G. Arnason, Kendra Zamzow, K. Zweifel, et al. "Contaminants at Arctic Formerly Used Defense Sites." *Journal of Local and Global Health Science* 2012, no. 1 (2012): 1–15.

Small, Gail. "War Stories: Environmental Justice in Indian Country." *Daybreak* 4, no. 2 (1994): 38–41.

Sonter, Laura J, Marie C Dade, James EM Watson, and Rick K Valenta. "Renewable Energy Production Will Exacerbate Mining Threats to Biodiversity." *Nature Communications* 11, no. 1 (2020): 4174.

Sovacool, Benjamin K, Saleem H Ali, Morgan Bazilian, Ben Radley, Benoit Nemery, Julia Okatz, and Dustin Mulvaney. "Sustainable Minerals and Metals for a Low-Carbon Future." *Science* 367, no. 6473 (2020): 30–33.

Spice, Anne. "Fighting Invasive Infrastructures: Indigenous Relations against Pipelines." *Environment and Society* 9, no. 1 (2018): 40–56.

Stark, Heidi Kiiwetinepinesiik. "Marked by Fire: Anishinaabe Articulations of Nationhood in Treaty Making with the United States and Canada." *The American Indian Quarterly* 36, no. 2 (2012): 119–49.

_____. "Respect, Responsibility, and Renewal: The Foundations of Anishinaabe Treaty Making with the United States and Canada." *American Indian Culture and Research Journal* 34, no. 2 (2010): 145–64.

Thompson, Darren. "Panel: Infrastructure Law Creating a "Gold Rush" for Tribal Energy Projects." *Native News Online*, October 15, 2022.

U.S. Civil Rights Commission. "A Quiet Crisis: Federal Funding and Unmet Needs in Indian Country," 2003.

U.S. Commission on Civil Rights. "Broken Promises: Continuing Federal Funding Shortfall for Native Americans," 2018.

Webster, Rebecca M. "This Land Can Sustain Us: Cooperative Land Use Planning on the Oneida Reservation." *Planning Theory & Practice* 17, no. 1 (2016): 9–34.

Wero, Shane, and Rena Martin. "Łeetso – Working and Living within the Monster: A Cultural Resources Study of Navajo Habitations within Former Uranium Mines in the Cove and Monument Valley Regions." *Kiva* 87, no. 3 (2021): 354–76.

Witgen, Michael. *An Infinity of Nations: How the Native New World Shaped Early North America*. Philadelphia, PA: University of Pennsylvania Press, 2011.

10 The Food Justice Movement

Justin Sean Myers

The cultivation of a food system that is both environmentally sustainable and socially just will require the creation of alliances between the food movement and the communities most harmed by current conditions. The food justice movement is laying the foundation for such coalition building.

(Alison Hope Alkon and Julian Agyeman,
Cultivating Food Justice)

The survival of Native America is fundamentally about the collective survival of human beings. The question of who gets to determine the destiny of the land, and of the people who live on it—those with the money or those who pray on the land—is a question that is alive throughout society.

(Winona LaDuke, *All Our Relations*)

The Food Justice Movement

The food justice movement in the United States is a term attached to numerous movements aiming to create a more just and sustainable food system and with it a more just and sustainable world.[1] It encompasses urban residents of color turning vacant land into community gardens to address inequities in food access, farmworkers, fast food workers, and restaurant workers fighting for higher wages and safer working conditions, and Indigenous nations working to restore traditional foodways. This chapter will tell the story of these movements within the movement and in so doing will outline the historical themes of the food justice movement: What it is, who it is, what it aims to realize, and how it pursues these aims.

What unites all these varied movements is that they seek to address inequities in wealth and power within the food system. The food justice movement has similarities to the traditional politics of the food movement, particularly the latter's critique of Big Ag and Big Food as detrimental to democracy, small farmers, people's bodies, and the planet.[2] However, the food justice movement shifts the focus away from a standalone

DOI: 10.4324/9781003214380-14

environmental sustainability politics to the fusion of sustainability and social justice, e.g., *just sustainabilities*, and from predominantly Anglo-American (white) small farmers to all people in the food system, particularly those who are working class and people of color.[3]

This shift is due to the food justice movement's focus on the role of white supremacy and institutional racism in shaping the inequities traversing the food system and how such inequities are interconnected with broader social, political, and economic relations that require a move in perspective beyond "just food."[4] This includes exploring the interconnections between food and racism in housing, education, employment, transportation, land-use planning, and the criminal justice, banking, and financial systems. With respect to Indigenous nations, the focus is pushed beyond white supremacy and institutional racism to explore how settler colonialism and the imposition of settler ecologies onto Indigenous nations is vital to understanding their loss of political sovereignty and why Indigenous peoples struggle with high rates of food-related health inequities, e.g., type II diabetes, hypertension, heart disease, and obesity.[5]

The food justice movement's shift from sustainability to just sustainability and from white farmers to all people in the food system is due to the movement's direct roots in civil rights, labor rights, immigrant rights, and environmental justice movements of the 1960s, not to mention their more radical Black Power, Chicano Power, and Red Power counterparts.[6] At the same time, the lineage of many of these movements can be traced back much farther in time to the Indigenous sovereignty and Black Freedom struggles that emerged alongside the creation of the United States.[7] The food justice movement has been shaped, in particular, by the environmental justice movement, which shifted the lens of environmental activism from protecting wildlife and "wild" places to addressing environmental inequities where people live, work, play, and pray.[8]

In focusing on the built environment, the environmental justice movement emphasized the unequal distribution of environmental burdens in working class communities and communities of color and how such communities experienced negative effects due to the disproportionate location of LULUs in their neighborhoods, including landfills, toxic waste dumps, chemical plants, mining facilities, agricultural pesticide spraying, slaughterhouses, oil refineries, and transportation corridors.[9] In emphasizing the disproportionate location of LULUs in communities of color, the movement employed the term "environmental racism" to underscore how zoning, urban planning, transportation systems, redlining, and systemic disinvestment all played a role in concentrating "environmental bads" in such communities.[10] The environmental justice movement also blended community mobilization and direct action with litigation, lobbying, and policy reform to move beyond the limited focus on distributive equity within the mainstream environmental movement toward prioritizing the realization of procedural equity.[11]

The influence of these longer and broader movements, especially the environmental justice movement, is apparent in how the food justice movement does not see food inequities as mere technical problems to be solved by philanthropic foundations, corporations, political officials, or affluent white folks.[12] Inequitable food relations are complex problems emerging from inequities in the distribution of power between Black, Indigenous, and people of color (BIPOC) communities and white communities, corporations, and the state. Given that inequitable food relations emerge from inequities in the distribution of power, food justice advocates contend that the communities who are experiencing inequities are those that need to mobilize to end them.[13] This bottom-up social change politics lies at the core of the food justice movement and its embrace of *just sustainabilities*, which incorporates procedural, substantive, and distributive justice.[14] Procedural justice underscores that marginalized communities need to have a voice in how the food system and the broader social system are organized. Substantive justice entails human rights and that people have a right to housing, food, clean air and water, and healthcare. Distributive justice emphasizes the equitable distribution of social and environmental benefits and burdens across social groups and ecological landscapes.

Given this tripartite conception of justice, advocates of the food justice movement fight not solely to reduce disproportionate burdens—such as pesticide exposure, food-based health inequities (such as diabetes, hypertension, heart disease, and obesity), and lack of access to a grocery store—but also realizing the right to food, democratizing control over the decision-making structures that shape economic, political, and food systems, as well as creating the conditions for the self-determination of communities of color. For Indigenous nations, this means going beyond the struggle for self-determination as an ethnoracial group to fight for their right to assert their political sovereignty as nations. Which is why Indigenous nations often use the language of food sovereignty rather than food justice given the importance of their nation-to-nation status with the United States.

Farming the City: Community Gardeners in East New York, Brooklyn

Think about the area where you live. How easy is it to obtain fresh veggies and fruits? Is the grocery store just a few blocks away? Is the supermarket a quick 5-minute drive? If you answered in the affirmative to any of these questions you are privileged, as you do not live in what public health scholars call a *food desert*—a low-income community with limited access to fresh fruits and vegetables because grocery stores or supermarkets are out of reach (more than 1 mile in an urban area and more than 10 miles in rural areas).[15]

According to the U.S. Department of Agriculture, 23.5 million people live in food deserts, and the majority of these people are located in urban areas.[16] Food deserts are not race-neutral places either, as they are disproportionately Black and Hispanic communities. At the same time, the food desert concept only explains what is absent, not what actually exists in these communities. The reality for many of these communities is that they are not food deserts so much as *food swamps*, "areas in which large relative amounts of energy-dense snack foods, inundate healthy food options."[17] These communities are often stocked full of bodegas, liquor stores, and fast food restaurants that provide cheap foods high in fats, oils, sugars, and salts and lacking in vitamins and minerals. In addition, if a grocery store is to be found in these communities, the produce sold is often of low quality, being damaged, spoiled, or past the sell-by date. The outcome of such inequitable food relations is that residents struggle with the hunger-obesity paradox, having higher than average rates of malnutrition as well as obesity, diabetes, and heart disease, problems that should be seen as processes of *structural violence*.[18]

Food justice activists and scholars emphasize how the food inequities residents experience in such communities are neither a natural occurrence nor the outcome of individual-level choices. They result from historical and structural processes that are rooted in institutional racism, segregation, and white supremacy. As such, the inequities residents experience are referred to not as food deserts or food swamps but as *food apartheid* because they are generated through state, corporate, and community actions. Actions that include the racialized practices of redlining, block-busting, urban renewal, and planned shrinkage, all of which disinvested in communities of color from the 1940s through the 1980s and were coterminous with federal programs for highway construction and single-family home ownership that built the suburbs as for "whites only." Coinciding with this shift in public investment was active resistance by white communities to the Civil Rights Movement and the racial integration of housing, education, and employment. All of this created food inequities for low-income communities of color as they pushed grocery stores to follow white flight (and higher disposable incomes) to the suburbs, a process that occurred alongside federal subsidies that incentivized the movement of fast-food into urban communities of color under the notion of "job creation."[19]

Residents in working class communities of color have not just meekly accepted these racialized investment practices and their production of food apartheid though, they have invested in turning vacant lots in their neighborhoods into community gardens and urban farms through the direct action of squatting and working with city officials and agencies to obtain lease rights or tenure rights to land as well as creating legislation to protect potential agricultural land from development. This is occurring in Atlanta, Baltimore, Boston, Cleveland, Los Angeles, Oakland, and Seattle, among other places. One of these communities is East New York, Brooklyn, a predominantly working class Black and Caribbean community.[20] Devastated by private and municipal disinvestment from the mid-to-late twentieth century, the community experienced the defunding and withdrawal of public services (education, transportation, healthcare, parks, trash services, and police and fire) and the bulldozing of entire neighborhoods with no plan by city government for their reinvestment and rebuilding. Household incomes collapsed, jobs disappeared, grocery stores closed down, and mass incarceration was put forth as the solution to mass poverty. Residents were left to fend for themselves and they soon began to convert the vacant land into community gardens and urban farms to feed themselves and the community.

Today, these efforts are stewarded by the food justice organization East New York Farms! (ENYF!). ENYF! emerged in 1998 out of a three-year asset-oriented participatory planning project and was tasked with the purpose of combating the community's social, economic, and ecological disinvestment through urban agriculture and scaling up of a fledgling community food economy. Its purpose was securing tenure to land for community gardeners, increasing food production in the community to address the lack of fresh affordable and culturally appropriate food in the community, and creating jobs for East New York's youth. Starting with 1 staff member, the organization now has 6 full-time staff and a 30-member youth program, manages several urban farms, networks with over 40 food-producing community gardens in East New York, and runs 2 farmers markets. East New York is now full of food production spaces that grow Black and Caribbean crops, including bitter melon, cucumbers, okra, callaloo, bush peas, Malabar spinach, and hot peppers. The Saturday farmers market is a community hub where in addition to fresh local fruits and vegetables, Caribbean vendors sell meat patties, sweet bread, and plantain chips as well as curried goat and chicken with sides of mac and cheese, rice and peas, and stewed greens. And on special weekends, the market is host to events that embrace and celebrate the people and culture of East New Yorkers, including hot pepper and bitter melon festivals, afro battles, poetry slams, and dancing contests. Such efforts have been leveraged to build alliances with other organizations in the community to contest gentrification and mass incarceration as well as demand more affordable housing and a voice in land-use planning and public investment decisions in East New York.

In this respect, ENYF! is reflective of numerous food justice organizations across the United States that are linking people, land, food, and community in ways to address racial and class inequities. Through such efforts, gardeners are able to, one, grow food as an independent income strategy and facilitate the growth of community-led economic development. Two, through embracing agroecology practices, gardeners are working to rebuild the health of the soil and revitalize a form of agriculture that works with, not against, ecological flows. Three, gardeners are able to revitalize their cultural and ethnoracial identities that were often disrupted through migration, a process that pushed residents to consume highly processed foods that are not culturally appropriate and therefore pressured them to assimilate to whiteness and the products of Big Food. Four, the production and consumption of fresh culturally appropriate foods empowers residents to combat the food-related health inequities residents experience due to their prior dependence on the products of Big Food.

Moreover, it is important to understand contemporary food justice organizations not as something inherently new but as the latest manifestation of a long-term practice of African, African American, and Afro-Caribbean people in the United States. Such efforts must be situated within the longer history of the Black Freedom Struggle, which has entailed the pursuit of land and economic self-reliance through agriculture for centuries. This can be traced back to the struggles over provisioning grounds by enslaved people in the domestic south, where they fought for land and time of their own to feed themselves in culturally appropriate ways as well as sell this produce on the market for independent income to gain a degree of freedom within a society organized around slavery.[21] Not to mention the history of runaway slaves in maroon communities that fled plantations and grew their own food to feed their freedom struggles.[22] Moreover, obtaining agricultural land by and for Black communities, often in a cooperative form, has long been framed as a way to build spaces of self-protection, create a foundation for the physiological health of Black people, and serve as an anchor institution for community-based economies outside of the white capitalist power structure.[23] Agricultural cooperatives were advocated by W.E.B. Du Bois in the early 1900s as well as Fanny Lou Hamer in the 1970s as a way to create the economic conditions for Black self-determination. While many of today's Black-led urban food justice organizations exist as community-benefit organizations and are not cooperatives, they do carry on the centuries-long struggle by Black America for land, food, and freedom.

Farmworkers: Fighting for Justice in the Fields

Behind only potatoes, tomatoes are the second most consumed vegetable in the United States. You may eat it in a taco or a burger, in spaghetti and pizza sauce, or diced up in a curry or salad. No matter how you eat your tomato, if you are eating that tomato in winter, odds are that it comes from Florida, which produces over 90 percent of winter tomatoes in the United States.

Despite the ubiquity of the tomato in our diet, have you spent much time thinking about the working conditions of tomato pickers? Probably not, but it is the inhumane conditions in the fields of Florida that have energized a national movement to improve the lives of farmworkers. The organization leading this movement is known as the Coalition of Immokalee Workers (CIW), named after the place where the pickers work, Immokalee, Florida. Prior to farmworker mobilization in the early 1900s, Immokalee was a place where workers had not received real wage increases since the 1970s, where workers lacked the right to overtime pay and the right to collectively bargain with their employers, where workers were beaten and kept under armed guard. This all began to change in 1993 when the CIW formed to counter what workers called "modern-day slavery" and utilized direct action in "three community-wide work stoppages" and a "month-long hunger strike" to improve their working conditions. These actions were amplified by the Anti-Slavery campaign that was organized by the CIW to "uncover, investigate and assist" federal officials in stopping human rights violations in the fields, a campaign that led to the prosecution of seven operations in the state of Florida that employed over 1,200 workers. These employers forced workers to pick food under the watch of armed guards, kept them under lock and key to prevent flight, assaulted, pistol-whipped, and shot those who refused to work or tried to escape and charged excessive prices for rent, food, drugs, cigarettes, and alcohol in order to keep them in perpetual servitude through debt-peonage.[24]

These successes increased the visibility and power of the CIW and led the organization to launch the Campaign for Fair Food in 2001.[25] This campaign reflected the CIW's embrace of a militant grassroots form of worker-based activism that does not rely on the state and its regulatory framework to improve working conditions but focuses instead on market-based social change through applying pressure on off-field corporate actors, such as institutional food purchasers (IFPs). These IFPs, including fast food companies, grocery stores, and food service companies, are able to shape the pricing of food commodities through their enormous buying power and therefore heavily influence the wages that farmers pay farmworkers. Consequently, these IFPs have contributed to the exploitative working conditions of farmworkers through putting downward pressure on farmers for lower agricultural prices so that they can keep food prices low for their consumers and increase their profit margins. Given that farmers are often price-takers, not price-makers, and dependent on access to IFPs for produce sales, their revenues are tied to meeting the price points IFPs are willing to pay. An outcome of this restructuring of power in the food system is that tomato pickers in Immokalee were earning "about half of what they earned thirty years ago" and had to pick close to "twice the amount a worker had to pick" 30 years ago to earn the minimum wage.[26]

These inequities are not just the result of concentrated power in IFPs however, they are also the direct outcome of the U.S. food system historically organized around unwaged and unfree slave labor or low-priced wage labor. Farmworkers across the country have historically been excluded from New Deal labor protections, e.g., collective bargaining rights, minimum wage laws, and overtime laws. This occurred because for President Franklin Delano Roosevelt (FDR) to get his New Deal for white industrial workers through congress he had to make concessions with Southern Democrats and Southwestern Republicans to exclude two forms of employment that were central to maintaining white supremacy in the South: farmworkers and domestics, both jobs largely occupied by Black and Chicanx workers.[27] This legacy still shapes farm work to this day as it is one of the least compensated jobs in the nation, one increasingly organized around a transient and undocumented labor force that does not have legal rights of citizenship and therefore experiences political and economic repercussions for fighting to improve their jobs, including losing their job and employer-provided housing as well as being deported.

Because farmworkers are often excluded from the right to collectively bargain and the state often works with farmers to secure a cheap pool of disposable and marginalized workers, farmworkers are increasingly trying to improve their conditions of employment not through challenging the state but by challenging major corporations. As a result, a central demand of CIW's Campaign for Fair Food was for a *penny-per-pound* increase for tomato pickers to be paid by IFPs. This may not sound like a lot but it means thousands of dollars for each tomato picker who is paid not by the hour but the piece. The going piece-rate in the early 2000s was 50 cents for every 32-lb bucket of tomatoes, which translated into an annual income of around $12,000.[28]

The CIW's penny-per-pound demand was leveraged by a media savvy strategy, which recognizes that the power of corporations today and their bottom lines are heavily influenced by their brand image, a fact that can also be leveraged by workers as a weakness. The CIW was able to do so through connecting the brand image of IFPs with labor rights abuses. This public shaming practice relied heavily on raising awareness about how the business practices of IFPs were playing a major role in the exploitation occurring in the fields. McDonald's, Sodexo, Kroger, and others, do not want you thinking about "modern-day slavery" when you bite into a double-quarter pounder with cheese, buy a slice of pizza at your university dining hall, or pick up a tomato in the grocery isle. Yet, this is precisely what the CIW was doing through its campaign. Central to the success of such a public image campaign was the formation of alliances with civil rights organizations, immigrant rights organizations, student organizations, and religious organizations that mobilized people to participate in tabling, flyering, protests, marches, and boycotts on behalf of the CIW's demands.

The CIW's Campaign for Fair Food follows in the footsteps of farmworker movements that have long targeted corporate brands, utilized the tactic of boycotts, and formed alliances with student, civil rights, and religious organizations. The CIW campaign is particularly reflective of the grape strikes and boycotts of the 1960s and 1970s in California led by Chicano and Filipino activists, including Cesar Chavez, Larry Itliong, and Dolores Huerta. The goal of these *secondary boycotts* was to shrink consumer demand for grapes and thereby reduce the volume of grapes purchased by major grocery stores in order to decrease the profits of grape growers. In doing so, the hope was that growers would sign labor contracts with the UFW that would pay the workers a higher wage, offer better unemployment and health benefits, secure protections against pesticide exposure, and enable union hiring halls over the exploitative labor contractor system.[29]

The 1960s and 1970s campaigns targeted Gallo's wine labels, DiGiorgio's S&W Foods and TreeSweet fruit juices, as well as the Safeway and A&P grocery store chains, among others. Protestant, Catholic, and Jewish organizations put pressure on public officials to back the strike and boycott, universities and Catholic schools suspended grape purchases, and city councils and mayors in New York, Detroit, Cleveland, and San Francisco endorsed the strikes and boycotts. Dock workers in European countries refused to unload grapes. The big grape growers lost hundreds of thousands of dollars and market sales shrank by over 20 percent. The losses pushed the growers to sign contracts with the UFW in 1970, a historic win, and the boycott was vital to farmworkers winning the right to collective bargain and obtain union contracts. Striking, by itself, was insufficient as growers often found strikebreakers (who were undocumented workers) to take their place and utilized the court system, law enforcement, and vigilantes to crush strikes. Boycotts brought external, and very public, pressure on the growers and harmed their bottom lines in ways that the workers by themselves could not.[30]

As of 2022, the CIW has been extremely successful in achieving its demands through the use of secondary boycotts. They have gotten the four largest fast food companies (Burger King, McDonald's, Subway, and Yum Brands) and the three largest food service providers (Compass Group, Aramark, and Sodexo) to sign Fair Food Agreements. Walmart, Whole Foods, Trader Joe's, Chipotle, The Fresh Market, Ahold, Giant, and Stop & Shop have also signed agreements. This momentum pushed the CIW to expand their organizing efforts and the Fair Food Campaign into tomato production in Georgia, South Carolina, North Carolina, Maryland, Virginia, and New Jersey, as well as strawberry and pepper production in Florida. The CIW is continuing these efforts today through boycotts against Publix and Wendy's, two companies that have refused to sign Fair Food Agreements for years.

Fast Food Workers: The Fight for $15

How often do you eat fast food? Everyday? Once a week? Once a month? Americans love their fast food and the ascent of the industry has long been framed as beneficial to society through its provision of affordable convenient food and entry-level jobs to teenagers and young adults. However, this rosy image would be challenged on November 29, 2012 when fast food workers in New York City engaged in a one-day strike to fight for better working conditions, higher wages, and the right to form and join unions. This one-day strike would become known as the birth of the Fight for $15 movement (FF15), a movement that has reframed the national debate around fast food around how raising the minimum wage is vital to combating poverty and inequality.[31]

This one-day strike in November was followed by walkouts with thousands of workers in July of 2013 and September of 2014. The September walkout was significant because it reflected an escalation in the movement's tactics, with employees not just walking out but engaging in more active forms of civil disobedience, including stopping automobile traffic and sitting on sidewalks. These actions were heavily supported by the Service Employees International Union (SEIU), through both material resources and leadership provision, and can be traced back to SEIU's 2010 Fight for a Fair Economy that sought "to build a grassroots movement by working closely together with local activists and community Groups."[32] In NYC, this meant FF15 emerged from existing relationships between SEIU and local social justice, labor rights, and immigrant rights organizations, including New York Communities for Change (NYCC), Jobs with Justice, Make the Road, and the Working Families Party, all of whom have provided resources and funds to the FF15.[33]

Alongside these mobilizations, organizations affiliated with the movement had been pushing politicians at the municipal and state level to support legislation that would increase the minimum wage to $15. The movement focused on increasing worker's wages through direct action and legislation rather than collective bargaining and union contracts because, legally, many fast food workers are considered employees of individual franchisees rather than their corporate chains, which would necessitate organizing each restaurant one-by-one rather than a single company-wide mobilization for unionization. Such efforts were not seen as an effective use of resources or politically viable since workers trying to unionize an individual store often experience employer harassment and are (illegally) fired and even if the workers were successful in unionizing, there is a strong possibility that the business would likely close up shop and lay off all the workers. To address this problem, the movement has pressured the National Labor Relations Board (NLRB) to recognize that employment and wage violations that occur at franchise locations are the responsibility of both franchise operators and their corporate chains, to limited success.

Since 2014, the movement for the FF15 has only grown in numbers, visibility, and power. That year is when the City of Seattle passed legislation for a $15 minimum wage. Then, in 2015, the movement was given national exposure and support through the presidential campaign of Senator Bernie Sanders, who actively called for a $15 minimum wage and spoke about the Fight for $15 at his rallies. As of 2018, the FF15 benefited 22 million workers to the tune of $68 billion in raises, which is "more than 14 times larger than the total raise under the last federal minimum wage increase, approved in 2007."[34] In 2019, the House of Representatives passed the Raise the Wage Act, which would increase the federal minimum wage of $7.25 each year until it reached $15 in 2025.[35] It would also slowly eliminate the separate tipped-minimum wage of $2.13. Such a bill would benefit up to 33 million workers, and 43 percent of working single mothers.[36] Unfortunately, it is stuck in the Senate as Democrats need the votes of ten Republicans to break a filibuster, votes they do not have. The movement has also had to fight against corporate-backed laws passed by Republican-led state legislatures that ban cities from raising their minimum wage above their state's minimum wage.[37]

Despite this opposition, $15 minimum wage legislation has passed in primarily Democratic states, including California, Connecticut, Florida, Hawai'i, Illinois, Maryland, Massachusetts, New Jersey, New York, and the District of Columbia. Many other states have passed minimum wage increases to $12–$14 as a result of the movement—either through legislation or ballot measures. The FF15's success since 2012 displays the tangible gains that can be achieved through a combination of direct action, coalition building, national media exposure, a political climate open to addressing inequality, strategic support by municipal and state politicians, and the benefits of national publicity via presidential campaigns. Overall, the changes in minimum wage legislation have meant a collective raise of $150 billion for over 26 million workers, with the average worker earning an additional $5,300 per year.[38] And on Labor Day in 2022, California Governor Gavin Newsom signed into law AB257 that created the Fast Food Council, which would consist of labor and management representatives that would set minimum standards for fast food workers, including wages, health and safety regulations, right to time off, and protections from discrimination and harassment. Beyond the immediate economic gains of increasing the minimum wage for workers, the long-term effects of the movement lie in its ability to shift the national debate as well as public policies away from supply-side economics and toward demand-side economics through emphasizing that a higher wage model is better at facilitating economic growth and more equitably distributing the benefits of that economic growth than the current low-wage/high-debt model.

Restaurant Workers: The Inequity of Living Off of Tips

Folks in the United States eat out a lot, not just at fast-food chains but also at casual dining restaurants with table service. From smaller mom-and-pop-style restaurants to the big chains, most folks in the United States eat at such establishments at least once a week, if not more. But as we finish our meal, wind down our conversations with family and friends, and tabulate the tip on our bill, do we ever think about the working conditions of the waitstaff serving us our food and why we even have a national custom of tipping?

If you start to investigate such questions, the answers you will find are not rosy. The base pay of many servers isn't even the regular minimum wage. Only seven states have the same minimum wage for tipped and non-tipped employees. In the other 43 states, tipped workers receive less than the minimum wage for non-tipped employees, with 18 states using the federal minimum tipped wage of only $2.13 per hour.[39] Tipping came over from England where it originated among the elite in the belief that it would ensure good service, tip being an acronym for "to insure promptitude."[40] Despite this aspirational belief, research finds little correlation between tipping and good service, the largest factor shaping tip size is bill size.[41] When tipping came to the United States after the Civil War, many folks opposed its existence as anti-democratic and anti-American, a recreation of the aristocratic master-servant relation.[42] People considered the receipt of a tip as a mark of stigma, a mark of degradation, so much so that seven states passed anti-tipping laws. But the practice was embraced by restaurants and railroad companies after the war, especially in the South, to shift the costs of a newly freed Black labor force away from employers and onto the general public. This practice, while preceding the New Deal, became institutionalized through the Federal Fair Labor Standards Act of 1938 that created a federal minimum wage, one that excluded workers who earned the minimum wage through tips, an exception that the restaurant and railroad industries fought hard for.[43]

It was not until 1966 and President Lyndon Baines Johnson's War on Poverty that the federal tipped minimum wage came into existence and was pegged at 50 percent of the federal non-tipped minimum wage.[44]

While this may seem like a small win, it was significant, as without any minimum wage protections, many servers were forced to live solely off of tips. Even with this win for servers, the real value of the federal tipped minimum wage has continued to decline under the pressure of the restaurant industry so that by 2022 the federal tipped minimum wage was worth only 28.4 percent of the non-tipped minimum wage.[45] This has occurred because in 1996 President Bill Clinton had to make concessions to Republicans to pass his legislation, much like FDR had to make concessions to Republicans and southern Democrats to pass his New Deal policies.[46] Clinton was pushing congress to raise the minimum wage to reduce poverty and increase consumer-buying power. In order to get their support for his legislation, he had to make a deal with the Republican-dominated Congress: Exclude the tipped minimum wage from the increase and therefore keep it frozen at $2.13. This was something that Republicans advocated for because they are often opposed to a minimum wage on ideological and economic grounds but it was also due to the immense lobbying power of the National Restaurant Association (NRA), which was opposed to an increase in the tipped minimum wage.[47]

By decoupling the tipped minimum wage from the non-tipped minimum wage, the restaurant industry has long been able to create a two-tiered wage structure where they are allowed to pay a subminimum wage and externalize the costs of their workforce onto customers and the general public, which has yielded increased and sometimes record profits for the major restaurant chains. The flipside is that servers have a poverty rate three times higher than the average worker, are twice as likely to utilize food stamps and Medicaid than the average worker, regularly experience wage theft, and lack employer-provided health benefits and paid sick days.[48] Such findings underscore that in many states tips are not a luxury for servers but constitute a significant portion of their salary and are the difference between survival and destitution. As these servers are disproportionately women and people of color the existence of the tipped minimum wage amplifies class, race, and gender inequities.

Despite the working conditions of waitstaff, this workforce has generally not mobilized collectively to challenge them, employing instead tactics of individual "exit"—where a worker quits one job to find another one—or individual resistance against their current employer through slowdowns, noncooperation, calling in sick, gossip, and taking their individual grievance to the manager. These are all valuable tools to cope with the day-to-day life of one's job, but they are not collective mobilizations that change the structure of employee-employer relations. This is where Restaurant Opportunities Center (ROC) comes in, as they have mobilized since 2011 to address conditions in the restaurant industry through building power outside of the traditional union form.[49] ROC emerged out of Al

Qaeda's attack on the World Trade Center on September 11, 2001, which killed 73 employees of Windows of the World and left the rest of the 300 employees without jobs. Windows of the World had been an extremely popular restaurant on the top floor of the north tower of the World Trade Center and its workforce was predominantly immigrants and unionized, represented by Local 100 of the Hotel Employees and Restaurant Employees Union (HERE). Six months later, ROC—New York (ROC—NY) emerged from efforts to support the Windows of the World workforce. Led by Fekkak Mamdouh and Saru Jayaraman, ROC—NY sought to improve the working conditions of restaurant employees across the country. Today, ROC United, the national level organization, has chapters in the Bay Area, Chicago, Los Angeles, Michigan, Minnesota, Mississippi, New Orleans, New York, Pennsylvania, and Washington D.C.

ROC emphasizes a three-pronged strategy of "research and policy, workplace justice, and promotion of the high road."[50] Rather than trying to recreate the union structure of HERE, ROC has instead opted for the model of an "independent worker center."[51] ROC's research wing blends surveys and interviews with local and national level statistics to generate data about the working conditions of restaurant employees that is then leveraged in campaigns for legislative action to win worker demands, including paid sick days, a higher minimum wage, stronger sexual harassment protections, and better health and safety regulations. The workplace justice wing prioritizes building worker-led power to address the inequities facing employees. This entails utilizing direct action, such as picket lines and boycotts, as well as litigation, to bring change to employer practices regarding wage theft and racial and gender discrimination. Such actions have resulted in changes in workplace policies as well as over $10 million dollars to workers for back wages and discrimination.[52]

Promotion of high road jobs occurs through ROC's Restaurants Advancing Industry Standards in Employment (RAISE) association that currently has over 900 high road employers. Such employers are committed to creating high road jobs for their employees, jobs that offer living wages, local hiring provisions, stable schedules, job security, and health and retirement benefits. These employers have also piloted ROC United's Racial Equity Menu toolkit, which aims at "desegregating the restaurant industry and opening up better-paying front-of-house positions to immigrants, people of color and women."[53] RAISE affiliated employers also hire workers who have graduated from ROC's COLORS Hospitality Opportunities for Workers (CHOW) Institute, which trains people of color, undocumented workers, and women to work as bartenders, managers, and fine dining servers. Overall, ROC has trained 5,000 workers for employment in high road jobs.[54]

The big action for 2022 is the Restaurant Workers Bill of Rights, which emerged on Labor Day following a 50-state 6-month long outreach to

ROC's 65,000 members collecting information on the needs of restaurant workers. The document is intended to serve as the organization's "advocacy North Star for the coming years" and calls for a thriving wage, paid leave, childcare, universal healthcare, protection from discrimination and harassment, and the ability to practice "democracy at and outside of work."[55]

Restoring the Salmon Runs: Indigenous Struggles for Food Sovereignty

The Klamath River was once the third largest salmon-producing river in the Western United States, behind only the Columbia and Sacramento rivers. Today, due to settler-imposed dams and diversions for agriculture in California and Oregon, the salmon runs are nearly nonexistent and their destruction has contributed to the ongoing genocide of the Karuk people. Yet, like other Indigenous nations across Turtle Island (North America), the Karuk are mobilizing and fighting to restore the salmon runs and with them the cultural, economic, ecological, and political sovereignty of their nation.[56]

The Karuk have lived with the salmon in the Klamath for thousands of years. Their stewardship of the river produced a magnificent bounty for the human and nonhuman life in the Klamath River basin. The river produced several salmons runs across the year, including "Spring and Fall Chinook, Coho, Humpback, Sockeye, and Chum, as well as steelhead and several species of lamprey."[57] This provided a stable supply of food throughout the year for the Karuk, a relationship that was quickly destroyed upon the arrival of settlers, miners, and the U.S. military in the 1850s, as they killed over 70 percent of the Karuk by the 1880s, forced them off their homeland, and confined them to reservations. Following a long line of bad-faith treaty negotiations, the U.S. congress refused to ratify the 1851 treaty between the Karuk and the federal government, which meant the Karuk had no title to land, no reservation, no fishing or hunting rights, and were not legally considered a tribe and thus had no political status as a nation from the point of view of the United States and therefore could not enter into political negotiations with the federal government, an issue that was not corrected until 1979.[58]

As a result of these processes of genocide and dispossession, 98 percent of the land in the Karuk homeland is now owned by U.S. Forest Service.[59] Moreover, the Karuk have been denied use rights to their occupied homeland by federal government officials who have deployed law enforcement to prevent them from fishing in the Klamath. The Forest Service has also denied the Karuk from being able to collect other "traditional forest foods, including deer, acorns, and mushrooms" as well as engage in management practices conducive to the stewardship of these foods, namely prescribed burning.[60] The destruction of the salmon runs and inability to forage and hunt for traditional foods has fueled a dramatic rise in food-related health inequities (e.g., type II diabetes and obesity) for the Karuk as they are forced to rely on a westernized diet of Big Food. The Karuk nation also struggles with maintaining their cultural lifeways as language, identity, gender, and social status are interconnected with the stewardship of traditional foodways. Thus, the loss of ownership of and access to their ancestral homeland has fueled social instability and mental health problems in family and community fabric for the Karuk.[61]

In this regard, Indigenous struggles are qualitatively different from other food justice movements documented in this chapter, as they are grounded in the destruction of their political sovereignty as nations, which was often done through the destruction of their food sovereignty, and underscores the role of settler colonialism in generating inequitable food relations for the Indigenous. This included General George Washington's scorched-earth strategy on the Haudenosaunee (Iroquois) in upstate New York in the 1770s and 1780s, where the military burned down entire communities and specifically targeted the food crops and food reserves of the Haudenosaunee to weaken their ability to resist and fight back against the United States.[62] These actions earned him the name "Town Destroyer." Such actions continued as the U.S. Army under Major General Phillip Sheridan waged a ruthless genocidal war on the Indigenous nations in the midwestern plains through the annihilation of the bison (buffalo) in the 1860s and 1870s.[63] Sheridan's goal was to force the Indigenous onto reservations to clear the land for settlers from the east to practice settler agriculture. Yet Indigenous nations on the plains would never give way to white settlers and their agriculture, nor settle down and practice settler agriculture themselves, as long as the bison existed. The Dakota and Lakota's (Sioux) way of life was organized around mobility with the bison, from which they obtained food, clothing, and housing material. Through their mass slaughter by settlers and the U.S. military, only a few hundred bison would remain several decades later and the Indigenous nations would be forced onto reservations a fraction of the size of their former lands. In Arizona, a similar event happened in the 1860s as the U.S. Army sought to force the Navajo off their land and onto Fort Sumner in order to clear the way for settlers traveling to the California Gold Rush.[64] Many Navajo fled to Canyon de Chelly to hide out but the army destroyed their crops, killed livestock, and burned down their peach orchards. Starving to death, the majority of the Navajo in Canyon de Chelly surrendered and endured the "Long Walk," where 9,000–10,000 Navajo were marched, under armed guard, to Fort Sumner.

In centering the role of settler colonialism and the U.S.'s scorched-earth policy on Indigenous food systems, Indigenous activists and scholars underscore that unlike in colonialism, where metropole countries extracted wealth from colonies, settler colonialism is where settlers come to create a new homeland that necessitates the erasure of Indigenous sovereignty.[65] Key here is that settler colonialism is a "structure not an event."[66] A structure that is reproduced materially through the inscription of "settler ecologies" onto Indigenous homelands in order to create a homeland for settlers, ecologies legitimated through the myth-making narratives of "feeding the world" and "taming the frontier."[67] In moving beyond locating inequitable food relations within the racialized political economy of capitalism, Indigenous food sovereignty movements underscore that efforts to rebuild and revitalize Indigenous food systems entail restoring their nation's political sovereignty, which requires not just Indigenous mobilization but the unsettling of settlers—their beliefs and values—so that settlers can work in solidarity with Indigenous decolonization struggles and movements to return land back to Indigenous communities.[68]

Akin to the experiences of other Indigenous nations, the imposition of settler ecologies onto the Karuk has been extremely detrimental, but they have also continued to fight such settler ecologies and their devastating effects through pushing for dam removal on the Klamath, with specific focus on four dams: Iron Gate, Copco 1, Copco 2, and J.C. Boyle, dams that are only used for electricity generation and not agriculture.[69] If successful, the dam removals would allow salmon to swim and spawn on 400 miles of the river that has been blocked for over a hundred years, ushering in the possibility of the revitalization of salmon migrations and with them the cultural and ecological health of the Karuk. Not only will dam removal open up spawning grounds to the salmon, but it should address low quality water and low flow water issues that have fueled disease outbreaks and mass salmon die-offs over the last 30 years.

These efforts have been pursued for decades by the Karuk, working with the Yurok and non-Indigenous settler organizations who advocate on behalf of river restoration and fish habitat restoration, e.g., American Rivers and California Trout.[70] These coalitions have lobbied the state governments of California and Oregon as well as the federal government to push the Bureau of Reclamation, which plays a large role in western water projects, to have to take into account the water and fishing rights of Indigenous nations in its day-to-day operations. This included pressuring the Federal Energy Regulatory Commissions (FERC) to implement dam removal requirements in the relicensing of hydroelectric dams on the Klamath operated by PacifiCorp. The Karuk also lobbied the Fish and Wildlife Service, the National Marine Fisheries Service, and the California Water Resources Control Board to implement protections for water quality and fish habitat in any relicensing agreement. The new license, if issued in 2006, would have lasted 50 years and this long timeline created a sense of urgency among the Karuk to mobilize even more on behalf of dam removal. To make their case and bring national and international media attention to the problem, the Karuk traveled, in 2004, to Scotland to protest at the shareholder meetings of Scottish Power, a multinational energy company that owned PacifiCorp. These mobilizations shifted to Omaha, Nebraska several years later when PacifiCorp was sold to Berkshire Hathaway Energy, owned by Warren Buffet.

Such mobilizations against state and corporate actors were effective in publicizing the issue and led to research studies and the creation of plans estimating the cost of dam removal and restoration versus fish ladders and dam maintenance. It became clear that it was more economical for PacifiCorp to dismantle the dams than keep them in operation while providing for salmon habitat. Emerging out of numerous meetings and studies was the Klamath Hydroelectric Settlement Agreement, which outlines the processes by which PacifiCorp's dam licenses are transferred to the Klamath River Renewal Corporation, an entity created for the explicit purpose of dam removal and river restoration. These plans are moving forward in 2022 as on August 26, FERC released its Final Environmental Impact Statement (EIS) for the Klamath River Renewal Project and recommended approval of the license transfer as well as the decommissioning and removal of the dams.[71] As the dams come down, the Karuk will once again be able to revitalize the salmon runs on the Klamath and with them regenerate their traditional cultural management practices of the river and with them access to their traditional foods. Such actions should help revitalize the social, cultural, physiological, and ecological health of the Karuk.

The Karuk are not alone in the struggle for Indigenous food sovereignty either, the movement exists across Turtle Island, as Indigenous peoples work to restore the ecological vibrancy of the land and with it their traditional food economies in order to facilitate economic development, strengthen cultural identity, and address food insecurity and diet-related diseases. At the White Earth Indian Reservation in Callaway, Minnesota, Ojibwe (Anishinaabe) people, through the nonprofit White Earth Land Recovery Project, are protecting waterways from destruction by mining and logging industries to enable the revitalization of wild rice cultivation and maple syrup harvesting. In Oklahoma, the Cherokee Nation Heirloom Garden and Native Plant Site operates a seedbank to provide Cherokees with native seeds that cannot be obtained commercially. The program started in 2006 and is extremely popular, distributing over 10,000 seed packets annually. The seeds offered include varieties of turkey gizzard beans, tobacco, sunchokes, basket gourds, jewel gourds, and dent corn, among others. In Arizona, the Tohono O'odham Nation, through the nonprofit Tohono O'odham Community Action (TOCA), are working on growing traditional indigenous foods, such as tepary beans, O'odham squash, yellow-meated watermelon, and O'odham corn, as well as revitalizing traditional harvesting practices for saguaro cactus fruit, cholla cactus bud, prickly pear fruit, wild spinach (amaranth), mesquite beans, and acorns. In South Dakota, the Rosebud Lakota (Sioux) tribe is working with the World Wildlife Fund to reinvigorate 28,000 acres of native grassland at the Wolakota Buffalo Range, which will be home to nearly 1,500 animals and operate as the largest Indigenous-owned bison herd on Turtle Island. As these examples underscore, there is no singular manifestation of Indigenous food sovereignty but a multitude of ways, as each expression is rooted in the land.

Conclusion: The Food Justice Movement

As this chapter should make clear, the food justice movement is not a monolith but one of multiplicity. It is a movement that emerges against an ongoing legacy of oppression and exploitation, one linking capitalism, white supremacy, and settler colonialism, a legacy that manifests itself in different forms of oppression against different groups in different locations, but one where those facing these injustices are organizing to build power in-and-through food to improve their quality of life and fight for a more just and sustainable world. This can take the form of workers throughout the food system mobilizing outside of the union form, building coalitions with civil rights, immigrant rights, religious, and student groups, and utilizing boycotts and legislation to force off-farm corporations to provide high road jobs to their workers. This can take the form of urban food justice organizations working with residents to build community food economies rooted in principles of social equity, racial justice, and environmental sustainability. Through gaining control over land and growing food by and for the community, these organizations aim to use food to build power to address systemic issues facing residents, including affordable housing, underfunded public education systems, and land-use planning. This can take the form of Indigenous nations fighting against settler colonialism and the destruction of their food systems through asserting their political sovereignty, regaining control over their ancestral homelands, and revitalizing their traditional foodways.

Across these divergent movements, there are also shared tactics and strategies that are fueling their success. Given their marginalized positions within the United States, each of these movements cannot rely on their own efforts to change the behaviors of state and corporate actors. To win, each of these movements has built coalitions with other movement organizations and utilized public media to flex external pressure on state and corporate actors to get them to acquiesce to their demands. With regard to workers in the food system, it is clear that they are mobilizing outside of the tradition union form due to their lack of legal rights and the current weakness of the traditional union movement as well as the power that can be unleashed through the use of secondary boycotts.

Notes

1 Alison Hope Alkon and Julian Agyeman, ed., *Cultivating Food Justice: Race, Class, and Sustainability* (Cambridge, MA: MIT Press, 2011); Robert Gottlieb and Anupama Joshi, *Food Justice* (Cambridge, MA: MIT Press, 2010).

2 Big Ag stands for the companies that control what and how food is produced in the United States, e.g., grains (Cargill, Archer Daniels Midland, Bunge), chemicals (Bayer, Syngenta, Corteva), pork (Smithfields, Seaboard Foods, Triumph), Chicken (Tyson, Perdue), and beef (Tyson, Cargill, National Beef). Big Food stands for the companies that control the food and beverage industries that shape what people consume, e.g., Hain Celestial, Post, General Mills, Unilever, Coca-Cola, McDonalds, Yum Brands (Taco Bell, KFC, Pizza Hut, WingStreet). In some instances, companies can be both Big Ag and Big Food, e.g., Tyson, which controls both how meat is produced and consumed.

3 Julian Agyeman, *Sustainable Communities and the Challenge of Environmental Justice* (New York: NYU Press, 2005).

4 Eric Holt-Giménez and Yi Wang, "Reform or Transformation? The Pivotal Role of Food Justice in the U.S. Food Movement," *Race/Ethnicity: Multidisciplinary Global Contexts* 5, no. 1 (2011): 83–102; Justin Sean Myers and Joshua Sbicca, "Bridging Good Food and Good Jobs," *Geoforum* 61 (2015): 17–26; Joshua Sbicca and Justin Sean Myers, "Food Justice Racial Projects: Fighting Racial Neoliberalism from the Bay to the Big Apple," *Environmental Sociology* 3, no. 1 (2017): 30–41.

5 Dina Gilio-Whitaker, *As Long as Grass Grows: The Indigenous Fight for Environmental Justice, from Colonization to Standing Rock* (Boston, MA: Beacon Press, 2020); Winona LaDuke, *All Our Relations* (Cambridge, MA: South End Press, 1999); Kari Marie Norgaard, *Salmon & Acorns Feed Our People: Colonialism, Nature & Social Action* (New Brunswick, NJ: Rutgers University Press, 2019).

6 Garrett M. Broad, *More Than Just Food: Food Justice and Community Change* (Berkeley: University of California Press, 2016); Gottlieb and Joshi, *Food Justice*; Gilio-Whitaker, *As Long as Grass Grows*.

7 LaDuke, *All Our Relations*; Monica White, *Freedom Farmers: Agricultural Resistance and the Black Freedom Movement* (Chapel Hill: The University of North Carolina Press, 2019).

8 Robert D. Bullard, *Confronting Environmental Racism: Voices from the Grassroots* (Cambridge, MA: South End Press, 1993).

9 Julie Sze, *Noxious New York: The Racial Politics of Urban Health and Environmental Justice* (Cambridge, MA: MIT Press, 2007).

10 Laura Pulido, "Rethinking Environmental Racism: White Privilege and Urban Development in Southern California," *Annals of the Association of American Geographers* 90, no. 1 (2000): 12–40.
    Laura Pulido, "Geographies of Race and Ethnicity 1: White Supremacy Vs White Privilege in Environmental Racism Research," *Progress in Human Geography* 39, no. 6 (2015): 809–17.

11 Luke W. Cole and Sheila R. Foster, *From the Ground Up: Environmental Racism and the Rise of the Environmental Justice Movement* (New York: New York University Press, 2001); Laura Pulido, *Environmentalism and Economic Justice: Two Chicano Struggles in the Southwest* (Tucson: University of Arizona Press, 1996).

12 Joshua Sbicca, *Food Justice Now!: Deepening the Roots of Social Struggle* (Minneapolis: University of Minnesota Press, 2018); Alison Alkon and Julie Guthman, ed., *The New Food Activism: Opposition, Cooperation, and Collective Action* (Berkeley: University of California Press, 2017).

13 I use food inequities and inequitable food relations interchangeably. In doing so, I aim to underscore how food inequities are embedded within broader classed, raced, and gendered relations of power, oppression, and resistance that shape the production, appropriation, and distribution of land, income, wealth, opportunity, and food.

14 Agyeman, *Sustainable Communities and the Challenge of Environmental Justice.*

15 U.S. Department of Agriculture, *Food Access Research Atlas: Documentation* (Washington, DC: United States Department of Agriculture, 2009).

16 *Ibid.*

17 Donald J. Rose, Chris M. Bodor, Janet C. Rice, Thomas A. Farley, and Paul L. Hutchinson, "Deserts in New Orleans? Illustrations of Urban Food Access and Implications for Policy," Paper presented at the *University of Michigan National Poverty Center/USDA Economic Research Service Conference Understanding the Economic Concepts and Characteristics of Food Access*, January 23, 2009, Washington, DC.

18 Justin Sean Myers, "Food and Hunger," in *Twenty Lessons in the Sociology of Food and Agriculture*, ed. Jason Konefal and Maki Hatanaka (New York: Oxford University Press, 2018), 223–38.

19 Justin Sean Myers, *Growing Gardens, Building Power: Food Justice and Urban Agriculture in Brooklyn* (New Brunswick, NJ: Rutgers University Press, 2023).

20 Myers, *Growing Gardens, Building Power.*

21 Eugene D. Genovese, *Roll, Jordan, Roll: The World the Slaves Made* (New York: Vintage Books, 1976).

22 Genovese, *Roll, Jordan, Roll.*

23 White, *Freedom Farmers.*

24 Coalition of Immokalee Workers, "About," *Coalition of Immokalee Workers,* https://ciw-online.org/about/

25 Gottlieb and Joshi, *Food Justice.*

26 Coalition of Immokalee Workers, "Facts and Figures on Florida Farmworkers," Coalition of Immokalee Workers, https://ciw-online.org/wp-content/uploads/12FactsFigures_2.pdf

27 Linda C. Majka and Theo J. Majka, *Farm Workers, Agribusiness, and the State* (Philadelphia, PA: Temple University Press, 1982).

28 Coalition of Immokalee Workers, "Facts and Figures on Florida Farmworkers."

29 Majka and Majka, *Farm Workers, Agribusiness, and the State.*

30 *Ibid.*

31 David Rolf, *The Fight for Fifteen: The Right Wage for a Working America* (New York: The New Press, 2016).

32 Maite Tapia, Tamara L. Lee, and Mikhail Filipovitch, "Supra-union and intersectional organizing: An examination of two prominent cases in the low-wage US restaurant industry," *Journal of Industrial Relations* 59, no. 4 (2017): 498.

33 Tapia et al., "Supra-union and intersectional organizing," 498–99.

34 National Employment Law Project, *Impact of the Fight for $15: $68 Billion in Raises, 22 Million Workers* (New York: National Employment Law Project, 2018), 1, https://s27147.pcdn.co/wp-content/uploads/Data-Brief-Impact-Fight-for-15-2018.pdf

35 David Cooper, *Raising the Federal Minimum Wage to $15 by 2024 Would Lift Pay for Nearly 40 million Workers* (Washington, DC: Economic Policy Institute, 2019), https://files.epi.org/pdf/160909.pdf

36 Cooper, *Raising the Federal Minimum Wage to $15 by 2024 Would Lift Pay for Nearly 40 million Workers*, 4.

37 Marni von Wilpert, *City governments are raising standards for working people—and state legislators are lowering them back down* (Washington, DC: Economic Policy Institute, 2017), https://files.epi.org/pdf/133463.pdf

38 Lathrop, Lester, and Wilson, *Quantifying the Impact of the Fight for $15*, 4.

39 ROC United. *2020 State of the Restaurant Workers: A Comprehensive Analysis of the U.S. Restaurant Workforce* (New York: ROC United, 2020) https://rocunited.org/wp-content/uploads/sites/7/2022/06/SORW_2020.pdf

40 Kerry Segrave, *Tipping: An American Social History of Gratuities* (Jefferson, NC: McFarland & Company, Inc., 2009); Ofer H. Azar, "The Economics of Tipping," *Journal of Economic Perspectives* 34, no. 2 (2020): 215–36.

41 Melanie Saltzman and Saskia De Melker, "Why Do We Tip?" *PBS*, March 28, 2016, https://www.pbs.org/wnet/chasing-the-dream/stories/why-do-we-tip/

42 Azar, "The Economics of Tipping."

43 Sylvia A. Allegretto and Steven C. Pitts, *To Work With Dignity: The Unfinished March Toward a Decent Minimum Wage* (Washington, DC: Economic Policy Institute, 2016), 4–5, https://files.epi.org/2013/Unfinished-March-Minimum-Wage.pdf

44 Saru Jayraman, *Forked: A New Standard for American Dining* (New York: Oxford University Press, 2019), 35.

45 Author's calculations based on federal tipped and federal non-tipped minimum wage in 2022.

46 Jayraman, *Forked*, 35–6.

47 The NRA's key members include Darden Restaurants (parent company of Red Lobster, Olive Garden and Capital Grille), Walt Disney, YUM! Brands (parent of Taco Bell, KFC and Pizza Hut), McDonald's, Marriott, Aramark, Sodexo, and Starbucks. See Restaurant Opportunities Centers United, *The Other NRA: Unmasking the Agenda of the National Restaurant Association* (New York: Restaurant Opportunities Centers United, 2014).

48 ROC United, *2020 State of the Restaurant Workers*, 2.

49 Tapia et al., "Supra-union and intersectional organizing."

50 José Olivia, "Restaurant Opportunities Center" in *Food Movements Unite: Strategies to Transform Our Food Systems*, ed. by Eric Holt-Giménez (Oakland, CA: Food First Books), 176.

51 Tapia et al., "Supra-union and intersectional organizing," 494.

52 Tapia et al., "Supra-union and intersectional organizing."

53 ROC United, "RAISE," ROC United, https://rocunited.org/raise/

54 ROC United, "History," ROC United, https://rocunited.org/history/

55 ROC United, "Restaurant Workers Bill of Rights," ROC United, https://rocunited.org/bill-of-rights/

56 Norgaard, *Salmon & Acorns Feed Our People*; Leontina M. Homel and Kari M. Norgaard, "Bring the Salmon Home! Karuk Challenges to Capitalist Incorporation," *Critical Sociology* 35, no. 3 (2009): 343–66.

57 Homel and Norgaard, "Bring the Salmon Home!," 351.
58 Homel and Norgaard, "Bring the Salmon Home!".
59 Cited by Homel and Norgaard, "Bring the Salmon Home!", 353.
60 Kari Marie Norgaard and Ron Reed, "Emotional Impacts of Environmental Decline: What can Native Cosmologies Teach Sociology About Emotions and Environmental Justice?" *Theory and Society* 46 (2017), 465.
61 Homel and Norgaard, "Bring the Salmon Home!"; Norgaard and Reed, "Emotional Impacts of Environmental Decline."
62 Barbara Alice Mann. *George Washington's War on Native America* (Westport, CT: Praeger 2005).
63 Nick Estes, *Our History is the Future* (New York: Verso, 2019), 89–132.
64 Susan Dolan, "How a Navajo Scientist Is Helping to Restore Traditional Peach Horticulture," *National Park Service*, https://www.nps.gov/articles/000/how-a-navajo-scientist-is-helping-to-restore-traditional-peach-horticulture.htm
65 Patrick Wolfe, "Settler Colonialism and the Elimination of the Native," *Journal of Genocide Research*, 8, *no.* 4 (2006): 387–409.
66 Wolfe, "Settler Colonialism and the Elimination of the Native," 388.
67 Kyle Powys Whyte, "Indigenous Food Sovereignty, Renewal, and US Settler Colonialism," in *Routledge Handbook of Food Ethics*, ed. By Mary C. Rawlinson and Caleb Ward (New York: Routledge, 2016), 354–65.
68 LaDuke, *All Our Relations*; Eve Tuck and K. Wayne Yang, "Decolonization is Not a Metaphor." *Decolonization: Indigeneity, Education & Society* 1, no. 1 (2012): 1–40.
69 S. Craig Tucker, "Bring the Salmon Home: Background Information for the Klamath River Dam Removal," Karuk Tribe, n.d., https://www.karuk.us/images/docs/press/press%20&%20campaigns/Bring%20the%20Salmon%20Home%20Fact%20Sheet.pdf
70 Tara Lohan, "Here's How the Largest Dam Removal Project in the U.S. Would Work," July 20, 2018, https://www.kqed.org/science/1927666/heres-how-the-largest-dam-removal-project-in-the-u-s-would-work
71 California Trout, "Major Milestone Met for Klamath Dams Removal," *California Trout*, August 26, 2022, https://caltrout.org/news/major-milestone-met-for-klamath-dams-removal

Bibliography

Agyeman, Julian. *Sustainable Communities and the Challenge of Environmental Justice*. New York: NYU Press, 2005.

Alkon, Alison Hope, and Julian Agyeman, eds. *Cultivating Food Justice: Race, Class, and Sustainability*. Cambridge, MA: MIT Press, 2011.

———, eds. *The New Food Activism: Opposition, Cooperation, and Collective Action*. Berkeley: University of California Press, 2017.

Azar, Ofer H. "The Economics of Tipping." *Journal of Economic Perspectives* 34, no. 2 (2020): 215–36.

Broad, Garrett M. *More Than Just Food: Food Justice and Community Change*. Berkeley: University of California Press, 2016.

Estes, Nick. *Our History Is the Future*. New York: Verso, 2019.

Genovese, Eugene D. *Roll, Jordan, Roll: The World the Slaves Made*. New York: Vintage Books, 1976.

Gilio-Whitaker, Dina. *As Long as Grass Grows: The Indigenous Fight for Environmental Justice, from Colonization to Standing Rock*. Boston, MA: Beacon Press, 2020.

Gottlieb, Robert, and Anupama Joshi. *Food Justice*. Cambridge, MA: MIT Press, 2010.

Homel, Leontina M., and Kari M. Norgaard. "Bring the Salmon Home! Karuk Challenges to Capitalist Incorporation." *Critical Sociology* 35, no. 3 (2009): 343–66.

Jayraman, Saru. *Forked: A New Standard for American Dining*. New York: Oxford University Press, 2019.

LaDuke, Winona. *All Our Relations*. Cambridge, MA: South End Press, 1999.

Majka, Linda C., and Theo J. Majka. *Farm Workers, Agribusiness, and the State*. Philadelphia, PA: Temple University Press, 1982.

Myers, Justin Sean. *Growing Gardens, Building Power: Food Justice and Urban Agriculture in Brooklyn*. New Brunswick, NJ: Rutgers University Press, 2023.

Norgaard, Kari Marie. *Salmon & Acorns Feed Our People: Colonialism, Nature & Social Action*. New Brunswick, NJ: Rutgers University Press, 2019.

Sbicca, Joshua. *Food Justice Now!: Deepening the Roots of Social* Struggle. Minneapolis: University of Minnesota Press, 2018.

Tapia, Maite, Tamara L Lee, and Mikhail Filipovitch. "Supra-Union and Intersectional Organizing: An Examination of Two Prominent Cases in the Low-Wage US Restaurant Industry." *Journal of Industrial Relations* 59, no. 4 (2017): 487–509.

White, Monica. *Freedom Farmers: Agricultural Resistance and the Black Freedom Movement*. Chapel Hill: The University of North Carolina Press, 2019.

Whyte, Kyle Powys. "Indigenous Food Sovereignty, Renewal, and US Settler Colonialism." In *Routledge Handbook of Food Ethics*, edited by Mary C. Rawlinson and Caleb Ward, 354–65. New York: Routledge, 2016.

Wolfe, Patrick. "Settler Colonialism and the Elimination of the Native." *Journal of Genocide Research* 8, no. 4 (2006): 387–409.

11 "We Are Missing Our Lessons to Teach You One"

Youth Activists on the Frontlines of Climate Justice

Jerusha Conner

September 2019 saw the world's largest ever mobilization for climate action when an estimated six million people around the world, including 650,000 demonstrators in the U.S., participated in climate strikes in the lead up to the U.N. Climate Action Summit in New York.[1] Seeking to capitalize on the momentum from these strikes, youth in the U.S. had grand plans for the 50th anniversary of Earth Day. In a jointly authored open letter from January 13, 2020, published in *MTV News*, leaders from ten different youth-led environmental and climate justice organizations proclaimed:

> Wednesday, April 22nd, the 50th anniversary of Earth Day, will be the launch of one of the most powerful civic actions for environmental protection in our history. It will provide an opportunity to listen to Indigenous peoples' wisdom, reflect on our connections to this earth, and serve as an invitation for everyone to make the decision to join us and commit to making climate change action a top priority. It will kick off three days of mass actions, including rallies, marches, strikes, teach-ins, and protests ... The three days of action are an open invitation to anyone who believes our generation, and all generations that come after us, deserve a future safe from climate catastrophe. The time to act is now. This must be the decade of climate action, and this must be the year it begins.

Then came the Covid-19 pandemic and its attendant shut-downs of schools and businesses.

Undaunted, the youth climate justice activists demonstrated their creativity and resilience as they turned to digital platforms, organizing a three-day live stream called Earth Day Live and calling for virtual strikes under the hashtags #resisttrumpocalypse, #divestfromdestructuion, #investinfrontlinesolutions, and #solidarityforsurvival. Explaining their decision to ask organizers to refrain from in-person mobilizations on Earth Day,

DOI: 10.4324/9781003214380-15

the US Youth Climate Strike Coalition cited their "responsibility, as the climate justice movement, to contain the spread and make the movement accessible for those who are most at risk, people who are immunocompromised and those with disabilities ... as they are disproportionately affected by the climate crisis." They continued, "In this critical election year, we vowed to take the youth-led climate movement to the next level. The Covid-19 epidemic presents a huge challenge, but we will rise to meet it. This is a defining moment that we, as young people, are uniquely prepared to tackle. It is our time to lead."[2] These statements illustrate several distinctive features of the youth climate justice movement: their concern with the intersection of identity and the climate crisis (e.g., Indigenous people; people with disabilities); their commitment to working in coalition; their digital media savvy; and their strategic leveraging of their own identity as youth.

Although in 2022, some observers wonder where all the youth climate activists have gone,[3] nine of ten of the organizations that issued the statements quoted above are still active, and new youth-led climate justice organizations continue to emerge. In-person protests and school strikes resumed in the fall of 2020, with youth-led climate protests occurring in 3,500 towns and cities in September 2020 alone.[4] In the U.S., youth climate activists organized a powerful electoral strategy in 2020 that helped win Joe Biden the presidency and both houses of Congress.[5] Since 2020, students from several colleges and universities, including Amherst College, the California State University system, the University of Montreal, and the University of Toronto, have won new commitments from their institutions to divest fully from fossil fuel companies and investment funds; and youth activists from more than 140 countries participated in the UN Climate Change Conference of Youth the week before the COP26 in Glasgow in 2021, including many delegates from Canada, Mexico, and the U.S., while 1,000,000 young people and their allies protested in the streets outside the conference.[6] Among the youth activists attending the 2021 and 2022 U.S.-based ACLU Summer Youth Advocacy Institute, 40% identified climate and environmental justice as a top issue area for their activism. Alongside youth, funders continue to show interest in the climate justice movement. In the spring of 2022, the Funders Collaborative on Youth Organizing announced a new fund to support 40 youth organizing groups working at the intersections of racial justice and climate justice.[7] These developments suggest that youth activism for climate justice, particularly in North America, has neither disappeared nor stagnated but continues to advance in key ways.

This chapter chronicles the rise of the youth-led climate justice movement and presents an overview of the current landscape of the movement in North America. I first trace its evolution from the modern environmental justice movement, focusing on the key moments, actors, and approaches that define it, as well as the factors that motivated young people to become climate justice activists. I then map the terrain of the movement, identifying four distinct types of youth-led climate justice organizations in North America. I conclude by discussing five features that I argue distinguish the contemporary youth-led climate justice movement at the present moment in time, some of which were introduced above: intersectional analysis; multi-solving; coalition work; digital organizing; and claims to youthfulness.

A Brief History of the Youth-Led Climate Justice Movement in North America

The history of the North American climate justice movement, which cannot be fully isolated from the global movement of which it is a part, is marked by the formation and growth of influential organizations, powerful direct actions, and various legal, institutional, and policy victories; however, it is also punctuated by moments of climate disaster, including most notably Hurricane Katrina in 2005. Both the immediate impacts of the storm, which fell disproportionately on low-income people of color, and the government's relief response – which again left the most marginalized with the least information, resources, and aid[8] – helped crystallize for many the issue of climate justice.[9] Other dramatic and devastating climate events, including Hurricane Sandy in 2012 and the 2018 Camp Fire, which destroyed the towns of Paradise and Concow, California, have served as catalysts for young people to join the climate justice movement, channeling the climate anxiety and grief they feel into productive action. Below, I elaborate on the constellation of factors that have motivated youth to become climate activists; however, I first sketch the emergence of the climate justice movement and then recount how youth staked out a leadership position within it, highlighting select examples.

The Rise of Climate Justice

Scholars David Schlosberg and Lisette Collins trace the birth of the climate justice movement to the Climate Justice Summit held during the COP6 meeting at The Hague in 2000, because it gave rise to the Environmental Justice and Climate Change Initiative a year later.[10] The Initiative, whose members included environmental justice and climate justice groups as well as other advocacy organizations, representing hundreds of communities, established ten principles of climate justice. In addition to calling for an end to fossil fuel exploration, a just transition to alternative energies, U.S. leadership, and international collaboration, the principles emphasized the importance of community participation, arguing "people must have a say in the decisions that affect their lives." The document also acknowledged that "low-income workers, people of color, and Indigenous peoples" will face the greatest impacts from climate change, and that future generations should be "taken into account" when formulating policy.[11] Two decades later, youth are demanding not just to be taken into consideration, but to have an active and vital voice in climate solutions. Nonetheless, the original 2002 principles helped establish an important foundation for the movement.

Scholsberg and Collins acknowledge that there was understanding of the disproportionate impacts of climate change on communities of color, particularly on African Americans in the U.S., prior to Hurricane Katrina, citing an influential report put out by the Black Congressional Caucus in 2004, entitled *African Americans and Climate Change: An Unequal Burden.* Nonetheless, they credit Hurricane Katrina with shifting how environmental justice scholars and advocates understood climate change. In the wake of Katrina, many in the environmental justice community "began to see climate change as another environmental condition that demonstrates the broader social injustice or poor and minority communities."[12] They point to the California Global Warming Solutions Act of 2006 as evidence of environmental justice activists' newfound embrace of climate change as a social justice issue. These activists successfully pressed lawmakers not only to avoid policies that would place undue burdens of risk and pollution on low-income or minoritized communities but also to include representatives from these communities on an advisory committee, thereby establishing a participatory mechanism consistent with the 2002 Principles of Environmental and Climate Justice. As more grassroots organizations took up the mantle of just climate policy, by 2014, Scholsberg and Collins argue, "climate change has become central to environmental justice organizing and discourse."[13]

The Role of Youth in the Environmental and Climate Justice Movements

Although most histories of the environmental justice movement do not include mention of young people and their role until much more recently,[14] youth have been engaged in environmental justice activism since the movement began, and youth environmental activism has a long history.[15] For example, sociologist Robert Bullard, who is considered the father of the environmental justice movement, locates the movement's origins in African American student protests over the drowning death of an eight-year-old girl in a garbage dump in Houston, Texas, in 1967.[16]

Among student activists in the late 1960s, environmentalism gained traction as an issue area. In 1969, students at the University of Michigan formed the Environmental Action for Survival (ENACT) organization, together with community leaders and faculty. In their charter, which can be seen in the Michigan in the World Environmentalism virtual exhibit (http://michiganintheworld.history.lsa.umich.edu/environmentalism/), they declared, "We are united as environmentalists, concerned will all facets of human survival on a fragile planet. Meeting this challenge requires immediate and direct action." In one of their first actions, the students of ENACT planned a four-day Teach-in on the Environment, repurposing a tactic pioneered a few years earlier on their campus to spread awareness and fuel opposition to the War in Vietnam. More than 13,000 people attended the teach-in's kickoff event on March 11, 1970. There, the keynote speaker, famed ecologist Robert Commoner, addressed disparities in exposure to pollution and toxicity faced by communities of color and explained that "the environmental movement need not come at the expense of other social movements, but rather as an extension of them." Another speaker at the kickoff rally, Ed Fabre, the student leader of Black Action Movement, drew direct connections between racism and environmental concerns in his speech, calling on environmentalists not to neglect race and other social justice issues in their advocacy.[17] Although not yet front and center, the seeds of environmental justice were clearly planted at this teach-in. A month after the teach-in on April 22, 1970, the first national teach-in on the environment, rebranded as Earth Day, occurred, attracting more than twenty million participants. Earth Day targeted both university and high school students, and since its advent, youth have always been engaged in this important day of action.

The early 1990s saw the creation of centers for the study of environmental justice at such institutions as University of Michigan, Xavier University, and Clark Atlanta University. Through the programing and coursework at these centers, students became not only formally trained in the concepts, theoretical frameworks, and research methods of environmental justice but also politicized. Some students credit the work of academic research centers like these with setting them on a path of environmental and climate justice activism and advocacy.[18]

During the 1990s, grassroots environmental justice community organizations also established links to campuses, working to educate and activate student allies. For example, in 1995, Chester Residents Concerned for Quality Living sponsored a retreat at Swarthmore College designed to teach students about environmental justice and the environmental injustices ravaging the city of Chester's predominantly Black and low-income residents. Chester (Pennsylvania) is home to one the largest trash incinerators in the U.S., and at the time, the Environmental Protection Agency conducted a study that found that 60% of children in the city had unacceptable levels of lead in their blood, and risks for cancer and respiratory conditions were also dangerously high.[19] As a result of the retreat at

Swarthmore, the Campus Coalition Concerning Chester (C4) was created. The C4 coalition eventually expanded to 15 campuses in four states. Students participated in C4 protests as well as community-based research to document and fight the environmental racism experienced by the city of Chester.

On the international stage, youth activists have been participating in U.N. climate negotiations since the 1992 Rio Earth Summit. At this gathering, a 12-year-old Canadian girl, Severn Suzuki, addressed world leaders, castigating them for continuing to "break" the planet and challenging them to take meaningful action.[20] But the arrival of the international youth climate movement did not come until 2005, when youth from many countries gathered in advance of COP11 under the banner, YOUNGO. It would be another three years before they were officially recognized as an observer constituency at COP by the UNFCCC Secretariat and member states, and it was not until 2011 that YOUNGO was formally confirmed to represent the voice of children and youth at the annual conference.

Meanwhile, 2011 marked two other key developments in the youth climate justice movement in the U.S. First, the lawsuit *Alec L. vs Gina McCarthy* was filed. Although it never advanced, it did lay the groundwork for legal arguments about how the government's failure to curb greenhouse gas emissions imperils children's rights to their future. Second, the organization 350.org, which environmentalist Bill McKibben founded with U.S. students in 2008, shifted its strategy and began to focus on blocking the development of Keystone XL Pipeline project, targeting the Obama administration. Scholars have referred to the Keystone XL campaign as the Selma or Stonewall for the climate movement and credit it and McKibben with fundamentally "transforming the U.S. wing of the climate movement."[21] Having taught a course on movement strategies in the spring of 2011, McKibben was inspired anew by the Civil Rights movement, and decided civil disobedience outside the White House was necessary to capture President Obama's attention and remind him that he needed the support of environmentalists to win reelection. In a significant win for the campaign to end the Keystone XL, 15,000 protestors, including many young people, showed up at the White House on November 6, 2011. Four days later, Obama announced he would be delaying his decision on the project. In March 2014, student members of 350.org organized a youth-led protest against the pipeline, which attracted more than 1,200 participants. They began with a rally at Georgetown, then marched to the home of John Kerry, where they staged a mock oil spill, and then marched to the White House, where they chained themselves to the fence. Nearly 400 youth protestors ended up arrested on that day.[22]

The organization 350.org also played a pivotal role in campus campaigns at this time. In the fall of 2012, McKibben and 350.org launched the "Do the Math tour," traveling on a biodiesel bus across the U.S. to do sold out shows in 21 different cities. The tour aimed to educate college students about how they could and why they should pressure their institutions to divest from the fossil fuel industry, explaining how current spending and burning trajectories added up to global catastrophe. McKibben argued:

> The logic of divestment couldn't be simpler: if it's wrong to wreck the climate, it's wrong to profit from that wreckage. The fossil fuel industry ... has five times as much carbon in its reserves as even the most conservative governments on earth say is safe to burn – but on the current course, it will be burned, tanking the planet. The hope is that divestment is one way to weaken those companies – financially, but even more politically. If institutions like colleges and churches turn them into pariahs, their two-decade old chokehold on politics in DC and other capitals will start to slip.[23]

In a blog from the tour, 350.org staff wrote, "Wow! We launched this new fossil fuel divestment campaign this November 7, and in less than a month, campaigns have sprung up on over 100 colleges and universities across the country. From big schools like the University of Michigan to small liberal arts colleges like Amherst, the idea of divestment is spreading like wildfire."[24] In 2012, Unity College in Maine became the first higher education institution in the U.S. to announce that it would divest its endowment from the fossil fuel industry. By April 2014, 11 other colleges had followed suit, and then in May, Stanford University announced its intention to "divest its holdings from 100 publicly traded coal companies."[25] Activist and writer Naomi Klein deemed this "the most significant victory in the youth climate movement to date."[26] By 2015, over 25 colleges and universities had announced commitments to divest, and fossil fuel divestment campaigns continued on approximately 400 campuses in the U.S. Scholars credit the "Do the Math tour" "with the rapid and widespread diffusion of the FFD [fossil fuel divestment] movement."[27]

Together with the advocacy organization Avaaz, 350.org also organized the September 2014 People's Climate March, marking yet another significant moment in the climate justice movement. An international day of action, with 2,646 events in more than 150 countries, the demonstration attracted an estimated 500,000 marchers in New York City, all demanding climate action in the lead up to the United Nations climate negotiations.[28] At the time, it was heralded as "the largest climate protest in history and largest social demonstration of the past decade."[29] Many children marched with their parents and students drove from as far away as Tulane University in New Orleans to participate.[30] Signs and costumes

Figure 11.1 Youth protestors at the 2014 People's Climate March in New York City. Photo by: Shadia Fayne Wood, Survival Media Agency.

at the march drew attention to the intersection of climate and social justice issues, as well as to the grave urgency of climate action. (See Figure 11.1.)

In an important win for the movement, in 2016 U.S. District Court Judge Anna Aiken issued a ruling in the case *Juliana vs. the U.S. government*, asserting that youth have a legitimate right to "a climate system capable of sustaining life."[31] The plaintiffs included 21 youth, between the ages of 12 and 23, including one Canadian youth. Their complaint, first filed in 2015, alleged that the government's failure to "preserve a habitable climate system for present and future generations" has violated the youngest generation's "fundamental constitutional rights to freedom from deprivation of life, liberty, and property; ... and Plaintiffs' rights as beneficiaries of the federal public trust."[32] Although the case has gone through several rounds of appeals as it has wound its way through the courts, and the youth plaintiffs are still awaiting rulings on motions at the time of this writing, the case has generated powerful legal arguments and rhetoric. For example, in a dissenting opinion in *Juliana* in January 2020, Judge Josephine Stanton wrote: "Plaintiffs' claims are based on science, specifically, an impending point of no return. If plaintiffs' fears, backed by the government's *own studies*, prove true, history will not judge us kindly. When the seas envelop our coastal cities, fires and droughts haunt our interiors, and storms ravage everything between, those remaining will ask: Why did so many do so little."[33]

The year 2016 also saw the rise of resistance to the Dakota Access Pipeline, when a group of young people from the Standing Rock Sioux Reservation established an encampment, where they led peaceful protests, religious ceremonies, and prayers, while blocking construction of the pipeline. Their resistance, chronicled under the hashtag #NoDAPL, inspired thousands of activists from around the world to join them in what became increasingly tense stand-offs over a period of several months, resulting in injury and mass arrests.[34] It was at Standing Rock, during this time, that the International Indigenous Youth Council was created.[35]

The Sunrise Movement, a youth-led climate justice organization that set its sights on building electoral influence, was formed in 2017. It played a key role in the 2018 U.S. midterms, helping to get out the vote for progressive candidates who committed to just climate policy action, including Alexandria Ocasio-Cortez, Ilhan Omar, Ayanna Pressley, and Rashida Tlaib: the freshmen Congresswomen would rise to prominence within the Democratic Party.[36] In November 2018 Sunrise activists staged an occupation of Nancy Pelosi's office, where they were joined briefly by Ocasio-Cortez, to demand action on the Green New Deal. This sit-in, which was widely covered by news media and went viral on social media, catapulted the Green New Deal into public discourse in the U.S. Through its creative organizing trainings, confrontational direct actions, and multi-faceted electoral strategy, Sunrise has emerged as a key player in the youth-led climate justice movement in the U.S. In interviews, Sunrise leaders told me and my colleagues that in their youth mobilizing strategy they were inspired by the efforts of a group of young people who had recently mounted a powerful challenge to the National Rifle Association and its grip on many politicians in the U.S.: the March for Our Lives movement.[37]

On Valentine's Day in 2018, a gunman entered Marjory Stoneman Douglas High School in Parkland, Florida and murdered 14 students and 3 staff members. Within hours after this horrific incident, students began to take charge of the narrative, rejecting politicians' thoughts and prayers and demanding gun reform and an end to the NRA. Their resolve inspired their peers around the country to organize nationwide walkouts to protest gun violence and call for school safety. The student-led walkouts on March 14, one month after the mass shooting, March 24, and April 20, on the anniversary of the Columbine High School massacre, mobilized thousands of young people around the U.S., marking one of the first times nationwide demonstrations were organized by high school students.

The walkouts caught the attention of a 15-year-old Swedish teenager, who had been growing increasingly despondent over the existential threat posed by the climate crisis. Greta Thunberg decided to adopt the tactic, launching her School Strike for Climate outside the Swedish parliament in August 2018. She committed to miss school every Friday, defying the rules and disrupting the social order, just as the young gun violence prevention activists had done in the U.S. Although school strikes for climate had been previously orchestrated by youth in England in 2010 and by youth on the first day of the Paris U.N. Climate Change Conference in 2015, the image of the Swedish teenager sitting by herself struck a chord, soon garnering widespread attention. It was not long before Greta was joined in her strike by other young people, and the movement began to spread, first to Belgium and then to Australia. In September 2018, #FridaysforFuture was launched to coordinate the growing movement. Greta's fiery indictments of world leaders at COP24 in 2018[38] and the World Economic Forum in 2019,[39] which went viral on social media, served as a call to action to the youngest generation to continue to speak truth to power. On March 15, 2019, in the Global Climate Strike for the future, nearly 1 million people demonstrated worldwide in about 2,200 events, across 125 countries. Another global student-led strike followed in May. In September, a global week of climate action was planned from the 20th to the 27th to coincide with the U.N. Climate Action Summit. Greta traveled by a wind-powered sailboat to attend the event. Protests were planned in 4,500 locations across 150 countries, and it is estimated that 4 million people participated in the September 20th strike worldwide, with another 2 million participating in the September 27th strikes. The surge of youth-led climate activism that swept the world in 2019 has been deemed a "watershed" moment in climate action[40] and Greta has been hailed as its figurehead.

Motivating Factors for Youth Involvement in Climate Justice Activism

For contemporary youth climate activists, there are several factors that propel them to activism. Results from a survey of nearly 300 U.S. youth engaged in environmental and climate justice activism revealed that the most common factor influencing them to become involved was "other activists," followed closely by something they had read, either online or in hand. Indeed, after their personal values and first-hand experiences of oppression, respondents rated other activists and books, articles, social media posts, or other texts as *the most influential* factors for their activism, well above the influence of family, friends, school, religion, and online influencers.[41]

The significance of "other activists" and personal experiences as key sources of motivation also emerged in an interview study colleagues and I conducted with 42 U.S.-based youth climate justice activists.[42] (All of the direct quotations below come from participants in this study.) In participants' accounts of why and how they joined the movement, four themes emerged: the Greta effect, the March for Our Lives effect, first-hand experiences of climate disaster, and a sense of responsibility from second-hand exposure.

The "Greta effect" surfaced as a powerful catalyst for some youth, especially among students in middle school and high school. One described Greta as "one of the real sparkers" of her motivation. Another recalled that her activism began "when Greta Thunberg started her school strike. It was like maybe half a year into that, when I was like 'Wow! She's my age, and she's fighting for an actually really good cause, and she's doing everything she can in her power." Greta's courage and commitment served to move many young people around the world to join her in demanding action from the world's leaders to secure their futures.

In addition to a Greta effect, we also identified a "March for Our Lives effect." Some of the young people who first dipped their toes into activist waters by participating in March for Our Lives walkouts and marches began to wade deeper with the rising climate justice movement. One young woman who was the lead organizer for her city's March for Our Lives chapter and a student ambassador with Ceasefire 365 remembered feeling "not as passionate about gun reform anymore and much more passionate about climate, and so I got involved in the climate movement." A Youth Climate Action Team (YCAT) leader observed, "A lot of climate organizers started at the March for Our Lives high point." They and another YCAT leader who got their start in activism by organizing their state or city's March for Our Lives sister marches eventually switched to climate activism because it felt more personal to them. One respondent deemed this transition the "March for Our Lives – climate pipeline." The March for Our Lives effect is also reflected in the shift made by Future Coalition, a youth-led organization that started as a coalition of youth-led gun violence prevention groups before becoming focused on youth-led environmental and climate justice organizations.

Another powerful motivator for several youth in our study was their first-hand experience of climate disaster. One young Sunrise organizer recounted his experience with a flash flood and having to jump into a canoe to be rescued from his apartment:

> It was really visible for me that it was not just one experience, but something that my generation, we're having to go through in any form: it could be air pollution, or flooding, or it can be drought, or any type of natural disaster. It would really knock on our door, so it made me feel like I really wanted to … It's almost all a matter of conscience, because there's something that I can see visibly that is coming, and I do have some capacity to do something with it. And the next step is just do it. So, that was how I was motivated.

Several participants explained how being personally impacted by climate disaster generated a sense of responsibility that goaded them to action. Another Sunrise participant recalled:

> I got involved when there were really serious wildfires in California, and the entire state was covered in smoke. You really couldn't escape it. There was nowhere to go, which was just a very intense experience for a lot of us, realizing that the climate crisis was no longer an abstract thing that was going to happen in the future, but it was something that was happening now, that was impacting us on a very emotional, intangible level. And it was very hard to get away from the feeling that like the whole world was going to burn up, and that there wasn't really ... like you couldn't move away and ignore it.

Shortly after this trauma, she saw the Sunrise sit-in in Nancy Pelosi's office and "knew immediately that this was the movement that I had been waiting for." Becoming active in the climate justice movement helped youth redirect the fear and pain they felt into productive, hopeful action.

Personal experiences of climate disaster were not the only source of galvanizing emotions for youth; indeed, witnessing climate destruction and its impacts on marginalized communities on the television or on social media channels served to activate several respondents. Some youth described intense feelings of "climate anxiety," "existential dread," and climate despair. Others spoke of the anger and frustration they felt witnessing injustice and knowing adults were doing little to redress it. Again, these emotions engendered in many youth a sense of "personal obligation" and responsibility to take action. One Sunrise respondent shared, "I just had a bunch of climate anxiety and was looking for a place to channel that. So, about a year and a half ago, I joined Sunrise in a more intentional way." Another respondent who suffered from debilitating climate anxiety discovered that attending meetings and being in organizing spaces "was the only thing that made me feel hopeful and energized by seeing all these people concerned with the same thing." Youth-led climate justice organizations held an important space for youth in our study not only to process their emotions related to climate change but also to experience more joyful and hopeful emotions, including a sense of community and connection and a sense of empowerment, as they learned how to take collective action to effect change.

Youth-led climate justice social movement organizations (SMOs) also served as a space in which youth could learn to think more critically about the issue of climate change and develop an intersectional analysis. Although some youth first joined the climate movement because of their love of nature or concern for animals like the polar bear, within youth-led climate justice SMOs, they quickly became politicized. As colleagues and I document,[43] through trainings, conversations with fellow activists, and organizational artifacts, such as platforms, vision statements, or guiding principles documents, youth learned about the "human toll" of the climate crisis. They learned to recognize the disproportionate impact of the crisis on marginalized populations, and they learned to forge critical solidarities with other social justice causes. They came to recognize the intersections of climate and capitalism and colonialism, and to see climate justice as inextricably linked to racial justice, Indigenous people's rights, immigrant rights, women's rights, and economic justice, among other issues. One YCAT member, for example, explained her motivation for staying involved in the movement: "The nature and the environment side [is] obviously really important, but for me, it was more like the humanity side and how my community is being impacted and how some people are being impacted more than others, and just being fed up with that injustice." Asked how she learned about these unjust impacts, she shared: "I was never really taught it, but especially as I became more involved, I became more aware of those injustices, and like a bunch of like intersectional things ... Understanding how certain communities, especially communities of color and low- income communities are affected more so, [came about] though a culmination of knowledge and learning as I went through my [activist] journey." The rhetoric and analysis that define the contemporary youth-led climate justice movement,[44] and activist spaces more broadly, proved influential in shaping many young climate activists' world view and critical consciousness.

Key Approaches of Youth-led Climate Justice Activism

When analyzing the approaches youth climate justice activists use, it is helpful to consider both their tactical and rhetorical repertoires, as both hold strategic power.

Tactical approaches. As evidenced in the abbreviated history above, youth climate justice activists draw on a wide range of tactics, including coalition building; electoral organizing; educating their peers and the public through webinars, forums, and other community-based events; testifying before Congress; penning op-eds and petitions; and pursuing legal recourse. They are perhaps best known, however, for their nonviolent direct actions. These direct actions include time-honored activist tactics, like marches and sit-ins, as well as more recent innovations, such as supergluing their hands to significant buildings or monuments,[45] die-ins, birddogging politicians, and going "on strike" from school. Importantly, when discussing their approach to direct actions, contemporary youth climate justice activists often explain how they seek to borrow and build on the tactics used in the past by prior generations of social justice activists. For example, the founders of Sunrise collectively studied the Civil Rights movement, the antiwar protests led by students in the Vietnam era, the LGBTQ movement, and the example of Mahatma Gandhi in India when they began building Sunrise.[46] Today's climate justice activists are students of the past, acutely aware of their role in helping to bend the arc of the moral universe toward justice, extending the legacy of their forerunners.[47]

In repurposing traditional activist tactics, today's youth climate justice activists adapt them to the current moment, using the tools of their day. The pairing of social media and direct action has proven particularly effective. For example, Sunrise engineered what it calls a "moment of the whirlwind" when a video of youth climate activists[48] confronting California State Senator Dianne Feinstein in her office over her refusal to support the Green New Deal went viral, after they posted it on Facebook Live.[49] In a moment of the whirlwind, an organization attracts new interest, and as one Sunrise leader explained to us in the Youth Movements and Voting Study, there is "so much momentum, so much energy, so much possibility" that the organization has to be careful to capitalize on it, lest the moment pass.[50]

In addition to staging direct actions, youth climate activists have embraced movement art, creating memes, videos, and other visual art installations. They refer to collective in-person art creation as "art builds," and they use art to amplify their campaign messages and inspire people both online and in person, at marches, protests, and art festivals. Highlighting the centrality of art in the movement, as part of its training for activists, Zero Hour features a one-hour module on "artful activism." The get-out-the-vote postcards that Sunrise members sent to 1 million first-time voters in 2020 featured 1 of 11 paintings created by members, depicting the power of voting. Youth vs. the Apocalypse (YVA), a group of predominantly low-income youth of color based in Oakland, California, uses Hip Hop as an organizing tool. YVA members have produced original songs and music videos, such as "No One is Disposable" and "Where's the Money At?" According to its website, the Toronto-based Youth Climate Collective (YCC) engages youth between the ages of 12 and 16 in "exploring the intersection of art and activism and the role of creativity in tackling the climate crisis, as we plan and activate green initiatives in our city."[51] Art and creativity have assumed a prominent role in the youth-led climate justice movement.

Rhetorical Approaches. With regard to their rhetorical approaches, youth climate justice activists again look to the past, this time borrowing strategies from the youth movement that immediately preceded theirs: the March for Our Lives. I argue that there are three primary rhetorical strategies deployed by young people in both movements.

First, just as March for Our Lives activists sought to trigger feelings of guilt and shame in adults by calling out adults' failure to protect them from harm and roundly critiquing their abdication of leadership, so too do young climate activists. This rhetorical strategy is effective not just because it humiliates adults for their dereliction of duty, but also because in coming "from the mouths' of babes," it draws attention to how the social order has been inverted. No longer are youth depending on adults to teach them, to look out for their best interests, and to exemplify leadership; now youth are stepping into the leadership vacuum to explain the science and spotlight solutions. Highlighting the painful irony of the situation, a popular sign at youth climate strikes, for example, reads: "We are missing our lessons to teach you one." At the same time that they are ready to lead, youth are quick to remind adults that they shouldn't have to. The disorienting and disquieting power of this disruption in the social order is perhaps best exemplified by Greta Thunberg's iconic lines:

> This is all wrong. I shouldn't be up here. I should be back in school on the other side of the ocean. Yet you all come to us young people for hope. How dare you! You have stolen my dreams and my childhood with your empty words.[52]

This same rhetorical tactic of shaming adults for their environmental and climate inaction was previously deployed by 12-year-old Severn Suzuki in her 1992 speech at the U.N. Earth Summit, when she repeatedly used the phrase, "I'm only a child, but I know...," to remind adult leaders of the scientific and moral knowledge they were recklessly ignoring. As she concluded her speech, she reflected:

> Parents should be able to comfort their children by saying, "Everything's going to be all right; it's not the end of the world, and we're – and we're doing the best we can." But I don't think you can say that to us anymore. Are we even on your list of priorities?[53]

Besides activating adult guilt and shame, youth activists in both the gun violence prevention and the climate justice movements seek to appeal to questions of morality. It is widely acknowledged in Western societies that youth are less jaded and cynical than their adult counterparts. Where adults, beholden to the status quo, see complexity that requires patience to navigate, youth see simplicity and urgency. While this difference of perspective is sometimes used to discredit youth as naïve, it is also at the root of the moral authority youth possess. They have the ability to cut through the noise and leverage a call for fairness and justice that is undeniable. This rhetorical strategy was on display when a co-founder and art director of

Zero Hour, Nadia Nazar, testified before Congress in 2019. She implored Congressional leaders to take action, saying: "The lives of my generation have been disregarded for far too long. You should put the interests of future generations first, not just because it is the right thing to do, but because many of us have the right to vote in just a couple of years."[54]

In addition to highlighting the moral imperative to act, Nazar's line above exemplifies the third rhetorical tactic deployed by youth in both the gun violence prevention and climate justice movement: attempting to stir fear in adult politicians by referencing their generation's coming power at the ballot box and beyond. At the March for Our Lives 2018 and 2022 rallies in DC, "Vote them out" was a resounding chant. In the leadup to the 2022 midterm elections, Sunrise has been relentless in its attacks against democratic politicians like Joe Manchin, Henry Cuellar, and Danny Davis, who profit and accept donations from the fossil fuel industry. Youth leaders also prefigure a time when they are in charge of decision-making, and older generations are at their mercy. As Greta Thunberg inveighed at the U.N. General Assembly in 2019, "*You're* failing us, but the young people are starting to understand your betrayal. The eyes of all future of all future generations are upon you, and if you choose to fail us, I say we will never forgive you."[55] This threat, delivered with searing emotional reverberation, has since been echoed on protest signs and has become a popular meme on social media. While youth certainly use other rhetorical strategies, such as humor and messages of love and unity, the three tactics discussed above hold particular potency because they can only be deployed by youth.

The Landscape of Youth-Led Climate Justice Activism

Type of Organizations

Sarah Pickard has argued that there are two main types of youth environmental movement organizations: formal and non-formal.[56] The former features a hierarchical leadership structure, membership fees, and limited participation from members and supporters, while the latter includes pressure groups, networks, and community-based organizations that tend to be leaderless or horizontally structured. While the distinction between formal and informal is useful, there is another dimension that further differentiates organizations within the movement: the scale of the problem they are targeting. Within the climate justice movement, there has long been a tension between place-based and global organizing, because local protests about local issues can be seen as disconnected from each other and from the global movement.[57]

The choices of where to focus one's organizing efforts are reflected among the four broad types of youth-led organizations that make up the movement: transnational, national, local, and campus-based. Each takes aim at a different set of challenges related to climate change. Transnational youth-led social movement organizations (YSMOs) focus on global climate change and seek to build the capacity of youth to pressure governments, multinational corporations, and umbrella organizations like the United Nations to establish agreements, policies, and practices that will slow climate change and mitigate its effects. Meanwhile, national YSMOs concentrate on national priorities and policies to reduce greenhouse gas emissions and reduce the consequences of climate change. While acknowledging the broader scope of the problem of climate change, local groups concern themselves with policies and projects at the state, city, or town level. Finally, campus-based groups pursue institutional change within their university, college, or higher education system designed to reduce reliance on fossil fuels and financial contributions to the industry, advance research and innovation in alternative energies and climate change mitigation efforts, and model more sustainable practices. Below, I offer snapshots of each type of YSMO, drawing on examples from the U.S. and Canada.

Examples of powerful transnational climate justice YSMOs include Polluters Out, a coalition of Indigenous and People of Color youth-led climate justice organizations around the world, and Fridays for Future, whose youth leaders in the U.S., Canada, and over 100 other countries coordinate the school strikes for climate inspired by Greta Thunberg. In the U.S., Zero Hour was founded by four teenagers in 2017 to provide training and resources for youth wanting to take action to address climate change. Since its founding, Zero Hour has established more than 50 chapters across 15 different countries. In addition to organizing direct actions, such as the Youth Climate March on July 21, 2018, Zero Hour has held a youth climate art festival, a youth climate lobbying day, and several webinars and rallies. Its members have testified before Congress.[58] As an organization, it has endorsed numerous legislative proposals, signed more than 80 letters to politicians, published a zine on pipelines, and produced an online seven-hour training series for aspiring climate activists around the world. Founded by Ana F. Gonzalez Guerrero and Dominique Souris in 2017 shortly after they had graduated from the University of Waterloo in Ontario, Canada, the Youth Climate Lab similarly offers youth-led capacity building programs, such as design labs, for youth around the world working to create policy solutions for climate justice. As of 2020, their projects had engaged nearly 1,400 youth climate leaders in more than 77 countries, resulting in 47 distinct youth-led climate solutions.[59]

At the national level, two prominent North American climate justice YSMOs are Climate Strike Canada (CSC) and the Sunrise Movement. CSC serves as an organizing hub, coordinating 17 local youth-led climate justice groups, sharing resources and tools and hosting events, such as online discussions and direct actions. On its website, the organization states, "Above everything else, we exist to give a voice to young people and to urge our government to end its legacy of colonization and exploitation and instead act to restore Indigenous rights and counter the climate crisis before the damage to our Earth becomes irreparable."[60] CSC sent a delegation to COP26 in Glasgow and issued a statement following the conference, decrying the gathering's lack of accessibility and calling on the Canadian government to meet its five demands, which began with justice for Indigenous peoples and ended with "achieve a 60% reduction of 2005 levels by 2030, and full decarbonization by 2050."[61] The Sunrise Movement, which was referenced above, was founded in 2017 by 12 young people who had all been involved in their various campus divestment campaigns as part of 350.org. Sunrise has embraced a very specific theory of change that focuses on building people power to create the political will to effect policy change. Shortly before their headline grabbing sit-it in Speaker of the House, Nancy Pelosi's office in November 2018, where they demanded climate action, they coalesced around the Green New Deal framework. In the 2020 and 2022 election cycles, they built a formidable electoral strategy to elect Green New Deal champions; during the 2020 campaign, they earned a seat at the table, shaping presidential candidate Joe Biden's climate policy, and in 2021 during negotiations for Biden's Build Back Better Act, which included ambitious climate provisions, they staged a 14-day hunger strike.

Falling under the umbrella of local climate justice YSMOs are statewide organizations, like the Palmetto Youth Movement, based in South Carolina, and YCAT, based in Wisconsin,[62] as well as city or community-based groups, such as YVA, based in Oakland, CA, and Climate Justice Montreal. Of course, many national organizations will have chapters or hubs that operate at the local level. Although they may enjoy a high degree of autonomy to focus on local issues, because they are still connected to a national organization with a particular theory of change, we would not consider these chapters to be examples of local YSMOs. Sustainabiliteens in Vancouver offers a good case of a local YSMO centered on climate justice. Founded in 2018 by a small group of high school students, Sustainabiliteens brings together high school-aged students from across Metro Vancouver to fight practices that contribute to climate change and press for local climate solutions. In addition to organizing the December 2018 school climate strikes and the 2019 Vancouver climate strike, which attracted thousands of participants, they have staged occupations of a British Columbia cabinet minister's office, as well as die-in's and funerals for their future. Asked about how they came together to found Sustainabiliteens, one of the founders told a journalist, "This isn't a movement that we started ... We're really following in the footsteps of Indigenous land defence, which has been going on for 500 years."[63] As discussed below, respect for both Indigenous peoples and the history of environmental and social justice activism are core features of many contemporary climate justice YSMOs.

The final type of youth-led climate justice organization is the campus-based student organization. These clubs, groups, associations, and organizations concentrate on educating peers, pressuring administrators and board members to adopt sustainable policies, and shifting institutional practices, such as ending on-campus recruiting by corporations associated with the fossil fuel industry. Students may also stage direct actions, including disrupting speakers who champion, support, or participate in extractive practices.[64] In some cases, these clubs may engage in off-campus activist campaigns to address policies or develop initiatives, such as a community garden, in the city or town in which their institution is located. Although some of these student groups may be associated with national networks, like 350.org, the Divestment Network, or Sunrise, others are home grown, unique to their specific campus. For instance, Rice University in Houston, Texas, offers two different student organizations that promote taking action to address climate change: Rice Climate Action and the Rice Initiative for Sustainable and Ethical Today (RISE Today). Occasionally, student-led climate justice organizations are associated with Environmental Justice Centers or Sustainability programs on campus.

While there may be overlap among the four types of organizations discussed above, differentiating them helps elucidate the varied terrain of the youth-led climate justice movement. Nonetheless, everyday youth activists rarely belong to only one social movement organization. In one study of youth climate activists from Sunrise, YCAT, and PYM, colleagues and I found that 65% reported belonging to two or more climate justice organizations.[65] Additionally, survey results from the 2021 ACLU youth advocacy institute showed that of the nearly 300 youth climate activists responding, 63% belonged to two or more movement organizations. Furthermore, whether through coalitions and collaborations or their own independent involvement, youth climate justice activists are commonly engaged in other social justice causes.[66] In the ACLU data, only 2% of climate and environmental justice respondents reported participating only in this movement. In fact, the average number of issues areas in which these climate justice activists reported active engagement was 8 (range 1–20; sd. 4.34).[67] This multi-issue engagement reflects a core tenet of environmental and climate justice, as the movement has long been associated with a wide range of struggles, including racial justice, women's rights, economic justice, and immigrant rights,[68] among others.

Features of the Movement

Despite diversity in the organizational structures, theories of change, and approaches adopted by contemporary climate justice YSMOs, there are cross-cutting threads that unify the broad youth climate justice movement. These include their commitment to intersectional analysis; their embrace of multi-solving, coalition building, and digital organizing; and their insistence on the importance of youth leadership.

Intersectionality

The language of intersectionality has become part of the everyday parlance of youth organizers in the climate justice movement. Guided by Crenshaw's theorizing,[69] the activists use the term to refer to different dimensions of identity that are linked to privilege or oppression. It is an intersectional lens that helps them to see how marginalized people (low-income, people of color, women and children, immigrants, Indigenous people, LGBTQ people, people with disabilities) who have contributed the least to climate change are most affected by its impacts. Youth climate activism reflects a shared commitment to centering the voices and experiences of those most adversely affected by climate disasters.

The young activists also use the term intersectional to characterize the ways in which climate change intersects with other social problems and climate justice intersects with other social justice causes. Discussing intersectionality and its role in the climate movement, a youth activist with Zero Hour, for example, told a journalist: "I find that there is not a single issue on this Earth that does not interconnect in one way or another ... The very same systems that are leading us to the brink of extinction are the same systems that create injustice everywhere."[70] Similarly, Diego Arreola, a 19-year-old climate justice activist from Mexico City explains, "Intersectional environmentalism is essential because the world is interconnected and no problem is isolated ... If we want to address poverty, racial justice, climate change, and economic inequality, we need to treat them as whole and find holistic solutions that address all the gaps in these realms."[71] This kind of analysis of the problem sets the stage for what has been called "a multi-solving" approach, highly favored by youth climate justice activists.

Multi-Solving

Multi-solving is the idea that multiple, interrelated social problems can be resolved through one comprehensive solution. For example, Sunrise supports the Green New Deal framework because it argues that millions of good jobs will need to be created in order to address the climate crisis and ensure a just transition to alternative energies. Sunrise credits itself with pushing President Joe Biden toward this analysis. Echoing a page from the Sunrise playbook, Biden tweeted, "When I hear *'climate,'* I think *jobs.* Good-paying, high-quality *jobs* that will help speed our transition to a green economy of the future and unleash sustainable growth."[72] A more localized example of multi-solving would be organizing a community-controlled food initiative to help address food insecurity, while also serving to enhance the social capital of local farmers, who may otherwise have limited reach.

Youth organizing, whether for climate justice or other issues, is also emblematic of multi-solving. It has long been hailed as a triple bottom line approach because it benefits the youth involved, their community, and the issue they set out to tackle. Indeed, youth organizing not only brings fresh solutions to bear on specific problems that youth have identified as pressing, but it also builds the civic capacity of a community, engaging the next generation in transformative leadership.

Coalitions and Critical Solidarities

In addition to multi-solving, the intersectional perspective informs youth climate justice activists' approaches to coalition building, both within and beyond the movement. As colleagues and I found in our study of Sunrise, it was imperative to the youth that their local hubs collaborate with community-based organizations, especially those led by Black, Brown, and Indigenous peoples, that had been working to address climate justice issues for years. They understood that they would be more effective if they sought to partner rather than compete with these organizations.[73]

Additionally, reflecting the fact that their members were typically engaged in a range of social justice movements, climate justice YSMOs embrace opportunities to support other movements. For example, as of July 2022, on its website, the first point of YVA's Points of Unity reads as follows:

> Climate change comes from the same system that approaches life saying that people of color, LGBTQ+ people, and those who are poor have no value in society except to make someone else rich. This system causes both climate chaos and those who aren't white and rich to struggle. When fighting climate change, we must find solutions to rebuild the entire system in order to get rid of the climate crisis. Based on this understanding, YVA stands in solidarity with movements for a sustainable world including but not limited to Black Lives Matter, Abolish ICE, LandBack and LGBTQ± Rights.[74]

Although coalition work is a hallmark of the current youth climate justice movement, young climate activists have been working in coalition for years. For example, the Canadian Youth Climate Coalition was founded in 2006, uniting 48 youth-led organizations across Canada.

Social Media Savvy

As digital natives, youth activists are sophisticated digital content creators, curators, and consumers. Within the youth climate justice movement, youth use social media to recruit, to mobilize their peers and resources for actions, to educate, to plan events, to pressure lawmakers and institutional agents, and to hold them accountable. A growing body of work examines the role and creative deployment of social media in youth climate activism.[75]

Leveraging Youth Identity

Finally, youth climate action is also distinguished by the salience of youth as identity category. Although Thew and colleagues have found that some of the European youth participating in U.N. climate negotiations in 2019 have moved away from emphasizing their own vulnerability as young people poised to inherit a dying planet in favor of highlighting the urgent needs of other marginalized populations, the scholars argue this is a mistake. "Whilst laudable, this impedes their mandate as representatives of younger generations."[76] As noted above, youth have capitalized on their identity as youth in their public speeches and direct actions, strategically leveraging their youth to strong rhetorical and political effect.

Just as the identity of youth matters in public-facing work, so too does it hold purchase in an organization's internal work to develop and carry out campaigns. Although intergenerational coalitions are understood by youth climate activists as critical to achieving their long-term goals, young people value having their own spaces to come together with peers to imagine and advance solutions. The sense of community and comradeship youth find in these organizations is a core reason they remain engaged in the struggle, even in the face of deep burnout.[77] I close with two quotations drawn from my research with young climate justice activists that underscore the importance of holding this youthful space for organizing:

> Many of us have had the experience of showing up at like a 350 or Sierra Club and finding that like the age skews about 50 years older than we are. And that's fine, but having a specifically youth space is really powerful. It draws in other young people and allow us to organize like our own people, basically.

> [What makes Sunrise different] is its emphasis on younger people pushing for change … It's different than a conservation organization, like the World Wildlife Fund, or some big, long-standing green,

like Greenpeace. Because they're so well known, they have to work within the political sphere, and they have that sort of pressure. I feel like Sunrise can really go against the status quo and take risks because they don't have that added pressure. They don't have to abide by any rules or anything like that.

Whether they are challenging the rules that have too long bound society in extractive relationships or formulating a radical alternative vision of a just, sustainable future that is unique to their generation, youth activists continue a long tradition of exercising significant leadership in the push for progress, both within and beyond the global climate justice movement.

Notes

1 Matthew Taylor, Jonathan Watts, and John Bartlett, "Climate Crisis: 6 Million People Join Latest Wave of Global Protests," *The Guardian*, September 27, 2019, https://www.theguardian.com/environment/2019/sep/27/climate-crisis-6-million-people-join-latest-wave-of-worldwide-protests

2 Future Coalition (@FutureCoalition), "In Light of the COVID-19 (Coronavirus) Pandemic, the US Youth #ClimateStrike Coalition is Pivoting Our #EarthDay Strikes Plans...," *Twitter*, March 13, 2020, https://twitter.com/futurecoalition/status/1238508271033614338?lang=cs

3 Eleanor Salter, "Where Have All the Young Climate Activists Gone?," *The Guardian*, June 9, 2020, https://www.theguardian.com/commentisfree/2022/jun/29/children-arent-the-future-where-have-all-the-young-climate-activists-gone?CMP=share_btn_tw

4 Elizabeth Cripps, *What Climate Justice Means and Why We Should Care* (London: Bloomsbury, 2022).

5 Hannah Miao, "Young Voters Helped Propel Biden to Victory," *CNBC*, November 13, 2020, https://www.cnbc.com/2020/11/13/election-young-voters-biden-democratic-party.html

6 William Brangham, "Why These Young People Came to the COP26 Climate Change Conference," *PBS*, November 11, 2021, https://www.pbs.org/newshour/world/why-these-young-people-came-to-the-cop26-climate-change-conference

7 Funders Collaborative on Youth Organizing (FCYO), "Apply to the Youth Organizing for Climate Action and Racial Equity (YO-CARE) Capacity-Building Fund!," accessed May 30, 2022, https://fcyo.org/uploads/resources/apply to-the-youth-organizing-for-climate-action-and-racial-equity-yo-care-capacity-building-fund-1_resource_625b9bc273c517342761cf6e.pdf

8 Robert Bullard and Beverly Wright, *Race, Place, and Environmental Justice after Hurricane Katrina* (Boulder, CO: Westview Press, 2009).

9 David Scholsberg and Lisette Collins, "From Environmental to Climate Justice: Climate Change and the Discourse of Environmental Justice," *Wires Climate Change* (2014), doi:10.1002/wcc.275

10 Ibid.

11 Environmental Justice and Climate Change Initiative, "10 Principles for Just Climate Change Policies in the U.S.," 2002, https://www.ejnet.org/ej/climate-justice.pdf

12 Shlosberg and Collins, "From Environmental to Climate Justice," 4.

13 Ibid, 5.
14 E.g., Edwardo Rhodes, *Environmental Justice in America* (Bloomington: Indiana University Press 2003); Dorcetta Taylor, "Introduction: The Evolution of Environmental Justice Activism, Research and Scholarship," *Environmental Practice* 13, no. 4 (December 2011): 280–301, doi:10.10170S1466046611000329
15 Benjamin Bowman and Sarah Pickard, "Peace, Protests, and Precarity: Making Conceptual Sense of Young People's Non-violent Dissent in a Period of Intersecting Crises," *Journal of Applied Youth Studies* 4 (2021): 493–510.
16 Luke Cole and Sheila Foster, *From the Ground Up* (New York: New York University Press, 2001).
17 Michigan in the World and the Environmental Justice HistoryLab, "Give Earth a Chance: Environmental Activism in Michigan," http://michiganintheworld. history.lsa.umich.edu/environmentalism/
18 Genevieve Pearthree, "Bridging the Divide: Activism and Academia in the Environmental Justice Movement" (Undergraduate honors thesis, University of Redlands, 2008).
19 Will Sullivan, "Too Much Pollution for One Place," *Nova*, August 23, 2017, https://www.pbs.org/wgbh/nova/article/too-much-pollution/
20 Severn Suzuki, Speech at U.N. Conference on Environement and Development, 1992, https://www.americanrhetoric.com/speeches/severnsuzukiunearthsummit. htm
21 George Hoberg, *The Resistance Dilemma: Place-Based Movements and the Climate Crisis* (Cambridge, MA: The MIT Press, 2021).
22 350.org, "Keystone XL Protest at the White House Leads to Mass Arrests," *350.org Press Release*, March 2, 2014, https://350.org/press-release/ keystone-xl-protest-at-the-white-house-leads-to-mass-arrests/
23 Bill McKibben, "The Fossil Fuel Resistance," *Rolling Stone*, April 11, 2013, http://www.rollingstone.com/politics/news/the-fossil-fuel-resistance-20130411
24 350.org, "Divestment Campaign Spreads to Over 100 Campuses," *350.org Blog*, November 29, 2012, https://math.350.org/news/
25 Jessica Grady-Benson and Brinda Satathy, "Fossil Fuel Divestment in US Higher Education," *Local Environment* (2015): 2, http://dx.doi.org/10.1080/ 13549839.2015.1009825
26 Hoberg, *The Resistance Dilemma.*
27 Grady-Benson and Satahy, "Fossil Fuel Divestment," 5.
28 Charlotte Alter, "Hundreds of Thousands Converge on New York to Demand Climate-Change Action," *Time Magazine,* September 21, 2014, https://time. com/3415162/peoples-climate-march-new-york-manhattan-demonstration/; Grady-Bensona and Satathy, "Fossil Fuel Divestment."
29 Alter, "Hundreds of Thousands Converge."
30 Ibid.
31 Andrew Gage, "Youth Are Leading the Climate Movement – In Court and on the Streets," *WCEL Blog*, October 15, 2019, https://www.wcel.org/blog/ youth-are-leading-climate-movement-in-court-and-streets
32 Youth Plantiffs' amended complaint, 2015, p. 50. https://www.ourchildrenstrust. org/court-orders-and-pleadings
33 Robinson Meyer, "A Climate Lawsuit Dissent That Changed My Mind," *The Atlantic*, January 22, 2020, https://www.theatlantic.com/science/archive/ 2020/01/read-fiery-dissent-childrens-climate-case/605296/
34 Ariande Montare, "Standing Rock: A Case Study in Civil Disobedience," *GP-Solo Magazine*, May/June, 2018, https://www.americanbar.org/groups/gpsolo/ publications/gp_solo/2018/may-june/standing-rock-case-study-civil-disobedience/#:

~:text=It%20would%20become%20the%20temporary,throughout%20 the%20spring%20and%20summer

35 International Indingneous Youth Council (IIYC), "Our Roots," https://indigenous youth.org/

36 Tessa Stuart, "Sunrise Movement, the Force Behind the Green New Deal, Ramps up Plans for 2020," *Rolling Stone,* May 1, 2019, https://www.rollingstone.com/ politics/politics-features/sunrise-movement-green-new-deal-2020-828766/

37 Jerusha Conner, Johnnie Lotesta, Tova Wang, and Kei Kawashima-Ginsberg, *The Role of Electoral Engagement in Youth Social Movements* (Medford, OR: CIRCLE, Tufts University, 2021), https://circle.tufts.edu/sites/default/ files/2021-09/Youth_Movements_Qual.pdf

38 John Sutter and Lawrence Davidson, "Teens Tell Climate Negotiators They Aren't Mature Enough," *CNN,* December 17, 2018, https://edition.cnn. com/2018/12/16/world/greta-thunberg-cop24/index.html

39 Greta Thunberg, Speech at the World Economic Forum, January 22, 2019, https:// fridaysforfuture.org/what-we-do/activist-speeches/#greta_speech_jan22_ 2019

40 Sarah Pickard, Benjamin Bowman, and Dena Arya, "We are Radical in Our Kindness: The Political Socialization, Motivations, Demands and Protest Actions of Young Environmental Activists in Britain," *Youth and Globalization* 2, no. 2 (2020): 251.

41 Jerusha Conner et al., "Burnout and Belonging: How the Costs and Benefits of Youth Activism Affect Youth Health and Wellbeing," *Youth* 3 (2023): 127–45. https://doi.org/10.3390/youth3010009

42 Conner, Lotesta, Wang, and Kawashima-Ginsberg, *The Role of Electoral Engagement.*

43 Jerusha Conner, Johnnie Lotesta, and Rachel Stannard, "Intersectional Politicization: A Facet of Youth Activists' Sociopolitical Development," *Journal of Community Psychology* 51, no. 3 (2023): 1345–64. https://doi.org/10.1002/ jcop.22941.

44 Harriet Thew, Lucie Middlemiss, and Jouni Paavola, "'Youth Is Not a Political Position': Exploring Justice Claims-Making in the UN Climate Change Negotiations," *Global Environmental Change* 61 (March 2020), https://doi. org/10.1016/j.gloenvcha.2020.102036

45 Sarah Pickard, "Young Environmental Activists are Doing It Themselves," *Political Insight* 10, no. 4 (December 2019): 4–7.

46 NBC News, "Inside the Sunrise Movement," *Video,* March 6, 2019, https:// www.nbcnews.com/video/inside-the-sunrise-movement-how-young-climate- activists-put-the-green-new-deal-on-the-map-145299104 3909,

47 Jerusha Conner, *The New Student Activists* (Baltimore: Johns Hopkins University Press, 2020).

48 Including youth from Youth vs the Apocalypse.

49 Guardian News, "Dianne Feinstein Rebuffs Youth Climate Activists' Calls for Green New Deal," *YouTube Video,* February 23, 2019, https://www.youtube. com/watch?v=jEPo34LCss8

50 Conner, Lotesta, Wang, and Kawashima-Ginsberg, *The Role of Electoral Engagement.*

51 Youth Climate Collective (YCC), "Using Creativity to Address the Climate Crisis," https://www.youthclimatecollective.ca/

52 Greta Thunberg, Speech to the UN Climate Action Summit, September 23, 2019, https://www.usatoday.com/story/news/2019/09/23/greta-thunberg-tells-un- summit-youth-notforgive-climate-inaction/2421335001/

53 Suzuki, Speech at U.N. Conference.
54 Emerald Pellot, "Nadia Nazar Is a Teen Activist Using Art to Organize for Climate Justice," *Yahoo News,* April 15, 2021, https://tinyurl.com/2p9he747
55 Greta Thunberg, "We'll Be Watching You," *NPR,* September 23, 2019, https://www.npr.org/2019/09/23/763452863/transcript-greta-thunbergs-speech-at-the-u-n-climate-action-summit
56 Sarah Pickard, "The Nature of Environmental Activism among Young People in Britain in the Early 21ˢᵗ Century," in *Political Ecology and Environmentalism in Britain,* ed. Brendan Prendiville and David Haigron (Newcastle upon Tyne: Cambridge Scholars, 2020), 89–109.
57 Hoberg, *The Resistance Dilemma.*
58 Sophie Hirsh, "4 Young Activist on Intersectionality in Climate Justice, Fighting from Home, and More," *Green Matters,* August 18, 2020, https://www.greenmatters.com/p/climate-activists-intersectionality
59 Youth Climate Lab (YCL). *Our Impact: Youth Climate Lab's First 3 and Next 3 Years* (2020), https://static1.squarespace.com/static/59ec036ef9a61ebf918040ac/t/5f0fac2954a6c178b01f0b0e/1594862645706/YCL-Our-Impact.pdf
60 Climate Strike Canada, "About," https://www.climatestrikecanada.org/about
61 Climate Strike Canada, "COP26 Statement," https://www.climatestrikecanada.org/cop26-statement
62 Conner, Lotesta, Wang, and Kawashima-Ginsberg, *The Role of Electoral Engagement.*
63 Olaniyn Olamide, "They Call Themselves Sustainabiliteens and They are Formidable," *The Tyee,* February 7, 2020, https://thetyee.ca/News/2020/02/07/Sustainabiliteens-Are-Formidable/
64 Jack Queen, "NYU Law Students Disrupt Paul Weiss Event over Exxon Win," *Law 360,* February 11, 2020, https://www.law360.com/articles/1243046/nyu-law-students-disrupt-paul-weiss-event-over-exxon-win
65 Conner, Lotesta, Wang, and Kawashima-Ginsberg, *The Role of Electoral Engagement.*
66 Grosse, Corrie, "Climate Justice Movement Building: Values and Cultures of Creation in Santa Barbara, California," *Social Sciences,* 8, no. 79 (2019): 1–26, doi:10.3390/socsci80300079
67 Conner et al., "Burnout and belonging."
68 David Pellow, *What is Critical Environmental Justice?* (Medford: Polity, 2018).
69 Kimberle Crenshaw, "Mapping the Margins: Intersectionality, Identity Politics, and Violence against Women of Color," *Stanford Law Review* 43, no. 6 (1991): 1241–1299.
70 Hirsh, "4 Young Activist on Intersectionality."
71 Sophie Hirsch, "How this 19-year old activist is fighting for intersectional environmental justice," *Green Matters,* November 15, 2021. *https://www.greenmatters.com/p/diego-arreola-intersectional-environmentalism*
72 Joe Biden (@JoeBiden), "When I hear 'climate,' I think jobs...," *Twitter,* June 15, 2022, https://twitter.com/joebiden/status/1537150703633850371
73 Conner, Stannard, and Lotesta, "Intersectional Politicization."
74 Youth vs Apocalypse, "Points of Unity," https://www.youthvsapocalypse.org/points-of-unity
75 Francesca Belotti, Stellamarina Donato, Arianna Bussoletti, and Francesca Comunello, "Youth Activism for Climate on and Beyond Social Media," *International Journal of Press Politics* 27, no. 3 (2022): 718–37, doi:10.1177/19401612211072776

76 Thew, Middlemiss, and Paavola, "'Youth is not a Political Position'," 1.
77 Conner, Lotesta, Wang, and Kawashima-Ginsberg, *The Role of Electoral Engagement.*

Bibliography

350.org. "Divestment Campaign Spreads to Over 100 Campuses." *350.org Blog*, November 29, 2012. https://math.350.org/news/
_____. "Keystone XL Protest at the White House Leads to Mass Arrests." *350.org Press Release*, March 2, 2014. https://350.org/press-release/keystone-xl-protest-at-the-white-house-leads-to-mass-arrests/
Alter, Charlotte. "Hundreds of Thousands Converge on New York to Demand Climate-Change Action." *Time Magazine*, September 21, 2014. https://time.com/3415162/peoples-climate-march-new-york-manhattan-demonstration/
Belotti, Francesca, Stellamarina Donato, Arianna Bussoletti, and Francesca Comunello. "Youth Activism for Climate on and Beyond Social Media." *International Journal of Press Politics* 27, no. 3 (2022): 718–37. doi:10.1177/19401612211072776
Biden, Joe (@JoeBiden). "When I hear 'climate,' I think jobs..." *Twitter*, June 15, 2022. https://twitter.com/joebiden/status/1537150703633850371
Bowman, Benjamin, and Sarah Pickard. "Peace, Protests, and Precarity: Making Conceptual Sense of Young People's Non-Violent Dissent in a Period of Intersecting Crises." *Journal of Applied Youth Studies* 4 (2021): 493–510.
Brangham, William. "Why These Young People Came to the COP26 Climate Change Conference." *PBS*, November 11, 2021. https://www.pbs.org/newshour/world/why-these-young-people-came-to-the-cop26-climate-change-conference
Bullard, Robert, and Beverly Wright. *Race, Place, and Environmental Justice after Hurricane Katrina.* Boulder, CO: Westview Press, 2009.
Climate Strike Canada. "About." Accessed July 8, 2022. https://www.climatestrikecanada.org/about
_____. "COP26 Statement." Accessed July 8, 2022. https://www.climatestrikecanada.org/cop26-statement
Cole, Luke, and Sheila Foster. *From the Ground up.* New York: New York University Press, 2001.
Conner, Jerusha. *The New Student Activists.* Baltimore: Johns Hopkins University Press, 2020.
_____, Johnnie Lotesta, and Rachel Stannard. "Intersectional Politicization: A Facet of Youth Activists' Sociopolitical Development." *Journal of Community Psychology* 51, no. 3 (2023): 1345–64. https://doi.org/10.1002/jcop.22941
_____, Johnnie Lotesta, Tova Wang, and Kei Kawashima-Ginsberg. *The Role of Electoral Engagement in Youth Social Movements.* Medford: CIRCLE, Tufts University, 2021. https://circle.tufts.edu/sites/default/files/2021-09/Youth_Movements_Qual.pdf
_____, Emily Greytak, Carly D. Evich, and Laura Wray-Lake. "Burnout and Belonging: How the Costs and Benefits of Youth Activism Affect Youth Health and Wellbeing," *Youth* 3 (2023): 127–45. https://doi.org/10.3390/youth3010009

Crenshaw, Kimberle. "Mapping the Margins: Intersectionality, Identity Politics, and Violence Against Women of Color." *Stanford Law Review* 43, no. 6 (1991): 1241–1299.

Cripps, Elizabeth. *What Climate Justice Means and Why We Should Care*. London: Bloomsbury, 2022.

Environmental Justice and Climate Change Initiative. *10 Principles for Just Climate Change Policies*. 2002. https://www.ejnet.org/ej/climatejustice.pdf

Funders Collaborative on Youth Organizing (FCYO). "Apply to the Youth Organizing for Climate Action and Racial Equity (YO-CARE) Capacity-Building Fund!" Accessed May 30, 2022. https://fcyo.org/resources/apply-to-the-youth-organizing-for-climate-action-and-racial-equity-yo-care-capacity-building-fund-1

Future Coalition (@FutureCoalition). "In Light of the COVID-19 (Coronavirus) Pandemic, the US Youth #ClimateStrike Coalition is Pivoting Our #EarthDay Strikes plans..." *Twitter*, March 13, 2020. https://twitter.com/futurecoalition/status/1238508271033614338

Gage, Andrew. "Youth Are Leading the Climate Movement – In Court and on the Streets." WCEL Blog, October 15, 2019. https://www.wcel.org/blog/youth-are-leading-climate-movement-in-court-and-streets

Grady-Benson, Jessica, and Brinda Satathy. "Fossil Fuel Divestment in US Higher Education." *Local Environment* (2015): 1–21. http://dx.doi.org/10.1080/1354 9839.2015.1009825

Grosse, Corrie. "Climate Justice Movement Building: Values and Cultures of Creation in Santa Barbara, California." *Social Sciences* 8, no. 79 (2019): 1–26.

Guardian News, "Dianne Feinstein Rebuffs Youth Climate Activists' Calls for Green New Deal." *YouTube Video*, February 23, 2019. https://youtube.com/watch?v=jEPo34LCss8

Hirsh, Sophie. "4 Young Climate Activist on Intersectionality in Climate Justice, Fighting From Home, and More." *Green Matters*, August 18, 2020. https://www.greenmatters.com/p/climate-activists-intersectionality

_____. "How This 19-Year Old Activist Is Fighting for Intersectional Environmental Justice." *Green Matters*, November 15, 2021. *https://www.greenmatters.com/p/diego-arreola-intersectional-environmentalism*

Hoberg, George. *The Resistance Dilemma: Place-Based Movements and the Climate Crisis*. Cambridge, MA: The MIT Press, 2021.

International Indigenous Youth Council (IIYC). "Our Roots." Accessed July 8, 2022. https://indigenousyouth.org/

McKibben, Bill. "The Fossil Fuel Resistance." *Rolling Stone*, April 11, 2013. http://www.rollingstone.com/politics/news/the-fossil-fuel-resistance-20130411

Meyer, Robinson. "A Climate Lawsuit Dissent That Changed My Mind." *The Atlantic*, January 22, 2020. https://www.theatlantic.com/science/archive/2020/01/read-fiery-dissent-childrens-climate-case/605296/

Miao, Hannah. "Young Voters Helped Propel Biden to Victory." *CNBC*, November 13, 2020. https://www.cnbc.com/2020/11/13/election-young-voters-biden-democratic-party.html

Michigan in the World and Environmental Justice HistoryLab. "Give Earth a Chance: Environmental Activism in Michigan." January 2018. http://michigan intheworld.history.lsa.umich.edu/environmentalism/

Montare, Ariande. "Standing Rock: A Case Study in Civil Disobedience." *GP-Solo Magazine*, May/June, 2018. https://www.americanbar.org/groups/gpsolo/publications/gp_solo/2018/may-june/standing-rock-case-study-civil-disobedience/#:~:text=It%20would%20become%20the%20temporary,throughout%20the%20spring%20and%20summer

NBC News. "Inside the Sunrise Movement." *Video*, March 6, 2019. https://www.nbcnews.com/video/inside-the-sunrise-movement-how-young-climate-activists-put-the-green-new-deal-on-the-map-145299104 3909

Olamide, Olaniyn. "They Call Themselves Sustainabiliteens and They are Formidable." *The Tyee*, February 7, 2020. https://thetyee.ca/News/2020/02/07/Sustainabiliteens-Are-Formidable/

Pearthree, Genevieve. "Bridging the Divide: Activism and Academia in the Environmental Justice Movement." Undergraduate honors thesis, University of Redlands, 2008.

Pellot, Emerald. "Nadia Nazar Is a Teen Activist Using Art to Organize for Climate Justice." *Yahoo News*, April 15, 2021. https://ca.style.yahoo.com/nadia-nazar-teen-activist-using-120000457.html?guccounter=1&guce_referrer=aHR0cHM6Ly93d3cuZ29vZ2xlLmNvbS8&guce_referrer_sig=AQAAADnFEfREeSSveaJGZ1oHdBHUlcKGGwU7S2Phz-UYK_WoASK7_tEckSNL1oHEVItxvx3Sk-8YJFFrHlti392OJhQrY_CY-6Gj9lK8ggZSbw3TdHHHBi6-yG4qZvisezaGt43XqFWzbOZu188FdClV9e11UEHV_xZAABHmcm_C0uNt

Pellow, David. *What Is Critical Environmental Justice?* Medford, OR: Polity, 2018.

Pickard, Sarah. "Young Environmental Activists Are Doing It Themselves." *Political Insight* 10, no. 4 (December 2019): 4–7.

_____. "The Nature of Environmental Activism among Young People in Britain in the Early 21st Century." In *Political Ecology and Environmentalism in Britain*, edited by Brendan Prendiville and David Haigron, 89–109. Newcastle upon Tyne: Cambridge Scholars, 2020.

_____, Benjamin Bowman, and Dena Arya. "We Are Radical in Our Kindness: The Political Socialization, Motivations, Demands and Protest Actions of Young Environmental Activists in Britain." *Youth and Globalization* 2, no. 2 (2020): 251–80.

Queen, Jack. "NYU Law Students Disrupt Paul Weiss Event over Exxon Win." *Law 360*, February 11, 2020. https://www.law360.com/articles/1243046/nyu-law-students-disrupt-paul-weiss-event-over-exxon-win

Rhodes, Edwardo. *Environmental Justice in America*. Bloomington: Indiana University Press, 2003.

Salter, Eleanor. "Where Have All the Young Climate Activists Gone?" *The Guardian*, June 9, 2020. https://www.theguardian.com/commentisfree/2022/jun/29/children-arent-the-future-where-have-all-the-young-climate-activists-gone?CMP=share_btn_tw

Scholsberg, David, and Lisette Collins. "From Environmental to Climate Justice: Climate Change and the Discourse of Environmental Justice." *Wires Climate Change* (2014). doi:10.1002/wcc.275

Suzuki, Severn. Speech at U.N. Conference on Environment and Development, 1992. https://www.americanrhetoric.com/speeches/severnsuzukiunearthsummit.htm

Stuart, Tessa. "Sunrise Movement, the Force behind the Green New Deal, Ramps up Plans for 2020." *Rolling Stone*, May 1, 2019. https://www.rollingstone.com/politics/politics-features/sunrise-movement-green-new-deal-2020-828766/

Sullivan, Will. "Too Much Pollution for One Place." *Nova*, August 23, 2017. https://www.pbs.org/wgbh/nova/article/too-much-pollution/

Sutter, John, and Lawrence Davidson. "Teens Tell Climate Negotiators They Aren't Mature Enough." *CNN*, December 17, 2018. https://edition.cnn.com/2018/12/16/world/greta-thunberg-cop24/index.html

Taylor, Dorcetta. "Introduction: The Evolution of Environmental Justice Activism, Research and Scholarship." *Environmental Practice* 13, no. 4 (December 2011): 280–301. doi:10.1017051466046611000329

Taylor, Matthew, Jonathan Watts, and John Bartlett. "Climate Crisis: 6 Million People Join Latest Wave of Global Protests." *The Guardian*, September 27, 2019. https://www.theguardian.com/environment/2019/sep/27/climate-crisis-6-million-people-join-latest-wave-of-worldwide-protests

Thew, Harriet, Lucie Middlemiss, and Jouni Paavola. "'Youth Is Not a Political Position': Exploring Justice Claims-Making in the UN Climate Change Negotiations." *Global Environmental Change* 61, (March 2020). https://doi.org/10.1016/j.gloenvcha.2020.102036

Thunberg, Greta. "Speech to the World Economic Forum." January 22, 2019. https://fridaysforfuture.org/what-we-do/activist-speeches/#greta_speech_jan22_2019

————. "Speech to the UN Climate Action Summit." September 23, 2019. https://www.usatoday.com/story/news/2019/09/23/greta-thunberg-tells-un-summit-youth-notforgive-climate-inaction/2421335001/

————. "We'll Be Watching You." *NPR*, September 23, 2019. https://www.npr.org/2019/09/23/763452863/transcript-greta-thunbergs-speech-at-the-u-n-climate-action-summit

Youth Climate Collective (YCC). "Using Creativity to Address the Climate Crisis." https://www.youthclimatecollective.ca/

Youth Climate Lab. *Our Impact: Youth Climate Lab's First 3 and Next 3*. 2020. https://static1.squarespace.com/static/59ec036ef9a61ebf918040ac/t/5f0fac2954a6c178b01f0b0e/1594862645706/YCL-Our-Impact.pdf

Youth vs Apocalypse. "Points of Unity." https://www.youthvsapocalypse.org/points-of-unity

Contributors

Jerusha Conner is a Professor of Education at Villanova University. Her research focuses on youth activism and organizing, student engagement, and student voice. She is the author of more than 60 journal articles and book chapters, 3 edited collections (*Political Activism in Post-Secondary Contexts*, 2022; *Contemporary Youth Activism*, 2016; and *Student Voice in American Education Policy*, 2015), and *The New Student Activists* (2020).

Rob Gioielli is a Professor of History and Environmental Studies at the University of Cincinnati. An urban and environmental historian, his research and teaching focus on how race and inequality have shaped social movements and sustainability in American cities. He is the author of *Environmental Activism and the Urban Crisis: Baltimore, St. Louis, Chicago*, and is currently writing an environmental history of white flight.

Elizabeth Grennan Browning is an Assistant Professor of History at the University of Oklahoma, and a former faculty fellow at Indiana University's Environmental Resilience Institute. She is the author of *Nature's Laboratory: Environmental Thought and Labor Radicalism in Chicago, 1886–1937* (Johns Hopkins University Press, 2022).

Zoltán Grossman is a Professor of Geography and Native Studies at The Evergreen State College in Olympia, Washington. He is a longtime community organizer and was a co-founder of the Midwest Treaty Network in Wisconsin. He earned his Ph.D. at the University of Wisconsin-Madison. He is a past co-chair of the Indigenous Peoples Specialty Group of the American Association of Geographers. Grossman was co-editor of *Asserting Native Resilience: Pacific Rim Indigenous Nations Face the Climate Crisis* (Oregon State University Press, 2012), and author of *Unlikely Alliances: Native Nations and White Communities Join to Defend Rural Land* (University of Washington Press, 2017).

Holly Miowak Guise (Iñupiaq) is an Assistant Professor of History at the University of New Mexico. Her research centers Indigenous oral histories with tribal and institutional archives. Her book manuscript, entitled "World War II in Alaska: Native Voices and History," is forthcoming from the University of Arizona Press. Her research has been supported by a Ford Foundation Postdoctoral Fellowship and an American Council of Learned Societies Fellowship. She administers an Indigenous oral history website: WorldWarIIAlaska.com.

A. J. Hudson is a recent graduate of Yale Law School (JD 2023), the Yale School of the Environment (MS 2019), and a doctoral candidate at the University of Oxford. Before graduate school, Hudson spent five years teaching in one of the most disenfranchised, polluted, and overpoliced neighborhoods in New York City. His scholarship, legal work, and activism all seek to redirect the body of traditional environmental law toward the objectives of civil rights, human rights, and climate justice. He is currently writing a legal history of coastal development in Mexico and the Caribbean in the context of the modern anti-tourism movements for environmental self-determination sweeping across postcolonial vacation destinations in Latin America and beyond.

Kyle Kajihiro is a Lecturer in the Departments of Ethnic Studies and Geography & Environment at the University of Hawai'i at Mānoa. His research examines U.S. imperial formation and militarization in Hawai'i and the Pacific and decolonial and demilitarization social movements that confront these processes. He works with a number of community organizations concerned with environmental justice, including Hawai'i Peace and Justice, Mālama Mākua, and the O'ahu Water Protectors.

Alessandro Morosin is an Assistant Professor of Sociology and Criminology at the University of La Verne. His scholarship focuses on environmental sociology, global capitalism, and participatory action research with communities that are resisting extractive megaprojects in Oaxaca, Mexico. In late 2016, he began his dissertation fieldwork in Oaxaca by meeting with Indigenous human rights defenders, NGOs, teachers, and environmental justice activists to document an emerging mobilization against proposed open-pit mines in the Isthmus of Tehuantepec. In it he analyzed how an Indigenous life-ethic motivates diverse networks of people to confront the longstanding federal plans for capitalist development in Mexico's "Isthmus Corridor," and the extractive logic that facilitates these *proyectos de muerte* ("projects of death"). Morosin has published in the *Journal of Political Ecology* and in *Latin American Perspectives* and contributed book chapters to a number of edited volumes. His monograph will critique the larger social and environmental crises in this strategic corridor of southern Mexico that are producing new mobilizations for class, ethnic, and gender justice.

Justin Sean Myers is an Associate Professor in the Department of Sociology at California State University, Fresno, and has published work on the politics of the food justice movement as well as the race and class tensions within the food movement. He is the author of *Growing Gardens, Building Power*, a book on urban agriculture and food justice in Brooklyn, New York.

Paul C. Rosier received his Ph.D. in American History from the University of Rochester in 1998. He currently serves as a Professor of History at Villanova University, where he teaches Native American History, American Environmental History, Global Environmental History, and 20th-Century American History. He also serves as the director of the Albert Lepage Center for History in the Public Interest at Villanova. He previously held the inaugural Mary M. Birle Chair in American History (2016–2022) and served as Department Chair (2013–2016). In 2001, he published *Rebirth of the Blackfeet Nation, 1912–1954*; in 2006, he co-edited *Echoes from the Poisoned Well: Global Memories of Environmental Injustice*; and in 2009, Harvard University Press published *Serving Their Country: American Indian Politics and Patriotism in the Twentieth Century*, which won the 2010 American Indian National Book Award. He has published numerous essays on Native American topics, including three articles in *The Journal of American History*. He is a founding member of the Editorial Board of the journal *Environmental Justice*. In addition to editing *Environmental Justice in North America*, he is completing a monograph on Native Americans' political history, entitled "Indigenous Citizens: Native Americans' Fight for Sovereignty, 1776–2020."

Lydia Schoeppner is a Senior Researcher at Narratives Inc., an environmental planning firm in Winnipeg. She is also a Faculty Fellow in Conflict Resolution Studies at Canadian Mennonite University in Winnipeg. Previously, she was a Teaching Assistant Professor in Conflict Resolution Studies at Menno Simons College, a college of Canadian Mennonite University at the University of Winnipeg. For her doctoral research in Peace and Conflict Studies (University of Manitoba, 2020) she collaborated with Inuit in Canada and Greenland to learn about Inuit conflict resolution mechanisms in the work of the Inuit Circumpolar Council and in two local Inuit communities (Pangnirtung in Nunavut/Canada and Maniitsoq in Greenland). Her article entitled "The role of international organizations in dealing with sovereignty conflicts in the Arctic" – published in the *Journal for Peace and Justice Studies* – received the Regional Studies Association's PhD Student Award for original and outstanding contribution to the field of regional studies in 2018. In spring 2022, she was a Visiting Research Fellow at the University of Notre Dame's Kroc Institute for International Peace Studies. Her current work and research focuses on environmental peacemaking with a focus on the role of storytelling and narration for conflict transformation.

Kyle Whyte is George Willis Pack Professor at the School for Environment and Sustainability, founding Faculty Director of the Tishman Center for Social Justice and the Environment, Principal Investigator of the Energy Equity Project, and Affiliate Professor of Native American Studies and Philosophy. Kyle is currently a U.S. Science Envoy, a chapter lead author on the National Climate Assessment, and serves on the White House Environmental Justice Advisory Council and the National Academies' Resilient America Roundtable. He is President of the Board of Directors of the Michigan Environmental Justice Coalition and the Pesticide Action Network North America. He is an enrolled member of the Citizen Potawatomi Nation.

Index